Natural Medicinal Plants

Edited by Hany A. El-Shemy

Published in London, United Kingdom

IntechOpen

Supporting open minds since 2005

Natural Medicinal Plants
http://dx.doi.org/10.5772/intechopen.91542
Edited by Hany A. El-Shemy

Contributors
Puneshwar Keshari, Biman Bhuyan, Prakash Rajak, Dipak Chetia, Dinithi C Peiris, Thilina Gunathilaka, Lakshika Rangee Keerthirathna, Motamarri V.N.L. Chaitanya, Firehiwot Belayneh, Heyam Saad Ali, Hailemikael Gebremariam Baye, Ramesh Kumar Ahirwar, Abdur Rauf, Yahya S. Al-Awthan, Naveed Muhammad, Muhammad Mukarram Shah, Saikat Mitra, Talha Bin Emran, Omar Bahattab, Mohammad S. Mubarak, Fongang Fotsing Yannick Stéphane, Bankeu Kezetas Jean Jules, Lenta Ndjakou Bruno, Gaber El-Saber Batiha, Iftikhar Ali, Pushpa Anantrao Karale, Shashikant Dhawale, Mahesh Antram Karale, Harsha Kashyap, Naveen Kumar Kaushik, Cheryl Sachdeva, Skunda Diliarosta, Rehani Ramadhani, Monica Prima Sari, Annisa Efendi, Muhammad Shoaib Amjad, Fozia Abasi, Huma Qureshi, Evelyn Asante-Kwatia, Abraham Yeboah Mensah, Lord Gyimah, Arnold Forkuo Donkor, Paula Vieira-Brock, Smita G. Bhat, Antony Justin, Sayani Banerjee, Chennu Manisha, Deepthi Murugan

First published in London, United Kingdom, 2022 by IntechOpen
IntechOpen is the global imprint of INTECHOPEN LIMITED, registered in England and Wales, registration number: 11086078, 5 Princes Gate Court, London, SW7 2QJ, United Kingdom
Printed in Croatia

British Library Cataloguing-in-Publication Data
A catalogue record for this book is available from the British Library

Additional hard and PDF copies can be obtained from orders@intechopen.com

Natural Medicinal Plants
Edited by Hany A. El-Shemy
p. cm.
Print ISBN 978-1-83969-275-8
Online ISBN 978-1-83969-276-5
eBook (PDF) ISBN 978-1-83969-277-2

For EU product safety concerns:
IN TECH d.o.o., Prolaz Marije Krucifikse Kozulić 3, 51000 Rijeka, Croatia, info@intechopen.com or visit our website at intechopen.com.

Meet the editor

Prof. Hany A. El-Shemy received a Ph.D. in Biochemistry from the University of Cairo, Egypt, and a Ph.D. in Genetic Engineering from the University of Hiroshima, Japan. He holds two patents and has written thirteen international books. He has also published more than 100 SCI journal papers and 55 conference presentations. Dr. El-Shemy was a technique committee member as well as chair of many international conferences. He has also served as editor for journals including *PLOS ONE*, *BMC Genomics*, and *Current Issues in Molecular Biology.* He has received several awards, including state prizes from the Academy of Science, Egypt (2004, 2012, and 2018), the Young Arab Researcher prize from the Shuman Foundation, Jordan (2005), and Cairo University Prizes (2007, 2010, and 2014). He served as an expert for the African Regional Center of Technology, Dakar, Senegal, as well as a visiting professor at Pan African University, African Union. He served as vice president of the Academy of Science and Technology, Egypt, from 2013 to 2014. Since 2014 he has been the dean of the Faculty of Agriculture, Cairo University. In 2018, he was elected a fellow of the African Academy of Science.

Contents

Preface

This book, *Natural Medicinal Plants*, includes sixteen chapters that address such topics as extraction of bioactive compounds from medicinal plants and herbs, medicinal plants and their pharmaceutical values, and usage of plants in traditional African medicine and in India. The book also provides scientific evidence on drugs developed from plants and their effectiveness for treating different illnesses, including COVID-19.

It is a useful resource for researchers, students, and other interested readers. It is also recommended for readers to take a look at a related book, *Natural Drugs from Plants.*

Hany A. El-Shemy
Faculty of Agriculture,
Biochemistry Department,
Cairo University,
Giza, Egypt

Chapter 1

Natural Products Altering GABAergic Transmission

Sayani Banerjee, Chennu Manisha, Deepthi Murugan and Antony Justin

Abstract

Gamma-amino butyric acid (GABA) is a major inhibitory neurotransmitter found in several regions of the brain and known to have various significant physiological roles as a potent bioactive compound. Malfunction of GABAergic neuronal signaling prompts to cause severe psychiatric symptoms in numerous mental disorders. Several drugs are available in clinical practice for neuropsychiatric disorders targeting through GABAergic pathway, with notable adverse effects. Interestingly, in recent years, researchers are focusing on natural compounds altering GABAergic neurotransmission for various psychiatric disorders due to its wide range of therapeutic efficacy and safety. The enormous variety of natural compounds, namely alkaloids, flavonoids, terpenoids, polyacetylenic alcohols, alkanes and fatty acids were reported to alter the GABAergic transmission through its receptors and or by influencing the transmission, synthesis and metabolism of GABA. Natural compounds are able to cross the blood brain barrier and influence the GABA functions in order to treat anxiety, mania, schizophrenia and cognitive disorders. Therefore, this current chapter describes on natural products which have the potential to alter the GABAergic neurotransmission and its therapeutical benefits in treating several neuropsychiatry disorders using various pharmacological methods.

Keywords: Natural products, GABA, agonist, metabolism, allosteric modulation, psychiatric disorders

1. Introduction

The ground breaking discovery of Gamma-aminobutyric acid (GABA) played an astonishing role in neural control theory in 1950's. In the human cortex GABA is the primary inhibitory neurotransmitter [1]. In the initial developmental stage of life, GABA functions as an excitatory element which influences many physiological processes like neuronal proliferation, neurogenesis, migration, differentiation and preliminary circuit building. After maturation of CNS, GABA acts as an inhibitory neurotransmitter which is controlled as chloride or cation transporter expression. GABA also plays a vital role in interstitial neurons development of white matter along with oligodendrocyte development. Whereas the basic fundamental cellular mechanisms are not well described though it is proven that a lot of neurological diseases are well involved through GABA dependant pathway which includes white matter abnormalities, including anoxic-ischemic injury, anxiety, insomnia and schizophrenia [2]. GABA receptors are majorly classified into two main types ionotropic $GABA_A$ and

$GABA_C$ receptors and the metabotropic $GABA_B$ receptor. $GABA_A$ acts by activating the fast-hyperpolarizing negative ion channel (Cl^-) and diffuse by the means of concentration gradient to hyperpolarize post synaptic mature neurons [3, 4]. Whereas another kind of ionotropic receptor was discovered $GABA_C$ with 3ρ subunits [5]. $GABA_B$ receptors consist of two subunits, $GABA_{B1}$and $GABA_{B2}$ which are responsible for slower inhibitory transmission. These receptor activations are coupled with K^+/Ca^+ channels through G-protein mediated secondary pathway [6].

Natural molecules with a wide range of chemical structures have been shown to have $GABA_A$ receptor modulating potential due to the structural heterogeneity of and more than one number of binding sites. It has different pharmacological effects depending on the mechanism of action, the binding site and the affinity of the compounds. These effects have been investigated using different *in vitro* and *in vivo* models [7–9].

The versatile binding nature of benzodiazepine binding site of GABA receptor allows multiple molecules to bind and modulate the functions of GABA in a very specific manner. So, this class of compounds are used for the treatment of anxiety, convulsion, insomnia by non-specifically modulating all five α subunits. This non selective nature of these compounds generates unwanted side effects like tolerance and dependence. Therefore, there is an immediate need for finding safe drugs, with increased anxiolytic and decreased sedative potential. In recent decades, various reports have been made on natural products with GABAergic activity and, different various methods have been used to describe the effects. Hence, this review aimed to collect the existing data and make the obtained results as comparable as possible, thus facilitating the discussion of structure–activity relationships [10].

1.1 Synthesis

GABA is mainly produced from α-decarboxylation of glutamate by the enzyme glutamic acid decarboxylase (GAD) and metabolized by the actions of GABA-transaminase (GABA-T) and succinic semialdehyde dehydrogenase (SSADH) into succinate respectively. Through the use of the pyridoxal-5′-phosphate-dependent interconversion steady state concentration of GABA is achieved in-vivo (apo-GAD). At least 50 percent of the total GAD present in the brain is apo-GAD [10]. Inorganic phosphate promotes the activation of GAD and blocked by aspartate, GABA and ATP. The ATP facilitates and stabilizes apo-GAD formation which further stimulates the development of GABA. At 37°c temperature apo-GAD has a half-life of few minutes without ATP. GAD mainly consists of two isoforms of distinct molecular weights (65 and 67 kD) which are the products of chromosomes 2 and 10 in humans.

After synthesis, GABA vesicular release has specific mechanisms. GABA is assembled using Mg_2^+ activated ATPase into vesicles. This method is energy-dependent and requires adenosine triphosphate and magnesium. Calcium-dependent GABA vesicular release appears to result in a temporary increase in the synaptic cleft's GABA concentration and the binding of the receptor to evoke action. Through the sodium and chloride reuptake mechanism of the GABA transporter (GAT) to the presynaptic neuron and surrounding glia, quick synapse removal takes place. GABA is then reused into metabolites that are eventually used for GABA resynthesis by breakdown. GABA-oxoglutarate transaminase, succinic semialdehyde dehydrogenase and glutamate decarboxylase (GAD) are three enzymes required for GABA metabolism and resynthesis. The deterioration of GABA to succine semi-aldehyde is catalyzed by the enzyme GABA oxoglutarate transaminase. The latter is then oxidized by means of succinic semialdehyde dehydrogenase into succinic acid. Ligands associated with these GABA procedures will regulate the action of GABA [11].

2. GABA receptor physiology and GABA ligands

In 1981, $GABA_A$ & $GABA_B$ subtypes of GABA were discovered by Hill and Bowery. $GABA_A$ was reported as chlorine sensitive ion channel which is allosterically modulated by barbiturates, benzodiazepines, neurosteroids and ethanol. Along with this $GABA_B$ receptors couple with Ca^+ and K^+ channels via G protein second messenger system. This receptor activation specifically happens through baclofen which is resistant through $GABA_A$ modulators [12].

As $GABA_B$ receptors are dimeric metabotropic in nature and the structure of pentameric $GABA_A$ receptors ideal for allosteric regulation. So research on these receptors is likely to develop novel therapy for the treatment of neurological and psychological disorders [13, 14] (**Figure 1**).

2.1 $GABA_A$ receptor

Among the three types of GABA receptors, the $GABA_A$ receptor is the best characterized one. For several selective ligands, this channel has numerous binding sites. One class of therapeutic drugs linked to this target are receptor modulators: benzodiazepines, non-benzodiazepines and barbiturates, most of which improve the effect of GABA by increasing the chloride channel opening [13].

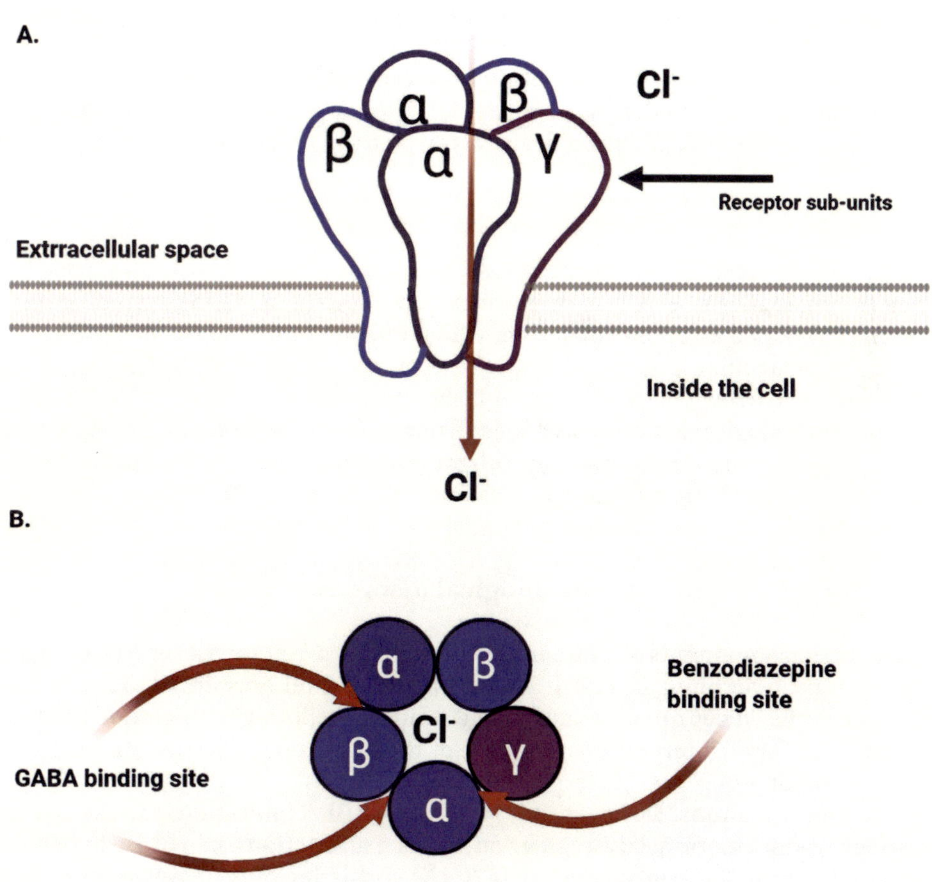

Figure 1.
A. The subunits of $GABA_A$ receptor and chlorine channel. B. Represents the barbiturate and benzodiazepine binding sites on GABA receptor subunits.

The $GABA_A$ agonist, muscimol, antagonist bicuculline and picrotoxin and inverse agonist FG 7142 are additional ligands which bind to the $GABA_A$ receptor. Some of these agents do not seem to have therapeutic benefits, but when used as pharmacological tools for the $GABA_A$ receptor they are the most significant ligands. Neuro-active steroids and partial benzodiazepine agonists (PBAs) are some newly discovered agents which are coming into recent considerations [14]. PBAs (e.g., bretazenil, imidazenil) are $GABA_A$ receptor activators, similar to benzodiazepines. Although they tend to have lower effectiveness compared to full benzodiazepine agonists, they give a more favorable side effect profile. Compared to other configurations found in more selective areas, this subtype is common throughout the brain [15].

2.2 $GABA_B$ receptor

$GABA_B$ receptors have seven membrane-spanning amino acid domains which are connected by a G-protein to its signaling pathway (K^+, Ca^{++} ion channels or adenylate cyclase). Presynaptic $GABA_B$ receptors are majorly coupled to calcium channels and their stimulation by the receptor results in decrease of calcium conductance and decline of GABA release. Thus, the receptors auto-regulates the discharge of $GABA_A$ and gives the $GABA_A$ system with negative feedback. On the other hand, Post-synaptic $GABA_B$ receptors are primarily linked to potassium channels and their stimulation led to increased conductance of K^+, hyperpolarization and decreased excitability of the neurons. The opening of T-type calcium channel is mainly associated with the actuation of $GABA_B$ receptor, resulting in calcium spiking activity that can contribute to absence seizure and is also included in signaling through the pathway of adenylate cyclase. It is therefore assumed that mediation of the $GABA_B$ receptor occurs through at least two distinct subtypes receptor [16–18].

2.3 $GABA_C$ receptor

The $GABA_C$ receptor, a subtype of $GABA_A$ receptor characterization started when the analogue of GABA cis-4-aminocrotonic acid (CACA) in cat spinal interneurons developed a depressant action, which was not inhibited by the $GABA_A$ antagonist bicuculline and varied from the depressant actions of the $GABA_B$ agonist baclofen. The $GABA_C$ receptor is distinguished from both $GABA_A$ and $GABA_B$ by their pharmacological actions. $GABA_C$ is structurally different from $GABA_A$ because $GABA_C$ is hetero oligomeric and homo oligomeric which means it composed of many subunits of the same subtype, it can be either r1 or r2 [19].

3. GABAergic system and neurological disorders

The main components of brain inhibitory circuits are networks of (GABAergic) interneurons in the amygdala [20]. This neurotransmitter is essential to maintain a balance between neuronal excitation and inhibition. Both glutamatergic neurons and the GABAergic interneurons compose of the basolateral nucleus (BLA). A relatively small group of GABAergic inhibitory neurons is closely regulated by Glutamatergic neurons. Devastation of GABAergic BLA inhibition, such as anxiety and depression, emotional dysregulation, and seizure actions, can cause hyperexcitability of the The central amygdala (CeA) consisting only of GABAergic neurons acts by converging inputs from the BLA as the primary output nucleus of the amygdala. In addition, the BLA, the central amygdala and all their associations play a key role in the regulation of the GABAergic system. As a result, these

GABAergic amygdala neurons are properly trained to perform a central role in the stress management. Nonetheless, even less is known about the association between the GABAergic amygdala inhibitory system and stress [21].

The sedative and hypnotic effects are mediated by α_1 subunit of $GABA_A$ receptors, whereas the anxiolytic effect is exhibited by the positive regulation of α_2 and (or) α_3 subunit of $GABA_A$ receptors. Furthermore, in learning and memory, the α_5 subunits play an important role. The cause of side effects, such as muscle relaxation or anterograde amnesia, is because of benzodiazepines, which are widely used in the treatment of anxiety, insomnia and seizures, functioning on various subunits (α_1, α_2, α_3 and α_5). Such drawbacks include the growth of resistance and dependence. The development of new and safer drugs with, for example, an efficient anxiolytic yet low sedative potential is therefore urgently warranted. Various studies have recently been performed on natural products with GABAergic involvement, and various types of approaches have been used to clarify the findings. Consequently, the purpose of this analysis is to gather current evidence and generate the findings obtained, thereby promoting the discussion of structure activity relationships [9].

In knock-out mice special kind of $GABA_B$ receptors are being introduced in mice that lack subunits of the $GABA_B$ receptor. In addition to psychiatric conditions, the phenotype of these mice shows evidence of $GABA_B$ receptor activation in epilepsy, sensorimotor gating, nociception and temperature control [22]. With almost the same behavioral phenotype as $GABA_{B1}$ Knockout mice whereas mice that lack the $GABA_{B2}$ subunit are currently developed. Some data suggest that these phenotypes underlie the lack of heteromeric $GABA_{B1}$ and $GABA_{B2}$ receptors. In order to evaluate the anxiolytic ability of other positive $GABA_B$ receptor modulators, further studies are required, but current evidence suggests that they may be a new category of anxiolytics with a higher side effect profile than benzodiazepines. The mechanisms involved in the anxiety activity impact of $GABA_B$ receptors are not well known. Future research should also focus on behavioral and electrophysiological approaches to the activation of $GABA_B$ receptors in major anxiety-related brain regions [23].

3.1 GABAergic system in schizophrenia

In late adolescence or early adulthood, schizophrenia is a mental health condition that commonly occurs. Its impact on speech, thinking, emotions and other areas of life can affect the social interactions and daily activities of people. In the presynaptic neuron, the carrier protein is available in GAT-1 and is mainly responsible for GABA reuptake in synapse. It plays a significant role in both phasic and tonic inhibition which is regulated by GABA. The synaptic potential of GABA is terminated by GAT-1 and it is managed by the duration and adequacy of GABAergic neurotransmission therefore, decreased GAT-1 levels demonstrate enhanced accessibility of GABA. In schizophrenia, numerous studies show decreased levels of mRNA encoding for the GAT-1 protein along with the decreased expression of GAD 67 mRNA. GAT-1 mRNA delivery is decreased and generally unchanged in most GABAergic neurons. GAT-1 mRNA concentration fluctuations are recognized in chandelier neurons. In schizophrenia, the thickness of immunoreactive GAT-1 cartridges is reduced, although axon terminal marker in other populations remains unaltered. Relatively low GAT-1 immunoreactive cartridge thickness indicates a significantly reduced GAT-1 protein correlated with a reduced level of GAT-1 mRNA. Therefore, in individuals with schizophrenia, the amount of GAT-1 protein-enclosing chandelier neurons decreased whereas the number of neurons comprising parvalbumin remained consistent. This outcome infers that the decreased degrees of GAT-1 mRNA are restricted to chandelier neurons.

The decline of GAD67 mRNA coding in the prefrontal dorsolateral cortex is the most predictable post-mortem finding in schizophrenia, which led to decrease in GAD67 levels of protein, despite the fact that this has been less widely considered. The schizophrenia-influenced subset tends to incorporate GABAergic neurons comprising parvalbumin. Expression of parvalbumin mRNA in schizophrenia is diminished in layer 3 and 4 of the prefrontal cortex (PFC). In the prefrontal cortex (PFC), the recent discovery indicates that the decreased articulation of GAD67 mRNA is unique for the GABA neuron subgroup [24].

Adequate histopathological data also suggests that, the association of GABAergic neurotransmission impairment with pathologies and cognitive dysfunctions of schizophrenia. The primary motor cortex (PMC), primary visual cortex (VC), anterior cingulate cortex (ACC) is distinguished by the similar GABAergic gene expression deficits as shown in the Dorsolateral prefrontal cortex, which includes selective parvalbumin-containing GABA neuron involvement. The greatest decreases in mRNA encoding levels for parvalbumin have been reported. In serious case reduction in the α_1 and δ subunits of GABA receptors, GAD67 mRNA, GAD65 mRNA and GAT-1 mRNA is displayed in the brain regions [10, 25].

3.2 GABAergic system in anxiety and depression

Both in animals and humans, depression and anxiety are most frequent causes of persistent stress. Two mechanisms are defined by anxiety models: fear processes are believed to be developed to allow us to change our emphasis on the first hint of risk and behavioral modification in order to prevent or eliminate an imminent or predicted overt danger [26].

The long-term potential activity strongly depends on the augmentation of GABA signaling which process through the $GABA_A$ receptors namely α_1 and α_2. The long-term potential response triggers are not only restricted to $GABA_A$ but also to the $GABA_B$ receptor. The $GABA_B$ receptor antagonists causing the long-term potential response on cortical along with thalamic centripetal synapses whereas the thalamic feed needs postsynaptic response from NMDA-receptor. The cortical actions controlled by pre-synaptic response on increased glutamate response by NMDA receptor independent activity, so activating GABA synapse thereby inducing $GABA_B$ receptor might help to arrest non associated long-term potential there by reducing agitation response [27].

By protruding to the central amygdala (CEA), CEA output neurons control the GABAergetic tone and form a spontaneous active neuron in lateral subdivisions. Aversive stimulus can reduce this inhibitory tone. CEA consists primarily of localized GABA neurons and the inhibition of GABA occurs through $GABA_A$ α_2 receptor. Therefore, for benzodiazepine-induced anxiolysis and anti-panic activity, CEA considered to be a significant target [28].

3.3 Epilepsy and GABAergic system

Epilepsy can be the consequence of disturbances in the homeostasis involving other neurotransmitters and neuromodulators, for example, glutamate, adenosine, norepinephrine, and acetylcholine. GABA receptor or transporter function alteration can allow the occurrence of seizure in the presence of normal GABA levels. Some data indicates that low occipital lobe GABA concentration (remote from the seizure focus) is a risk factor for seizure recurrence. Low GABA levels predispose but may not be sufficient for seizures to become clinically effective [28, 29].

In case of adults, status epilepticus induces a complete re-organization of the networks, with cell death, axonal growth leading to an increased glutamatergic drive. This, in turn, will decrease the threshold of seizure generation and thus contribute to seizure generation. Somatostatin innervates the dendrites of the principal cells in the hippocampus and triggers a chemical imbalance between excitatory and inhibitory neurotransmitters which leads to a reduction of the inhibitory strength that is necessary but not sufficient to generate ongoing seizures. An additional important factor is the persistent increase of the intracellular chloride concentration that leads to a long-lasting shift in the depolarizing direction of the actions of GABA that will also contribute to seizure generation [30, 31].

4. Natural products and GABA

Due to the different binding sites present on GABA (A) receptor, various receptor modulating compounds have been identified and depending on the mode of action, the affected binding site, and the compounds' affinity. Radioligand binding assays have been confirmed the capacity of the ligand for the displacement of a molecule from its binding site. Various studies helped us understand the link between modulation of the receptor and associated effects, such as anxiolytic, sedative, and anticonvulsive properties (**Figure 2**; **Table 1**).

Radioligand binding assays are simple but influential tool for reviewing receptors. They mainly analyze the interactions of hormones, neurotransmitters, growth factors, and related drugs with the receptors, studies of receptor interactions with second messenger systems, along with the characterization of regulatory changes in receptor number, subcellular distribution, and physiological function. So these assays are widely used in a numerous disciplines, including pharmacology, physiology, biochemistry, immunology, and cell biology. The fundamentals of the radioligand binding assay are fairly simple. The receptor of interest is incubated with an appropriate radioligand for a suitable period of time and then the radioactivity bound to the receptor is determined. There are three major types of experiments:

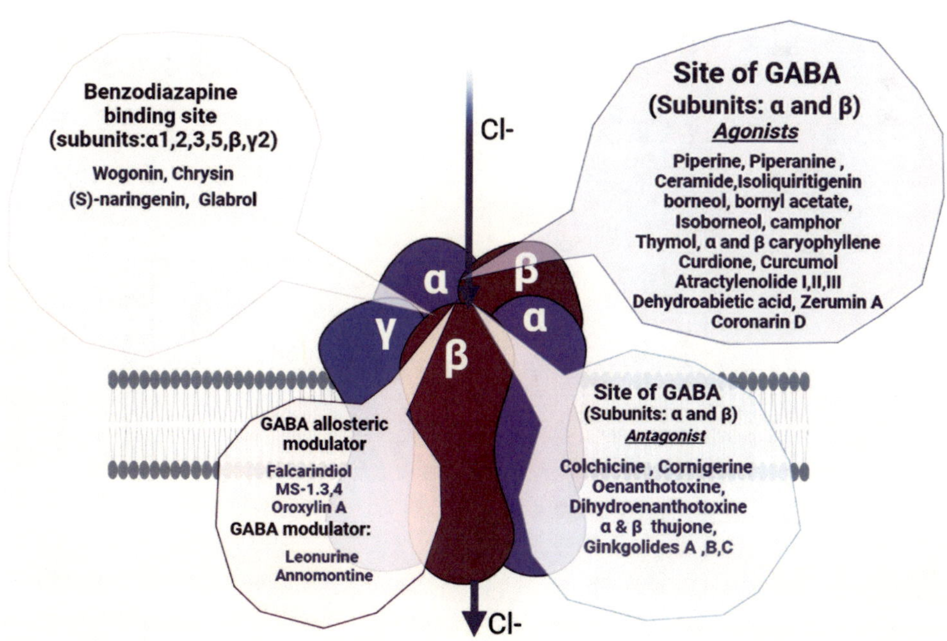

Figure 2.
Action of natural products on the specific site of GABA receptor.

Class of Compound	Source	Mechanism	Active Componud	In-Vivo/ In-Vitro	Reference
Alkaloids					
	Colchicum autumnale, Colchicum szovitsii	GABA competitive antagonist	Colchicine Cornigerine	Electrophysiological studies, Radio ligand binding assay	[32, 33]
	Leonurus japonicus	GABA modulator	Leonurine	Radio ligand binding assay	[34]
	Piper nigrum	GABA (A) receptor agonist	Piperine piperanine	Electrophysiological studies	[35]
	Annona purpurea	GABA modulator	Annomontine	Elevated plus maze test	[36]
	Aconitum leucostomum	non-competitive GABA(A) receptor antagonist	Songorine	Electrophysiological studies, Radio ligand binding assay	[37]
Alkanes					
	Sarcophytum auritum	GABA agonist	Ceramide	Elevated plus maze test & Light dark test	[38]
	Oenanthe fistulosa	GABA (A) receptor antagonist	Oenanthotoxine, Dihydroenanthotoxine	Electrophysiological study	[39]
	Oenanthe crocata	GABA (A) allosteric modulator	Falcarindiol	Electrophysiological study	[40]
	Cussonia zimmermannii	GABA (A) allosteric modulator	MS-1 MS-2 MS-4	Electrophysiological studies	[41]
Flavones					
	Scutellaria baicalensis	Allosterically blocks GABA-mediated receptor	Oroxylin A	Electrophysiological study, Radio ligand binding assay	[42]
	Scutellaria baicalensis	Acts at benzodiazepine site of GABAA receptor	Wogonin	Radio ligand binding assay, Elevated plus maze test and hole board test	[43]
	Scutellaria baicalensis	Partial agonist of central benzodiazepine receptor	Chrysin	Elevated plus maze and hole board test	[44]

DOI: http://dx.doi.org/10.5772/intechopen.99500

Class of Compound	Source	Mechanism	Active Componud	In-Vivo/ In-Vitro	Reference
Flavanes					
	Mentha aquatic	Acts at benzodiazepine site of GABA receptor	(S)-naringenin	Radio ligand binding assay	[45]
	Glycyrrhiza glabra	Acts at benzodiazepine site of GABA receptor	Glabrol	Radio ligand binding assay	[46]
Isoflavanoids and chalcones					
	Adenocarpus cinncinatus	GABA (A) receptor agonist	2′,4′,7-trihydroxy-8-(3-methylbut-2-en-1-yl) isoflavone	Electrophysiology study	[47]
	Angelica dahurica	GABA (A) receptor agonist	Isoliquiritigenin	Electrophysiology study, Radio ligand binding assay	[48]
Monoterpenes					
	Valeriana officinalis, Matricaria chamomilla, Lavandula officinalis	GABA (A) receptor agonist	(+)borneol, (−)-borneol, (−)-bornyl acetate, Isoborneol, camphor	Electrophysiology study	[49]
	Artemisia absinthium	GABA (A) receptor antagonist	α- thujone, β- thujone	Electrophysiology study; Radio ligand binding assays	[50]
	Thymus vulgaris	GABA (A) receptor agonist	Thymol	Electrophysiology study	[51]
Sesquiterpenes					
	Sideritis hyssopifolia	GABA (A) receptor agonist	α caryophyllene, β caryophyllene	Electrophysiology study	[52]
	Rhizoma curcumae oil	GABA (A) receptor agonist	Curdione, Curcumol	Electrophysiology study	[53]

Class of Compound	Source	Mechanism	Active Componud	In-Vivo/ In-Vitro	Reference
	Atractylodes macrocephala	GABA (A) receptor agonist	Atractylenolide I Atractylenolide II Atractylenolide III	Electrophysiology study	[47, 54]
	Illicium anisatum	non-competitive antagonist of $GABA_A$ receptor	Anisatin	Electrophysiology study	[55]
Diterpenes					
	Boswellia serrata	GABA (A) receptor agonist	Dehydroabietic acid	Electrophysiology study	[47]
	Curcuma kwangsiensis	GABA (A) receptor agonist	Zerumin A Coronarin D	Electrophysiology study	[56]
	Ginkgo biloba	GABA (A) receptor antagonist	Ginkgolides A Ginkgolides B Ginkgolides C	Electrophysiology study	[57, 58]
Triterpenes					
	Actaea racemosa	GABA (A) receptor agonist	Cimigenol-3-O-β-D-xylopyranoside 25-O-acetylcimigenol-3-O-α-L-arabinopyranoside	Electrophysiology study	[59]
	A. racemosa	GABA (A) receptor agonist	23-O-acetylshengmanol-3-O-β-D-xylopyranoside	Electrophysiology study, Elevated Plus maze test and open field test	[59, 60]

Table 1.
Natural compounds altering GABAergic transmission.

saturation, kinetic, and inhibition. A saturation curve can be made by considering amount of receptor as constant and concentration of radioligand as variable. From this type of experiment the receptor density and the affinity of the receptor for the radioligand can be estimated. If the amount of receptor and radioligand is constant and the time is the variable, then kinetic data which are obtained from forward and reverse rate constants can be assessed. If the amount of a competing nonradioactive drug included in the incubation is the only variable, then the affinity (Ki) of that drug for the receptor identified by the radioligand can be estimated [61].

In Xenopus oocytes assay, xenopus oocytes are the immature egg cells of the South African clawed frog *Xenopus laevis*, these have striking appearance with two colors, the light colored vegetal pole and the dark animal pole, where the nucleus is found. TEVC recording (This method is a type of patch-clamp electrophysiology method to inject current into a cell with one electrode and recording the change in voltage with the other electrode) is used to measure membrane potential of the oocyte is clamped at a constant value. Two electrodes spear the oocytes in which one intracellular microelectrode measures the membrane potential (voltage electrode) and the second one (intracellular microelectrode) controls the current. This is also called current electrode and uses as a feedback circuit to pass sufficient current to the oocyte for maintaining the voltage clamp. The current which is flowing through the current electrode can provide a measurement of the of chloride specific ion channels [62].

4.1 Alkaloids

Radioligand binding assay using [^{35}S] TBPS and [^{3}H] flunitrazepam analyzed the weak partial agonistic activity of Colchicine and (−) cornigerine along with six other colchicinoids from *Colchicum brachyphyllum*. These two molecules displayed 25% of the action of 10 μM allopregnanolone, but the (−) activity of colchicines was none. (−) colchicines acted as a GABA competitive antagonis [32, 33]. GABA modulation is also recorded in proto alkaloid leonurine, belonging to an East Asian herb called *Leonurus japonicus*, indicated for anxiety, depression, nervousness, and insomnia. The molecule showed half-maximal inhibitory concentration (IC_{50}) values of 15 μg/mL and 123 μg/mL, respectively in a radio ligand assay with [^{3}H] gabazine and [^{3}H] flumazenil [34].

Piperine and piperanine belonging to the class of piperidine-alkaloids were investigated in the immature egg cell of *Xenopus laevis*. The binding site of the molecule was confirmed to be a benzodiazepine binding site as Flumazenil (5 mg/kg) [35].

A β-carboline named annomontine also shows GABA dependant activity which was separated from the plant *Annona purpurea* [36]. In the EPM test, the compound increased the time spent in the open arms and the open arm entries at 10 and 30 mg/kg, but not the total arm entries. These effects are controlled via the benzodiazepine binding site as it was confirmed with antagonist Flumazenil at a dose of 3 mg/kg. Another alkaloid, a non-competitive GABA(A) receptor antagonist is diterpene alkaloid and this was separated from the plant *Aconitum leucostomum* having an IC_{50} 19.6 μM. Radioligand studies with the help of [^{3}H] muscimol produced an IC_{50} value of 7.06 μM and a Kd value of 6.31 nM. Specific binding site specificity was shown by β-carbolines and picroacridine alkaloids as they bind to the benzodiazepine binding site Benzodiazepine binding sites can at least be excluded for piperidine alkaloids and protoberberine alkaloids. This holds for songorine in GABA/muscimol binding site as well [37].

Three colchicinoids displayed unspecific binding with weak action on both benzodiazepine and TBPS/bicuculline binding sites. Colchicine is the antagonist, but androbiphenyline and cornigerine are partial agonists. Protoalkaloid Leonurine shows binding to various sites, with decreased affinities to, GABA/muscimol and the benzodiazepine binding site. Protoberberine type 2 alkaloids were able to modulate GABA(A) receptors, but unsaturated type 1 alkaloids displayed no effects.

4.2 Alkanes

The odor substance, 1-octen-3-ol is part of the $GABA_A$ sensory receptor modification research and has a stimulation rate of 295 ± 50 percent at a particular concentration of 300 μM and 1 μM GABA [52]. Ceramide (N-[(2S,3R,4E,6E)-1,3-dihydroxyhenicose-4,6-dien-2-yl] tridecanamide) separated from the Red Sea soft coral *Sarcophytum auritum*. This works as GABA agonist and produce anxiolytic effect in animal models evaluated by EPM test in which the animal spent more time in open arms and time spent in light in light/dark test whereas all its action can be altered by the action of bicuculline (GABA antagonist) [38, 52].

Two polyacetylenes extracted from *Oenanthe fistulosa*, Oenanthotoxine and dihydroenanthotoxine, provided major inhibitory activity on GABA receptors with IC50 values of 0.835 μM and 1.29 μM respectively on GABA (A) receptors The potential explanation for the indications which include water drop worm intoxication (Oenanthe crocata) and facial muscle contractions which is due to the inhibition of GABAergic reactions. The substance that allosterically regulate GABA-binding, non-competitively inhibits ion channel and also eradicates the desensitization of the receptor was constituted by modes of action of oenanthotoxin [39].

At a very low concentration the component falcarindiol obtained from *Oenanthe crocata* effectively regulates GABAergic currents. This component at low concentration (1 μM for falcarindiol and 0.1 μM for falcarinol) promotes the ion currents caused by GABA, on the other hand at higher concentration it inhibits the action of GABA. Moreover, the sedative but not convulsive result in animals is triggered by the low-dose effect, whereas the large doses in insect herbivores act as insecticides [40].

The three polyacetylenes MS-1, MS-2, and MS-4 were obtained from *Cussonia zimmermannii* with recorded GABA(A) modifying activity [41].

However, the potency and/or affinity were demonstrated in the small micro molar range, but that varies significantly in terms of toxicity. Two structural characteristics (groups of allyl and terminal hydroxyl) that are present in five (most) poisonous natural products produced toxicity. It suggests that the terminal hydroxyl class is vital for the toxicity. Further, both the oenanthotoxins and dihydrooenanthotoxins require the allyl hydroxyl group but are highly toxic. On the other hand falcarinol and falcarindiol, which have an allylic class but not the final hydroxy group, showed decreased toxicity. None of the two "toxic characteristics" are present in the last three polyacetylenes group and are also not documented to display inhibitory behavior consistent with this theory. It would be necessary to investigate whether hydrolyzation has led to GABA (A) receptor antagonism because MS-4 has a terminal acetyloxy-group [38, 52].

4.3 Flavones

The substance Oroxylin A, allosterically to block GABA-mediated receptor by its action on chloride currents, and thus it describes the results of a previous in vivo study in which the substance exhibited antagonistic diazepam-induced effects [42, 43].

Wogonin was considered for the induction of GABA-induced chloride currents by using electrophysiological methods where it shown a stimulation of 57% at a concentration of 30 μM in the presence of 1 μM GABA where at 3 μM half maximal stimulation was noticed. It was also tested pharmacologically at a dose of 7.5, 15 and

30 mg/kg by using Elevated plus maze and hole board test. The wogonin showed anxiolytic effects. These data recommend that wogonin yielded anxiolytic by positive allosteric modulation of the $GABA_A$ receptor complex through benzodiazepine site interaction [43].

The chrysin is from *Scutellaria baicalensis* class which was separated from *Passiflora caerulea* [44, 63]. Chrysin was testified as partial agonist of central benzodiazepine receptors which reduced anxiety and does not induced sedative and muscle relaxation. The pharmacological effect of chrysin was observed in mice at 1 mg/kg in Elevated plus maze test. The anxiolytic effect was observed by increasing the number of entries and time spent in the open arm. The horizontal wire test showed a decreased percentage of animals grasping the horizontal wire, while in the hole board test an increase in time spent head-dipping at 3 mg/kg was observed, but no sedative effects at doses of 3 and 6 mg/kg [44].

Flavone compounds like wogoninnn and chrysin shows diazepam like anxiolytic effect whereas Oroxylin A antagonizing the effects provoked by diazepam.

4.4 Flavanes

(S) naringenin was isolated from the ethanol extract of *leaves Mentha aquatica* and evaluated against [^{3}H] flumazenil which exhibits an IC50 value of 26 mM. This compound can effectively modulate GABA function [45].

Glabrol, is the prenylated flavanone its three Diels-Alder type derivatives, sanggenon C, D, and G and were obtained from the root extract of *Morusalba*. All three molecules, with EC50 values in the range from 13.4 to 16.7 μM, increased chloride-induced GABA by over 700 percent (100 μM) [46].

In particular, two 8-lavandulyl-flavanones produced GABA-induced chloride impulses to potentiate by about 600 percent compared to the third 8-lavandulyl-flavanonol which is substantially less active.

The compounds like (S) naringenin, glabrol and 8-lavandulyl-flavanones acts at benzodiazepine site of GABA receptor which was analyzed using radio ligand binding assay.

4.5 Isoflavanoids and chalcones

Adenocarpus cinncinatus considered as the source of 2′, 4′, 7-trihydroxy-8-(3-methylbut-2-en-1-yl) isoflavone. Its stimulatory effect exhibits a uplift of GABA-induced chloride currents [64]. At a concentration of 30 nM, the substance increased GABA-induced chloride currents by 135 percent with a maximum potentiation of 581 percent at a level of 100 μM.

Isoliquiritigenin increased GABA-induced currents by of 151% at a dose of 10 M with a patch-clamp method on dorsal raphe neurons [48].

The *Sophora flavescens lavandulyl chalcone* is Kuraridine, which potentiates GABA-induced chloride currents by 719.7 percent at a dose of 10 M with a maximal activation rate of 891.5 percent [56].

The findings for isoflavonoids and chalcones are consistent with the results of the last two sections: isoflavone genistein blocks chloride currents in the same way as its flavone equivalents apigenin. The binding of [^{3}H] flunitrazepam inhibits chalcone isoliquiritigenin, furthermore the prenylated types show a marked ability of more than 500 percent (95.97) to around 900 percent.

In these compounds the substitution of one hydroxy and one methoxy group in both aromatic rings shows better potency. Overall, all of these compounds shown GABA (A) receptor agonist type action.

4.6 Terpenes

4.6.1 Monoterpenes

(+) borneol, (−) borneol, (−) bornyl acetate, is borneol, and camphor acting on GABA(A) receptors which were stated in *oocytes of Xenopus laevis*. With the increased stimulation reported for (+)-borneol and (−)-borneol, all other substances resulted in a marked maximum potentiation of GABA-induced chloride currents. EC50 values were, however, in the large micromolar range with the smallest score reported for bornyl acetate (111.2 μM) [49, 65].

In a radioligand binding assay measured on α and β thujone against [^{3}H] EBOB, where the substances displayed IC50 values of 13 and 37 μM. The β-thujone was identified as a non-competitive antagonist with an IC50 value of 21 μM in additional electrophysiological studies. Studies have confirmed these molecules acts by allosteric decrease of GABA-induced chloride currents. α-thujone has been reported in a survey on GABAergic miniature inhibitory currents to decrease their frequency and amplitude and to moderately influence their kinetics. The study concluded that alpha-thujone had gating receptor activity as this substance decreased the amplitude of current reactions to exogenous GABA and influenced their initiation, desensitization, and neutralization [50]. Epoxy-carvone was studied using MES, PTZ, and picrotoxin-induced seizure models for its anticonvulsant properties [66].

In *Xenopus oocytes*, thymol an aromatic monoterpene is known from a variety of Thymus species, was examined on $\alpha_1 \beta_2 \gamma_3$ where chloride-induced GABA-currents increased by 416% at a concentration of 100 μM [51].

Isopulegol has been tested in-vivo for its anxiolytic ability. In the hole board test and Elevated plus maze [EPM] test at a concentration of 25 and 50 mg/kg, the isopulegol has been shown to raise the number of head dips in the hole board test which specifies anxiolytic effect in which the number of open arm entries along with the time spent in the arm was also increased in EPM test. In the EPM test of isopulegol results reduced the animal's aversion to the open arms as well as promoted the exploration which specifies anxiolytic effect [67].

In an anxiolytic-like behavioral study, the (+)-limonene epoxide at various doses of 25, 50, and 75 mg/kg showed an improvement in open arms inputs and time spent in open arms in the EPM test and decrease in the number of crossing, grooming, and rearing is found in the open field test, further implying the sedative effects of the drug [55]. The anxiolytic effect was reported by a follow-up study in which the compound demonstrated a decrease in the number of buried marbles in the buried marble test at a dose of 25, 50, and 75 mg/kg [68]. In several studies, Carvacryl acetate was also tested for anxiolytic and sedative effects. The EPM test shows that the compound increased the number of open arm entries at a dose of 100 mg/kg and the time spent in the open arm at doses from 25 to 100 mg/kg. In case of the light/dark test it increased the number and time spent in the light area at doses from 25 to 100 mg/kg. In the buried marbles test reduction of buried marbles number was observed at doses from 25 to 100 mg/kg, but no co-ordination impairment in the Rotarod test and no decrease in locomotor activity is observed in the open field test were measured at the same doses [69].

A few monoterpenes have been studied for their GABA receptor modulation action and the highest potential of chloride channel opening was observed for bicyclic alcohols, like (+) and (−)-borneol whereas isoborneol showed distinct potentiation. Oxidation of the hydroxy-group or the presence of an exocyclic methylene group causes decrease in the activity. The only monocyclic monoterpenes positive receptor modulation was observed by thymol.

4.6.2 Sesquiterpenes

Two monoterpenoid moieties namely α caryophyllene and β caryophyllene belonging to *Sideritis sp*. that displayed medium modulation of GABA-mediated chloride channels (117 and 115%, respectively).

Curdione and curcumol were extracted from the oil of *Curcumaerhizoma* and were tested on $GABA_A$ receptors expressed. The molecules increased GABA-mediated chloride channel activity with 133 and 175.7%, respectively at a concentration of 50 μM. The EC50 value of Curcumol was found to be 34.4 μM and the highest activity of 251% was found at 300 μM [53].

The highest induction of GABA-mediated chloride channel of around 400% was found in (+) cuparenol and (+)-dihydrocuparenic acid.

At 300 μM, when Atractylenoids I, II and III from *Atractylodesmacrocephala* was tested on GABA (A) receptors highest stimulation of 96 to 166% was observed with an EC50 value of 12, 70 and 99 μM, respectively [54, 70].

Anisatin is oxygenated sesquiterpene lactone separated from *Illicium anisatum* is a potent noncompetitive antagonist of $GABA_A$ receptor that has an activity similar to picrotoxin [71]. Studies demonstrated that anisatin at 1 μM decreased chloride currents created by 30 μM GABA to 41.7%. The IC50 value was measured with 1.10 μM along with an IC50 value of 0.42 μM for picrotoxinin, which is the active compound of picrotoxin was obtained. An indication that anisatin binds to the picrostatin site of the receptor was shown in a radioligand binding assay with an IC50 value of 0.43 μM against [^{3}H] EBOB. One very potent sesquiterpene is xenovulene A, which was separated from the fungus Acremonium striatum (now classified as Sarocladium striatum) [72].

As a result of the structural differences of the sesquiterpenes only restricted conclusions on their structure–activity relationship can be drawn. Reduction of the acidic function to an alcoholic function does not change the activity whereas the change of the isopropenyl-function of compound to a plane isopropanyl-moiety leads to a significant loss of activity.

4.6.3 Diterpenes

In this Section 14 diterpenes which are having the actions on GABA are discussed. Miltirone, a Salvia miltiorrhiza tanshinone, was assessed against [^{3}H] flunitrazepam with an IC50 value of 0.3 M in a radioligand-binding analysis [73].

Dehydroabietic acid has been segregated and examined in *Xenopus laevisocytes* from *Boswellia thurifera*, now known as *Boswellia serrata* [47]. GABA-induced chloride currents were enhanced by the substance by 397.5 percent at 100 μM and displayed an EC50 value of 8.7 μM. Isopimaric and sandaropimaric acid were extracted and examined from *Biota Orientalis*, currently known as Platycladusorientalis, in the Xenopus oocyte assay [74]. The substances showed a maximum stimulation effect of 425.2 and 855.7 percent of GABA-induced chloride currents at 500 μM and EC50 values of 141.6 and 33.2 μM, respectively.

Two diterpenes of phyllocladane namely 17-dihydroxyphyllocladane-3-one and 16,17,18-trihydroxyphyllocladane-3-one types were obtained from *Aloysia virgata* and assessed for GABA(A) affinity to [^{3}H] flumazenil with inhibitory constant [Ki] of 111 and 56 μM. Both compounds were studied in vivo, with compound, 17-dihydroxyphyllocladane-3-one which exhibits increased locomotor activity at a dose of 1 mg/kg in the locomotor activity test and increased rearing at 0.3 and 1 mg/kg in the hole board test. Compound 16,17,18-trihydroxyphyllocladane-3-one increased the number of head dips at 0.3 and 3 mg/kg, the number of rears at a dose of 1 mg/kg and the time spent head-dipping at a dose of 3 mg/kg. The compound at a dose of 1 mg/kg increased the number of open arm entries in the EPM test and the time spent in the light area as well as the number of transitions in the light/dark test [75].

Two diterpenes of type labdane, cerumin A and coronarin D, were obtained from Curcuma kwangsiensis [57]. In the Xenopus oocyte assay, substances at a 300 M concentration stimulated GABA-induced chloride currents by 309.4 and 211.0 percent, with EC50 values of 24.9 and 35.7 M. Ginkgolides A B and C which are diterpene trilactones of *Ginkgo biloba*, are moderately active $GABA_A$ receptor antagonists with Ki values of 14.5, 12.7 and 16.3 M in *Xenopus laevis oocytes* [58].

Some results suggest that compounds like 7-methoxyrosmanol and galdosol increases 10-fold receptor affinity by an oxo-group at 7^{nth} position instead of methoxy group. On the other hand, for compounds isopimaric acid and sandaropimaric acid, the change from the 7th to the 8th position of the double bond and thus to the C-ring of the substance doubles the maximum stimulatory effects and significantly decreases the EC50 value There are no clear variations in the inhibitory action of bilobalide and ginkgolide A-C in their IC50 values or in their ability to inhibit chloride current induced by GABA. Therefore, all these diterpenes works as GABA receptor agonists which help in chloride current flow.

4.6.4 Triterpenes

Asiatic acid was separated from *Centella Asiatica* and its anxiolytic effects were analyzed in the EPM test. The compound displayed no action on the open arm time but reduced the motile time and the highest speed at 30 mg/kg. These actions were blocked by flumazenil [76].

Ginsenoside C, is a glycoside isolated from *Panax ginseng* was tested on GABA(A) receptors expressed in xenopus laevis oocytes which were found to potentiate GABA-induced chloride currents with an EC_{50} value of 53.2 μM [77].

Four cycloartane glycosides actein, cimigenol-3-O-β-D-xylopyranoside 25-O-acetylcimigenol-3-O-α-L-arabinopyranoside, 23-O-acetylshengmanol-3-O-β D-xylopyranoside were extracted from *Actaea racemosa* root systems (black cohosh) and assessed in *Xenopus laevis* oocytes for their capacity for GABA-induced chloride currents [74]. Substances like actein, cimigenol-3-O-β-D-xylopyranoside and 25-O-acetylcimigenol-3-O-α-L-arabinopyranoside which are isolated constituents from rhizomes of Actaea rasemosa exhibited potentiation of GABA-induced chloride currents in the range of 256 to 378 percent at a concentration of 300 M, while 23-O-acetylshengmanol-3-O-D-xylopyranosides reported stimulation of 1947 percent and were also shown to generate small chloride currents due to lack of GABA. The EC50 values for the four glycosides were estimated from 26 to 36 μM. The pentose moiety cleavage led to a substantial decline in anxiety-related behavior (particularly for substance 23-O-acetylshengmanol-3-O-D-xylopyranosides. This compound was used in several in-vivo studies for the examination of its anxiolytic and sedative properties. It increased the number of open entries at 0.6 mg/kg in the EPM test whereas reduced stress-induced hyperthermia at doses of 0.2, 0.6, 2 and 6 mg/kg. In the open field test, this compound reduced the distance traveled at doses of 6, 20 and 60 mg/kg and also increased the time spent in the centre at a dose of 60 mg/kg, while the number of entries into the centre was reduced [60].

The discussion of the structure–activity of triterpenes is not influenced by the lack of comparable structures (scaffolds) compared to the last two subsections, but by the variety of test systems used for their analysis. However, it is possible to compare at least some of the known triterpenes from ginseng and black cohosh. Electrophysiological data showed lower EC50 levels for the three ginseng triterpenes ginsenoside C. Unfortunately, the maximum chloride current stimulation values were only observed for the two aglycones and were recorded to be 54.1 and 23.3 percent, respectively (at a concentration of 100 M). It can be concluded that the receptor modulation of the glycoside would be of significant concern after examining substance 23-O-acetylshengmanol-3-O-D-xylopyranosides, where the xylose moiety cleavage changed the potentiation of GABA-induced chloride currents from 1692 percent to 64 percent (100 M) and thus into the range of ginseng aglycon. Both compounds 23-O-acetylshengmanol-3-O-D-xylopyranosides and ginsenoside C disclose a four-ring structure with a side chain linked to ring D when contrasting their scaffolds. The prenylate and oxyprenylate side chains have enhanced activity, which is reminiscent of the structure-action-relationship of coumarins. The ginsenoside side-chain will stand for the prenyl moiety in the case of the triterpenes under consideration and that of substance23-O-acetylshengmanol-3-O-D-xylopyranosides for the more active epoxylated form. However, this molecule has additional characteristics that may contribute to its pronounced effect, such as keto-function at position 16 or acetyloxy-group at C-23, which both differentiate the compound from the other slightly less active cycloartanoids [59].

The neurosteroid binding site would be most obvious and consistent with the fact that neurosteroids are the most effective natural $GABA_A$ receptor modulators and, in the absence of GABA, are also capable of evoking chloride currents [78]. However, the hydroxy group at position 3 and the keto group at position 17 or 20 are considered to be important for neurosteroid binding activity. As far as the structure of compound is concerned, the keto group may well lead to the binding of the receptor in position 16 instead of position 17, but the fact that the role of the compound almost vanishes with the xylose moiety does not support this theory unless the binding of the neurosteroid site can be improved by the residue of sugar instead of the hydroxy group in position 3. Barbiturates, on the other hand, are also known to activate GABA(A) receptors directly at higher concentrations and the site of barbiturate binding is thought to be similar to that of neurosteroids [79].

5. Conclusion

Natural products with GABA receptors activity were identified in the literatures and discussed in this chapter. Depending on the number of related compounds and test systems used, it was possible to draw in the vicinity of conclusions regarding their structure–activity relationships. As most of the studies examined flavones, and these studies mainly applied radio ligand binding assays, substitution patterns responsible for increased receptor affinity could be associated with one flavone even with diazepam-like Ki values. As far as receptor regulation is concerned, flavones are either non-competitive antagonists or partial agonists. However, certain compounds also exhibited anxiolytic or anticonvulsant effects. Other phenolic compounds addressed in this study were, for example, coumarins, where prenylated compounds demonstrated higher stimulation of the receptor. The association of phenyl residues and pronounced receptor modulation has also been observed for flavanes, isoflavonoids, and chalcones and may be of interest to the production of GABA(A) receptor modulators. Besides, the structural features required for the positive or negative regulation of the polyacetylene and monoterpene receptors as well as the effect of deglycosylation on certain triterpenes have been highlighted. Very few studies have been found on the subtype-specificity of natural products. One example is the enhanced modulation of isopimar and sandaropimaric acid receptors after the exchange of α1-subunit for α2 or α3-subunits. Neolignane honokiol must also be stated in this sense, although the effect was more dependent on the GABA(A) receptor subunits. Data obtained from recorded in vivo studies may be helpful in this regard, as many compounds have been known to exhibit anxiolytic effects without exhibiting sedative or muscle relaxant properties.

Author details

Sayani Banerjee, Chennu Manisha, Deepthi Murugan and Antony Justin*
Department of Pharmacology, JSS College of Pharmacy, JSS Academy of Higher Education and Research, Ooty, Nilgiris, Tamilnadu, India

*Address all correspondence to: justin@jssuni.edu.in

References

[1] Jones EG. Gabaergic neurons and their role in cortical plasticity in primates. Cereb Cortex. 1993; 361 – 372.

[2] Wu C, Sun D. GABA receptors in brain development, function, and injury. Metab Brain Dis. 2015; 30(2): 367-79.

[3] Ozoe Y. GABAA Receptor Channels. J Pestic Sci. 1996; 21(2): 217-22.

[4] Blednov YA, Benavidez JM, Black M, Leiter CR, Osterndorff-Kahanek E, Johnson D, et al. GABAA receptors containing p1 subunits contribute to in vivo effects of ethanol in mice. PLoS One. 2014; 9(1): e85525.

[5] Boue-Grabot E, Roudbaraki M, Bascles L, Tramu G, Bloch B, Garret M. Expression of GABA receptor ρ subunits in rat brain. J Neurochem. 1998; 70(3): 899-907.

[6] Luján R, Shigemoto R, López-Bendito G. Glutamate and GABA receptor signalling in the developing brain. Neuroscience. 2005; 130(3): 567-80.

[7] Johnston G a R, Johnston G a R. Receptor Channel Pharmacology. Receptor. 2006; 1867-85.

[8] Rudolph U, Möhler H. GABA-based therapeutic approaches: GABAA receptor subtype functions. Curr Opin Pharmacol. 2006; 6: 18-23.

[9] Nilsson J, Sterner O. Modulation of GABAA Receptors by Natural Products and the Development of Novel Synthetic Ligands for the Benzodiazepine Binding Site. Curr Drug Targets. 2011; 12(11): 1674-88.

[10] Erlander MG, Tillakaratne NJK, Feldblum S, Patel N, Tobin AJ. Two genes encode distinct glutamate decarboxylases. Neuron. 1991; 7(1): 91-100.

[11] Keynan S, Kanner BI. γ-Aminobutyric Acid Transport in Reconstituted Preparations from Rat Brain: Coupled Sodium and Chloride Fluxes. Biochemistry. 1988; 27(1): 12-7.

[12] Hill DR, Bowery NG. 3H-baclofen and 3H-GABA bind to bicuculline-insensitive GABAB sites in rat brain. Nature. 1981; 290: 149-52.

[13] Marksitzer R, Benke D, Fritschy JM, Trzeciak A, Bannwarth W, Mohler H. GABAa-receptors: Drug binding profile and distribution of receptors containing the α2-subunit in situ. J Recept Signal Transduct. 1993; 13: 467-77.

[14] Mondadori C, Jaekel J, Preiswerk G. CGP 36742: The first orally active GABAB blocker improves the cognitive performance of mice, rats, and rhesus monkeys. Behav Neural Biol. 1993; 60(1): 62-8.

[15] Herb A, Wisden W, Lüddens H, Puia G, Vicini S, Seeburg PH. The third γ subunit of the γ-aminobutyric acid type a receptor family. Proc Natl Acad Sci U S A. 1992; 89(4): 1433-7.

[16] Cunningham MD, Enna SJ. Evidence for pharmacologically distinct GABAB receptors associated with cAMP production in rat brain. Brain Res. 1996; 720: 220-4.

[17] Hashimoto T, Kuriyama K. In vivo evidence that GABA(B) receptors are negatively coupled to adenylate cyclase in rat striatum. J Neurochem. 1997; 69(1): 365-70.

[18] Scott RH. Modulation of Neuronal T-Type Calcium Channel Currents. 1990; 38(2): 285-94.

[19] Bormann J, Feigenspan A. GABAc receptors. Trends Neurosci. 1995; 18(12): 515-9.

[20] Huang ZJ, Di Cristo G, Ango F. Development of GABA innervation in the cerebral and cerebellar cortices. Nat Rev Neurosci. 2007; 8(9): 673-86.

[21] Jie F, Yin G, Yang W, Yang M, Gao S, Lv J, et al. Stress in regulation of GABA amygdala system and relevance to neuropsychiatric diseases. Front Neurosci. 2018;12.

[22] Henderson C, Wijetunge L, Healy M, Therapeutics S, Patrick T, Centre W, et al. Reversal of disease phenotypes in the Fragile X mouse model by selective activation of GABA-B receptors. 2012; 4(152): 1-42.

[23] Heaney CF, Kinney JW. Role of GABAB receptors in learning and memory and neurological disorders. Neurosci Biobehav Rev. 2016; 63: 1-28.

[24] Pierri JN, Chaudry AS, Woo TUW, Lewis DA. Alterations in chandelier neuron axon terminals in the prefrontal cortex of schizophrenic subjects. Am J Psychiatry. 1999; 156(11): 1709-19.

[25] Mizukami K, Ishikawa M, Hidaka S, Iwakiri M, Sasaki M, Iritani S. Immunohistochemical localization of GABAB receptor in the entorhinal cortex and inferior temporal cortex of schizophrenic brain. Prog Neuro-Psychopharmacology Biol Psychiatry. 2002; 26(2): 393-6.

[26] Baldwin DS, Anderson IM, Nutt DJ, Allgulander C, Bandelow B, Den Boer JA, et al. Evidence-based pharmacological treatment of anxiety disorders, post-traumatic stress disorder and obsessive-compulsive disorder: A revision of the 2005 guidelines from the British Association for Psychopharmacology. J Psychopharmacol. 2014; 28(5): 403-39.

[27] Thoeringer CK, Ripke S, Unschuld PG, Lucae S, Ising M, Bettecken T, et al. The GABA transporter 1 (SLC6A1): A novel candidate gene for anxiety disorders. J Neural Transm. 2009; 116(6): 649-57.

[28] Haubensak W, Kunwar PS, Cai H, Ciocchi S, Wall NR, Ponnusamy R, et al. Genetic dissection of an amygdala microcircuit that gates conditioned fear. Nature. 2010; 468: 270-6.

[29] Hyder F, Petroff OAC, Mattson RH, Rothman DL. Localized 1H NMR measurements of 2-pyrrolidinone in human brain in vivo. Magn Reson Med. 1999; 41(5): 889-96.

[30] Petroff OAC, Rothman DL, Behar KL, Mattson RH. Low brain GABA level is associated with poor seizure control. Ann Neurol. 1996; 40(6): 908-11.

[31] Petroff OAC, Pleban LA, Spencer DD. Symbiosis between in vivo and in vitro NMR spectroscopy: The creatine, N-acetylaspartate, glutamate, and GABA content of the epileptic human brain. Magn Reson Imaging. 1995; 13(8): 1197-211.

[32] Bueno OF, Leidenheimer NJ. Colchicine inhibits GABA(A) receptors independently of microtubule depolymerization. Neuropharmacology. 1998; 37(3): 383-90.

[33] Alali FQ, El-Elimat T, Li C, Qandil A, Alkofahi A, Tawaha K, et al. New colchicinoids from a native Jordanian meadow saffron, Colchicum brachyphyllum: Isolation of the first naturally occurring dextrorotatory colchicinoid. J Nat Prod. 2005; 68(2): 173-8.

[34] Rauwald HW, Savtschenko A, Merten A, Rusch C, Appel K, Kuchta K. GABAA Receptor Binding Assays of Standardized *Leonurus cardiaca* and *Leonurus japonicus* Extracts as Well as Their Isolated Constituents. Planta Med. 2015; 81: 1103-10.

[35] Zaugg J, Baburin I, Strommer B, Kim HJ, Hering S, Hamburger M.

HPLC-based activity profiling: discovery of piperine as a positive GABAA receptor modulator targeting a benzodiazepine-independent binding site. J Nat Prod. 2010; 73(2): 185-91.

[36] Del Carmen Rejón-Orantes J, González-Esquinca AR, De La Mora MP, Roldan Roldan G, Cortes D. Annomontine, an alkaloid isolated from *annona purpurea*, has anxiolytic-like effects in the elevated plus-maze. Planta Med. 2011; 77(4): 322-7.

[37] Zhao XY, Wang Y, Li Y, Chen XQ, Yang HH, Yue JM, et al. Songorine, a diterpenoid alkaloid of the genus Aconitum, is a novel GABAA receptor antagonist in rat brain. Neurosci Lett. 2003; 337(1): 33-6.

[38] Eltahawy NA, Ibrahim AK, Radwan MM, Zaitone SA, Gomaa M, Elsohly MA, et al. Mechanism of action of antiepileptic ceramide from Red Sea soft coral Sarcophyton auritum. Bioorganic Med Chem Lett. 2015; 25(24): 5819-24.

[39] Appendino G, Pollastro F, Verotta L, Ballero M, Romano A, Wyrembek P, et al. Polyacetylenes from Sardinian *Oenanthe fistulosa*: A molecular clue to risus sardonicus. J Nat Prod. 2009; 72(5): 962-5.

[40] Wyrembek P, Negri R, Kaczor P, Czyżewska M, Appendino G, Mozrzymas JW. Falcarindiol allosterically modulates gabaergic currents in cultured rat hippocampal neurons. J Nat Prod. 2012; 75(4): 610-6.

[41] Baur R, Simmen U, Senn M, Séquin U, Sigel E. Novel plant substances acting as β subunit isoform-selective positive allosteric modulators of GABAA receptors. Mol Pharmacol. 2005; 68(3): 787-92.

[42] Kim DH, Jeon SJ, Son KH, Jung JW, Lee S, Yoon BH, et al. The ameliorating effect of oroxylin A on scopolamine-induced memory impairment in mice. Neurobiol Learn Mem. 2007; 87(4): 536-46.

[43] Wang H, Hui KM, Chen Y, Xu S, Wong JTF, Xue H. Structure-activity relationships of flavonoids, isolated from Scutellaria baicalensis, binding to benzodiazepine site of GABAA receptor complex. Planta Med. 2002; 68(12): 1059-62.

[44] Wolfman C, Viola H, Paladini A, Dajas F, Medina JH. Possible anxiolytic effects of chrysin, a central benzodiazepine receptor ligand isolated from Passiflora Coerulea. Pharmacol Biochem Behav. 1994; 47(1): 1-4.

[45] Jäger AK, Almqvist JP, Vangsøe SAK, Stafford GI, Adsersen A, Van Staden J. Compounds from *Mentha aquatica* with affinity to the GABA-benzodiazepine receptor. South African J Bot. 2007; 73(4): 518-21.

[46] Cho S, Park JH, Pae AN, Han D, Kim D, Cho NC, et al. Hypnotic effects and GABAergic mechanism of licorice (*Glycyrrhiza glabra*) ethanol extract and its major flavonoid constituent glabrol. Bioorganic Med Chem. 2012; 20(11): 3493-501.

[47] Rueda DC, Raith M, De Mieri M, Schöffmann A, Hering S, Hamburger M. Identification of dehydroabietc acid from Boswellia thurifera resin as a positive GABAA receptor modulator. Fitoterapia. 2014; 99(1): 28-34.

[48] Cho S, Kim S, Jin Z, Yang H, Han D, Baek NI, et al. Isoliquiritigenin, a chalcone compound, is a positive allosteric modulator of GABA A receptors and shows hypnotic effects. Biochem Biophys Res Commun. 2011; 413(4): 637-42.

[49] Granger RE, Campbell EL, Johnston GAR. (+)- And (−)-borneol: Efficacious positive modulators of GABA action at human recombinant

α1β2γ2L GABAA receptors. Biochem Pharmacol. 2005; 69(7): 1101-11.

[50] Höld KM, Sirisoma NS, Ikeda T, Narahashi T, Casida JE. α-Thujone (the active component of absinthe): γ-aminobutyric acid type A receptor modulation and metabolic detoxification. Proc Natl Acad Sci U S A. 2000; 97(8): 3826-31.

[51] Priestley CM, Williamson EM, Wafford KA, Sattelle DB. Thymol, a constituent of thyme essential oil, is a positive allosteric modulator of human GABA A receptors and a homo-oligomeric GABA receptor from *Drosophila melanogaster*. Br J Pharmacol. 2003; 140(8): 1363-72.

[52] Kessler A, Sahin-Nadeem H, Lummis SCR, Weigel I, Pischetsrieder M, Buettner A, et al. GABAA receptor modulation by terpenoids from Sideritis extracts. Mol Nutr Food Res. 2014; 58(4): 851-62.

[53] Ding J, Wang JJ, Huang C, Wang L, Deng S, Xu T Le, et al. Curcumol from Rhizoma Curcumae suppresses epileptic seizure by facilitation of GABA(A) receptors. Neuropharmacology. 2014; 81: 244-55.

[54] Singhuber J, Baburin I, Kählig H, Urban E, Kopp B, Hering S. GABA A receptor modulators from Chinese herbal medicines traditionally applied against insomnia and anxiety. Phytomedicine. 2012; 19: 334-40.

[55] De Almeida AAC, Costa JP, De Carvalho RBF, De Sousa DP, De Freitas RM. Evaluation of acute toxicity of a natural compound (+)-limonene epoxide and its anxiolytic-like action. Brain Res. 2012; 1448: 56-62.

[56] Schramm A, Ebrahimi SN, Raith M, Zaugg J, Rueda DC, Hering S, et al. Phytochemical profiling of Curcuma kwangsiensis rhizome extract, and identification of labdane diterpenoids as positive GABAA receptor modulators. Phytochemistry. 2013; 96: 318-29.

[57] Huang SH, Duke RK, Chebib M, Sasaki K, Wada K, Johnston GAR. Ginkgolides, diterpene trilactones of *Ginkgo biloba*, as antagonists at recombinant α1β2γ2L GABA A receptors. Eur J Pharmacol. 2004; 494(2-3): 131-8.

[58] Ivic L, Sands TTJ, Fishkin N, Nakanishi K, Kriegstein AR, Strømgaard K. Terpene trilactones from *Ginkgo biloba* are antagonists of cortical glycine and GABA(A) receptors. J Biol Chem. 2003; 278(49): 49279-85.

[59] Cicek SS, Khom S, Taferner B, Hering S, Stuppner H. Bioactivity-guided isolation of GABAa receptor modulating constituents from the rhizomes of *Actaea racemosa*. J Nat Prod. 2010; 73(12): 2024-8.

[60] Strommer B, Khom S, Kastenberger I, Cicek SS, Stuppner H, Schwarzer C, et al. A cycloartane glycoside derived from *actaea racemosa* L. Modulates GABAAreceptors and induces pronounced sedation in mice. J Pharmacol Exp Ther. 2014; 351(2): 234-42.

[61] Toews L, David B. Radioligand binding methods: practical guide and tips. Am Physiol Soc. 2018; 421-9.

[62] Kvist T, Hansen KB, Bräuner-Osborne H. The use of Xenopus oocytes in drug screening. Expert Opin Drug Discov. 2011; 6(2): 141-53.

[63] Medina JH, Paladini AC, Wolfman C, de Stein ML, Calvo D, Diaz LE, et al. Chrysin (5,7-di-OH-flavone), a naturally-occurring ligand for benzodiazepine receptors, with anticonvulsant properties. Biochem Pharmacol. 1990; 40(10): 2227-31.

[64] Çiçek SS. Structure-Dependent Activity of Natural GABA(A) Receptor Modulators. Molecules. 2018; 23(7).

[65] Quintans-Júnior L, Guimarães A, Araújol B, Oliveira G, Santana M, Moreira F, et al. African journal of biotechnology. African J Biotechnol. 2002; 9(39): 6566-72.

[66] Czyzewska MM, Mozrzymas JW. Monoterpene α-thujone exerts a differential inhibitory action on GABAA receptors implicated in phasic and tonic GABAergic inhibition. Eur J Pharmacol. 2013; 702: 38-43.

[67] Silva MIG, de Aquino Neto MR, Teixeira Neto PF, Moura BA, do Amaral JF, de Sousa DP, et al. Central nervous system activity of acute administration of isopulegol in mice. Pharmacol Biochem Behav. 2007; 88(2): 141-7.

[68] De Almeida AAC, De Carvalho RBF, Silva OA, De Sousa DP, De Freitas RM. Potential antioxidant and anxiolytic effects of (+)-limonene epoxide in mice after marble-burying test. Pharmacol Biochem Behav. 2014; 118: 69-78.

[69] Pires LF, Costa LM, Silva OA, De Almeida AAC, Cerqueira GS, De Sousa DP, et al. Anxiolytic-like effects of carvacryl acetate, a derivative of carvacrol, in mice. Pharmacol Biochem Behav. 2013; 112: 42-8.

[70] Rueda DC. Ethnomedicine-based discovery and characterization of plant-derived GABA A receptor modulators with new scaffolds. 2014.

[71] Ikeda T, Ozoe Y, Okuyama E, Nagata K, Honda H, Shono T, et al. Anisatin modulation of the γ-aminobutyric acid receptor-channel in rat dorsal root ganglion neurons. Br J Pharmacol. 1999; 127(7): 1567-76.

[72] Andäng M, Hjerling-Leffler J, Moliner A, Lundgren TK, Castelo-Branco G, Nanou E, et al. Histone H2AX-dependent GABAA receptor regulation of stem cell proliferation. Nature. 2008; 451(7177): 460-4.

[73] Lee CM, Wong HNC, Chui KY, Choang TF, Hon PM, Chang HM. Miltirone, a central benzodiazepine receptor partial agonist from a Chinese medicinal herb Salvia Miltiorrhiza. Neurosci Lett. 1991; 127(2): 237-41.

[74] Zaugg J, Khom S, Eigenmann D, Baburin I, Hamburger M, Hering S. Diterpenes from *Biota orientalis* That Decrease Locomotor Activity in Mice. J Nat Prod. 2011; 74: 1764-1772.

[75] Wasowski C, Marder M. Central nervous system activities of two diterpenes isolated from Aloysia virgata. Phytomedicine. 2011; 18(5): 393-401.

[76] Ceremuga TE, Valdivieso D, Kenner C, Lucia A, Lathrop K, Stailey O, et al. Evaluation of the anxiolytic and antidepressant effects of asiatic acid, a compound from Gotu kola or *Centella asiatica*, in the male Sprague Dawley rat. AANA J. 2015; 83(2): 91-98.

[77] Choi SE, Choi S, Lee JH, Whiting PJ, Lee SM, Nah SY. Effects of ginsenosides on GABAA receptor channels expressed in Xenopus oocytes. Arch Pharm Res. 2003; 26(1): 28-33.

[78] Hosie AM, Wilkins ME, Smart TG. Neurosteroid binding sites on GABAA receptors. Pharmacol Ther. 2007; 116(1): 7-19.

[79] Bowery NG, Smart TG. GABA and glycine as neurotransmitters: A brief history. Br J Pharmacol. 2006; 147: 109-19.

Chapter 2

The Ghanaian Flora as a Potential Source of Anthelmintic and Anti-Schistosomal Agents

Evelyn Asante-Kwatia, Abraham Yeboah Mensah, Lord Gyimah and Arnold Donkor Forkuo

Abstract

Parasitic infections including schistosomiasis and soil transmitted helminthiasis are the most commonly encountered Neglected Tropical Diseases (NTDs) in the world. These diseases remain a major public health concern affecting millions of people especially those living in poor regions where access to effective conventional health care is a challenge. Interventions to control these infections in endemic areas have not been successful due to the high cost of drugs, limited availability as well as inequity of access to preventive chemotherapies. Another problem is the development resistance to the limited number of recommended medications due to their intensive use in both human and live-stock. There is an increasing awareness of the potential of natural products as chemotherapeutic agents to combat parasitic infections. Natural products may offer an unlimited source of chemically diverse drug molecules which may be safe, efficient, less toxic, less expensive and readily available for use especially in low-income countries. The Ghanaian flora provides such a ready source for new therapeutic interventions for the local population. Several researches have provided evidence of the anti-parasitic activity of Ghanaian medicinal plants. This chapter provides a review with special focus on medicinal plants collected from Ghana with anthelmintic and anti-schistosomal activity. Evidence of pharmacological activities of crude extracts, fractions and bioactive phytoconstituents as well as possible mechanisms of action where investigated are discussed.

Keywords: schistosomiasis, worms, parasite, helminth, Ghana, herbal medicine

1. Introduction

Neglected tropical diseases (NTDs) include a collection of chronic, disabling, and physically disfiguring infectious diseases that usually affect dwellers of poor rural populations in tropical and sub-tropical countries of the world [1]. Apart from their negative impact on the health of victims, NTDs exert an immense socio-economic burden on the society as a result of the social stigma and physical disabilities associated with them. These interrelated negative outcomes perpetuate a cycle of poverty and unproductivity resulting in a consistent decline in economic growth [2]. As a major element of the Millennium Development Goals (MDGs), much effort is being put in for the elimination of the NTDs [3].

Among the NTDs, helminth infections especially soil-transmitted helminthiasis (STHs) and schistosomiasis are among the most prevalent afflictions of humans [4]. About 2 billion people are estimated to suffer from helminth infections worldwide, out of whom 300 million suffer from severe morbidity [5]. The negative impact of helminth infections on human growth and development (including cognitive development in childhood and nutritional status), pregnancy and work performance cannot be overemphasized. Though considered as acute health problems in some developed parts of the world, chronic parasitic infections are common and recurrent in poor communities and usually result in long-lasting complications making them a significant health threat to the populations who are continuously at risk for infection [6].

Over the years, many highly effective chemotherapeutic agents have been developed for treating helminth infections. Unfortunately in the setting of rural poverty where these diseases are mostly prevalent, access to healthcare facilities and the cost of medications are a challenge [7, 8]. Additionally, environmental factors and unavoidable domestic or occupational exposures, strongly favor the process of re-infection even after a successful therapy [9, 10]. Given that these infections also require lengthy treatment regimens with related costs which cannot be afforded by the affected victims, many patients seek for alternative treatment options especially the use of herbal medicines which are readily available and less expensive [9, 11].

Herbal extracts have been used in traditional medicines since ancient times for the effective treatment of human diseases [12]. Ethnobotanical studies in various regions of the world have documented medicinal plants used for the treatment of various parasitic infections. Scientific investigations of selected plants have also revealed remarkable activity of medicinal plants against specific human parasites [13, 14]. In Ghana, numerous medicinal plants play an important role in the healthcare system of rural communities. The Ghanaian flora provides a ready source for new therapeutic interventions for the local population [15–17]. This chapter provides a review with special focus on medicinal plants collected from Ghana with anthelmintic and anti-schistosomal activity.

1.1 Soil transmitted helminthiasis (STH)-the disease burden and current chemotherapy

Soil transmitted helminth (STH) infections are a group of infections which are acquired by the ingestion of, or contact with, soil containing infectious worm eggs or larvae [18]. STHs have been reported as the most common parasitic infections encountered in humans with an estimation of more than 1 billion people infected with at least one or more helminth parasites. They constitute an important global health challenge in resource deprived parts of the world and are prevalent in areas of poor sanitary conditions [19].

The main species of clinical importance are the intestinal roundworm (*Ascaris lumbricoides*), the whipworm (*Trichuris trichiura*) and the hookworms (*Necator americanus* and *Ancylostoma duodenale*) [18]. Common symptoms of intestinal helminthiasis include abdominal pains, nausea, itching and diarrhea and in severe cases, anemia, pneumonia, eosinophilia and malnutrition. School-aged children and preschool children are the most vulnerable group who harbor the greatest numbers of intestinal worms. As a result, they experience growth stunting and diminished physical fitness as well as impaired memory and cognition [20]. Although helminth infections are not known to be lethal as compared to other infections, they are recurrent among poor people and pose an enormous impact on the socio-economic status of the society affected [21].

Anthelmintics are a group of antiparasitic drugs that expel worms and other internal parasites out of the body by either stunting or killing them. For the treatment of STHs, the benzimidazoles specifically albendazole and mebendazole are the current treatment drugs of choice [19]. The main challenge with these anthelminthics is the development of resistance due to the intensive use of drugs in both human and live-stock [22]. With few new drugs evolving against helminth infections over the years, the fight against these parasites could become a losing battle, thus the need to search for new alternatives.

1.2 Schistosomiasis—the disease burden and current chemotherapy

Schistosomiasis, widely known as bilharzia, is caused by infection with blood flukes of the genus *Schistosoma* which is transmitted through contact with infected fresh-water snail vectors. Schistosomiasis is reported to be the 2nd leading endemic parasitic disease in the world after malaria. The disease affects more than 240 million people in tropical and subtropical areas with about 90% cases reported from sub Saharan Africa [23, 24].

Five species of the schistosome parasite namely: *Schistosoma mansoni*, *Schistosoma haematobium*, *Schisosoma japonicum*, *Schistosoma mekongi*, and *Schistosoma intercalatum* usually affect humans [25]. In sub-Saharan Africa the main burden of disease is usually attributed to *S. mansoni* and *S. haematobium* which cause intestinal and urinary schistosomiasis respectively [10]. The infection is mainly characterized by painful bloody urination in urinary schistosomiasis or blood stained diarrhea in intestinal schistosomiasis. Long term effects include liver fibrosis, renal failure, cancer of the bladder, infertility and increased risk of contracting HIV. In children, schistosomiasis results in malnutrition, growth retardation, cognitive defects and chronic anemia [6, 26].

For the eradication of schistosomiasis, control programmes have been based on preventive chemotherapy. The WHO endorsed and advocated for mass drug administration (MDA) especially among school children utilizing a single oral dose of 40 mg/kg praziquantel [27]. Unfortunately, the unavailability of the drugs due to cost, poor drug coverage, inequity of access to chemotherapy and non-compliance to therapy due to adverse side effects have impeded the progress of this approach [7, 28]. The expansion of preventive chemotherapy has also raised concerns about the potential development of resistance to praziquantel (PZQ) which remains the only commercially readily available drug for the control of schistosomiasis [29]. Some studies have reported low cure rates of PZQ attributing this to possible mutation of the schistosome parasite as well as inactivity of PZQ against early stages of the worms [30, 31]. It is thus not a satisfactory situation to have only one single effective treatment. Ideally, other anti-schistosome drugs should be developed so that the classical strategy of avoiding development of resistance could be followed.

1.3 Methods used in this review for identifying medicinal plants with anthelminthic and anti-schistosomal activities

Reported anthelmintic and anti-schistosomal activities of medicinal plants collected from various parts of Ghana were obtained from electronic databases including PubMed, SciFinder and Google Scholar. The inclusion criteria were that: (i) plants should be used in Ghanaian traditional medicine for treatment of worm infestations or expulsion of worms and schistosomiasis (urinary and intestinal) or other condition characterized by the symptoms of the above diseases (ii) plant should have been investigated for anthelmintic or anti-schistosomal (cercarididal) activity using one or more validated *in vitro* or *in vivo* models (iii) the right botanical names, plant parts used, types of extracts prepared, active constituents and mechanisms of action if investigated were mentioned. Consideration was also given to plants with significant activity differences with reference to control groups.

2. Plants with anthelminthic activity identified from Ghana

The anthelmintic activity of plant extracts was mostly studied by evaluating their effect on worms after direct exposure for a period of time. Earthworms including *Pheretima posthuma*, *Lumbricus terrestris*, *Eudrilus eugeniae* and *Caenorhabditis elegans* were employed as target organisms due to their anatomical and physiological similarity to the human intestinal round worm, ease of availability, adaptability to laboratory conditions and ease of handling.

2.1 *Alcornea cordifolia* (Schumach & Thonn) Müll. Arg. Euphorbiaceae

Alchornea cordifolia, commonly called the Christmas bush is a straggling, laxly branched evergreen dioecious shrub growing up to about 8 m tall. It is locally known as '*agyama*' in the Ghanaian Akan language and an essential medicinal plant in traditional medicine. Various parts of the plant are used to treat jaundice, diarrhea, rheumatic pains, malaria, fever, wounds, colds, asthma, amoebic dysentery and worm infections. Other literatures report its use in the treatment of urinary and gastrointestinal infections, leprosy, yaws, filariasis as an antidote to snake venom [32].

The anthelminthic potency of the petroleum ether, chloroform and methanol extracts of *A. cordifolia* leaves were investigated by evaluating its effect on the gross motility and mortality of earthworms (*Pheretima posthuma*). The extracts displayed significant ($p < 0.001$) concentration-dependent anthelminthic activity at concentration range of 0.75 to 12.00 mg/mL. At the highest concentration, worm paralysis was effectuated between 10 and 26 mins whiles death occurred between 57 to 93 min. The effect of the extracts in reducing the paralysis and death times of the worms was significantly higher than the effect on albendazole-treated worms [33].

2.2 *Alstonei boonei* De Wild (Apocynaceae)

Alstonia boonei is an indigenous African tree mostly distributed in the evergreen rain forest of tropical West Africa. In Ghana it is locally called '*Nyame dua*' meaning God's tree in the Akan language. In the western coastal regions of Africa, this plant is well known for its extensive use in traditional medicine for treating rheumatism, general body pains, worm infestation and diabetes. A cold infusion of the fresh or dried bark is used as a vermifuge to expel intestinal worms and other intestinal parasites in children [34].

The methanol extracts (50–150 mg/mL) of the stem bark and roots of *A. boonei* were investigated *in vitro* for anthelmintic effects against the adult Indian earthworm, *Pheretima posthuma* by direct exposure of worms to the extracts. The stem bark extract exhibited a concentration dependent anthelmintic activity causing paralysis of worms within 15–55 mins and death within approximately 100 mins which was significant ($p < 0.01$) compared to the untreated group. The stem bark extract had a better anthelmintic effect than the root bark [35].

In another study, the aqueous and ethanolic stem bark extracts (50–200 mg/mL) of *A. boonei* demonstrated significant anthelmintic activity against *Lubricus terretris*.

While worms in the untreated group saw no paralysis or death after 120 mins of exposure, the extract-treated worms were paralyzed within 8–16 minutes of exposure and died within approximately 21–27 minutes of exposure [36].

2.3 *Azadirachta indica* A. Juss. (Meliaceae)

Azadirachta indica, commonly known as neem, is a fast-growing and long lived evergreen tree which grows up to about 15 m tall with long, spreading branches that form a dense, large rounded crown. The plant is a multipurpose medicinal plant which also provides food and timber and is widely distributed in several regions of Asia and Africa. It is well known for its insecticidal and insect-repelling property. Various parts of the plants are reported to be used for the treatment of many ailments in traditional medicine including malaria, fever, upper respiratory tract infections, wound healing, sexually transmitted infections and skin diseases [37].

The anthelmintic activity of the ethanolic extract of *A. indica* seeds was investigated *in vivo* using albino rats (*Rattus norvegicus*) infected with helminth species including: *Hymenolepsis diminuta*, *Enterbius vermicularis* and hookworm. The rats were treated with the alcoholic extracts (20–60%) over a 3-week period and fecal samples were examined for eggs. The extract treated groups showed declining levels of egg count by the 3rd week and complete elimination of worms by the end of 21 days when treated with 40–60% of neem seed extract. Weight loss and death were however recorded at 60% concentration of extract raising some concern about the toxicity of the seed extract [38].

2.4 *Carica papaya* Linn. (Caricaceae)

The pawpaw tree is well known for its nutritional and medicinal values. The leaf decoction is used as a galactogogue and in the treatment of tonsillitis, ulcerative stomatitis, hemorrhoids, asthma, urinary tract infections, as poultice for sores and gingivitis and in the treatment of helminth infections. The roots are used as antidote to various poisons. The fruits are used to treat indigestion, chronic diarrhea, ringworm infections, bleeding piles, and amoebic dysentery [39]. Almost all parts of the plant are documented to be used for managing helminth infections. In Ghana, 74% traditional healers used this plant for treating helminth infections [40].

In a comparative assessment of the anthelminthic activity of various parts of the plant, the hydroethanolic extracts of the leaves, stem bark, and seeds of *Carica papaya* were tested against *P. posthuma* as the target organism. The results indicated that all crude extracts prepared were more effective than albendazole in reducing paralysis ($p < 0.0001$) and death times ($p < 0.0001$) of worms. Extracts from the seeds at 2.5 mg/mL were the most effective causing worm paralysis and death at 9.26 ± 0.03 and 20.12 ± 0.01 mins respectively. This was more potent than the standard anthelmintic albendazole at the same concentration which gave paralysis and death times of 19.45 ± 0.57 and 31.43 ± 0.28 mins respectively [41].

2.5 *Combretum mucronatum* Schumach & Thonn. (Combretaceae)

Ethnopharmacological reports from parts of Ghana revealed the extensive use of the leaves of *Combretum mucronatum* for treatment of human and livestock helminth infection [40]. The leaves from this plant species is monographed in the Ghana Herbal Pharmacopeia for the treatment of infections with worms [42].

In a previous study, the alcoholic leaf extract of *C. mucronatum* was assayed *in vitro* for anthelmintic activity against free-living nematode, *Caenorhabditis elegans* using levamisole as a positive control. The extract demonstrated anthelmintic activity with a worm survival rate of 89.2% at 0.1 mg/mL and 58.1% at 1 mg/mL [40].

In another study, fractions and purified compounds from *C. mucronatum* leaves were tested *in vitro* for their anthelmintic activity against *C. elegans*. Unsubstituted oligomeric proanthocyanidins (PACs) mainly composed of epicatechin units were identified as the active compounds of the hydroethanolic leaf extracts. The compounds demonstrated a dose-dependent anthelmintic activity ranging from 1 to 1000 mM and activity was found to increase with increasing molecular size. The anthelmintic activity was suggested to be by interaction of the PACs with some unidentified proteins of the target organism [43]. Further, the mechanism of anthelmintic activity of the PACs was determined by transcriptome analysis. PACs were found to interact with proteins within the worm's intestinal membrane as well as enzymes and peptides to elicit anthelmintic effects [44]. Another proposed mechanism was interaction of the tannins with cuticular proteins, particularly proline-rich collagen in the worm cuticle [45].

2.6 *Cyperus difformis* Linn. (Cyperaceae)

Cyperus difformis is an annual plant with smooth leaves and fibrous reddish roots. It is native to the subtropical and tropical areas but also distributed and widespread in South Europe, Asia and Americas. It is regarded as one of the world's commonest weeds found growing in wet swampy soils among rice plantation. It is very common in Ghana and traditionally used for the management of scorpion bites and malaria [46].

The anthelmintic and helminth resistance modifying activities of methanol extract of *C. difformis* was investigated against the adult Indian worm, *P. posthuma* using albendazole, mebendazole and levamisole as reference anthelmintics. The extract exhibited a concentration dependent anthelmintic activity against *P. posthuma* with significant ($p < 0.001$) paralysis and death times of 66.67 ± 1.8 and 140.7 ± 2.3 mins respectively at extract concentration of 20 mg/mL [47].

Further the extract at 1, 2 and 5 mg/mL significantly potentiated the activity of albendazole, mebendazole and levamisole against the test organism. In the presence of 2 mg/mL of the extract the paralysis and death times of albendazole (8 mg/mL) against *P. posthuma* were reduced from 41.33 ± 0.33 and 106.67 ± 0.88 min respectively to 33.33 ± 0.88 and 85.67 ± 1.2 min, respectively. Similar results were obtained for mebendazole and levamisole [47].

2.7 *Garcinia cola* Heckel (Guttiferae)

Garcinia cola also known as "bitter cola" is a valuable medicinal plant in African traditional medicine widely accepted for its numerous medicinal properties. It is usually called the wonder plant due to the usefulness of every part of the plant. The seeds are chewed as an aphrodisiac and used to cure cough, dysentery and upper respiratory tract infections [48, 49]. The latex from the stem is used against sexually transmitted infections and applied externally to heal wounds. The sap is used in curing parasitic diseases. Chewing sticks produced from the stems are used as masticatory for nervous alertness and for treating coughs and throat infections [50].

In a previous study, the methanol stem bark extract of *G. cola* (1—50 mg/mL) demonstrated a concentration dependent anthelmintic activity, decreasing paralytic and death times of *P. posthuma* with increasing extract concentrations. At 50 mg/mL, the extract had a paralytic time of 39.29 ± 0.12 min and death time of 54.29 ± 0.01 [51].

2.8 *Morinda lucida* Benth. (Rubiaceae)

Morinda lucida is an evergreen shrub growing from about 3 m to 18 m tall. It has a dense crown with slim, crooked branches. The plant is occasionally grown in home gardens. It is locally called '*konkroma*' in the Ghanaian Akan language. It is a multipurpose species yielding dyes, timber, fuel and traditional medicines. The plant is reported to be used in managing diabetes, hypertension, dysentery, stomach-ache, leprosy and gonorrhea. Traditionally, the stems are used to treat piles while the leaves are used to treat fever. A decoction of the bark or leaf is used in the treatment of jaundice and against itch and ringworm. The leaves and twigs are sold as a medicinal tonic for young children [52].

In a previous study, the methanol stem bark extract of *M. lucida* (10–50 mg/mL) reduced worm motility and caused death of the adult Indian earth worm, *P. posthuma* with a paralytic time of 18.17 ± 0.03 min and death time of 24.34 ± 0.21 min at 50 mg/mL [51].

2.9 *Moringa oleifera* Lam. (Moringaceae)

Moringa oleifera is a fast growing perennial evergreen or deciduous plant which grows up to a maximum height of 7–12 m. It has an open crown of drooping fragile branches bearing feathery foliage of opposite pinnate leaves, a crooked bole and dark gray stem bark. *M. oleifera* has been naturalized in many tropical and subtropical regions of the world including Africa, Arabia, South Asia, South America and India where it is commonly referred to as horseradish tree and drumstick tree [53]. Various parts of the plant are used in traditional medicine to treat various diseases including skin infections, anemia, asthma, bronchitis, catarrh, chest congestion, cholera, diabetes, hypertension and many other illnesses [54].

The foliage of *M. oleifera* was investigated for anthelmintic activity in wild caught *Achatina achatina* Linnaeus (edible snails). After feeding the snails on the foliage for 10 weeks, the proportion of parasitic infection in the treated group was estimated using dissecting and microscopic techniques. At the end of the treatment period, 96% of snails in the untreated group were observed to have their kidneys infected with roundworms as opposed to 24% of snails in the treated group. The percentage prevalence of parasitic infection in the treated and control groups was significantly different ($p < 0.0001$). Similar results were recorded for the infection of the lungs highlighting the anthelmintic value of *M. oleifera* in the control of worm infection in edible snails [55].

2.10 *Ocimum basilicum* Linn (Lamiaceae)

Ocimum basilicum is a tender-growing aromatic annual herb indigenous to West Africa and India. It is commonly called basil or sweet basil and locally known in the Ghanaian Akan language as '*Nunum*'. The herb is ubiquitously known for its therapeutic potentials in African folk medicine. In Ghana, basil is used in its fresh form as spice and flavoring in soups and sauces due to its strong spicy aroma. The whole plant is used to treat worm infestation, inflammation, pain, diarrhea, gastrointestinal infections and eye-related diseases [56].

In vitro anthelmintic activity of the hexane and ethanolic extracts of the fruits of *O. basilicum* was investigated against *Eudrilus eugeniae*. At a concentration range of 0.25–5 mg/mL, the extracts displayed a concentration dependent anthelmintic activity which was observed to be significantly ($p < 0.001$) higher compared to mebendazole-treated worms. At 5 mg/mL, paralysis was observed at 11.85 ± 0.71,

27.90 ± 0.42 and 94.04 ± 2.57 mins for the ethanol extract, hexane extracts and mebendazole-treat worms respectively. Similarly, death of worms was recorded at 24.74 ± 0.42, 85.18 ± 0.07 and 522.77 ± 1.53 mins respectively for the ethanol extract, hexane extracts and mebendazole [57].

2.11 *Paullinia pinnata* L. (Euphorbiaceaae)

Paullinia pinnata is a woody climber growing in tropical regions worldwide. In Ghana, it is locally called '*toantini*' in the Akan language. Preparations from the whole plant is used to treat dysentery. The mashed roots are used as poultice to heal chronic wounds and to treat leprosy. The root decoction is also used to cure coughs, pneumonia, gonorrhea, fractures, bacterial infections and abscesses. It is popularly known for its aphrodisiac property and used to treat erectyle dysfunction [58]. In addition, extracts of leaves and roots have been described for the treatment of helminth infestations particularly ancylostomiasis [40].

The hydroethanolic extract of the roots of *P. pinnata* was investigated in an *in vitro* mortality assay against the free-living nematode *Caenorhabditis elegans* as well as the larval stages of the parasitic helminths: *Ancylostoma caninum*, *Haemonchus contortus*, *Toxocara cati* and *Trichuris vulpis*. From the assay, the extract showed lethal activity against *T. cati* (LC_{50} = 112 µg/mL), *T. vulpis* (LC_{50} = 17 µg/mL), and *C. elegans* (LC_{50} = 2.5 of mg/mL), but not against *A. caninum*. Additionally, the effects of the extract on egg hatching and larval migration of the sheep parasite, *Haemonchus contortus* were investigated *in vitro*, but no inhibitory activity was observed [59].

In another study, the 70% aqueous acetone extract, solvent fractions and isolated compounds from the roots of *P. pinnata* were investigated for anthelmintic against *C. elegans*. From the results, the ethyl acetate fraction showed the highest anthelmintic effects with an LC_{50} of 1.1 mg/mL followed by the crude extract (LC_{50} = 1.9 mg/mL) and the aqueous fraction (LC_{50} = 2.9 mg/mL). Oligomeric proanthocyanidins were identified as the main active compounds. A mortality rate of at least 70% was observed for all proanthocyanidin containing fractions at 1 mg/mL [60].

2.12 *Plumbago zeylanica* Linn. (Plumbaginaceae)

Plumbago zeylanica is a perennial shrub with semi woody stems and numerous branches. It is a valuable medicinal plant widely used in Africa and Asia for the treatments of common ailments like hemorrhoids, diarrhea, leprosy, arthritic pains, toothache and as aphrodisiac and wound healing [61].

In a previous, observations were made for the time taken for different solvent extracts of the leaves of *P. zeylanicum* at concentrations of 300, 100 and 30 mg/mL to paralyze and kill *Pheretima posthuma*. The ethyl acetate extracts showed significant ($p < 0.0001$) concentration-dependent anthelminthic activity with the highest effect at 300 mg/mL causing paralysis at 7.39 ± 0.94 min and death at 11.81 ± 1.10 min. The methanol extract at 300 mg/mL demonstrated slightly lower anthelmintic effect with paralysis at 17.23 ± 1.68 min and death at 21.83 ± 2.60 min [62].

2.13 *Rauwolfia vomitoria* Afzel. (Apocynaceae)

Rauwolfia vomitoria commonly called the African Snakeroot or African Serpent root is a small tree or shrub that grows up to about 20 m tall in tropical Africa. It is locally called '*kakapenpen*' in the Asante dialect of Ghana. In traditional medicine, the plant is recorded to be used in the treatment of convulsions, malaria fever, insomnia, arthritis, pain, high blood pressure, diabetes, stomach problems and as an emetic. The leaves are applied topically for skin infections, swelling and snake bites. It is placed in the rectum for the expulsion of worms and for dysmenorrhea [46].

The leaves and stem bark of *R. vomitoria* demonstrated significant ($p < 0.001$) anthelmintic activity against the Indian adult earthworm *P. posthuma*. The methanol extracts of the stem bark caused paralysis of worms at 11.17 ± 0.088 min and reduced the death time to 21.67 ± 0.733 similar to the effect of albendazole at 10 mg/mL which had a worm death time of 21.03 ± 0.258 min [63].

2.14 *Sclerocarya birrea* (A. Rich) Hochst (Anacardiaceae)

Sclerocarya birrea is a dioecious small to medium sized tree growing up to about 20 m high and 1.2 m in diameter. The plant is distributed from Gambia, Ghana and Nigeria in West Africa, across Cameroon in Central Africa, to Ethiopia and Sudan in East Africa and to South Africa, usually found growing in open farm lands and natural vegetation [64]. The stem-bark, roots and leaves are used to treat several ailments including diabetes mellitus, diarrhea, dysentery, proctitis, ulcers, inflammation, arthritis, hypertension, skin diseases, and malaria [65].

The anthelmintic activity of the aqueous and ethanolic extracts of the roots of *S. birrea* were evaluated against earth worms. The extracts displayed significant ($p < 0.001$) concentration-dependent anthelmintic activity at 12.00 to 0.1875 mg/mL. The observed effect was higher compared to albendazole-treated worms [66].

2.15 *Vernonia amygdalina* Del. (Asteraceae)

Vernonia amygdalina is tropical shrub which grows up to about 3 m high. The plant is distributed throughout tropical Africa and has been domesticated in some parts of West Africa including Nigeria and Ghana where it is commonly called the bitter leaf. It is a highly valuable vegetable in West and Central Africa which is consumed as part of various dishes. In traditional medicine the leaf decoction is used to treat fever, malaria, diarrhea, dysentery, hepatitis and cough, as a laxative and as a fertility inducer [67]. The root extracts are also used for treating malaria and gastrointestinal disorders. One of the most common medicinal uses of *V. amygdalina* is as a treatment against intestinal worms including nematode infections [68]. The use of the leave decoctions against intestinal worms, especially pinworms was confirmed in an ethnobotanical survey in the Ashanti Region of Ghana [40].

In a previous study, the anthelmintic activity *of V. amygdalina* leaves were investigated against *Lumbricus terrestris* (earth worm). Unlike the negative control groups which remained alive and active after 6 hours of exposure to normal saline, all worms treated with the aqueous and ethanol leaf extracts (50–200 mg/mL) of *V. amygdalina* were noted to be paralyzed within 4.05 ± 1.06 to 59.94 ± 8.25 and 3.56 ± 0.37 to 33.18 ± 12.4 mins respectively ($p < 0.0001$). The effect was concentration dependent [36].

In another study, the stem bark extracts (ethanol and chloroform extracts) of *V. amygdalina* were observed to produce a synergistic anthelmintic effect when combined with the seeds of *Carica papaya* [69].

2.16 *Voacanga Africana* Stapf. (Apocynaceae)

Voacanga africana is a small tree or shrub, reaching up to 6 m tall in height with a low widely spreading crown. In Ghana, it is locally known as '*ofruma*' in the Asante language. Various plant parts are used medicinally throughout its distribution area [70]. The leaf decoction is used to treat dysentery, diarrhea, cutaneous and subcutaneous parasitic infections, leprosy, oedema, gout, paralysis and convulsion. The stem bark or roots decoctions are used as wound healing agents and used to treat boils, malaria, sexually transmitted diseases like gonorrhea, and skin diseases such as eczema and scabies. They are also taken to treat cardiovascular diseases and rheumatoid arthritis. The leaf latex is put in the teeth to treat dental caries or dripped in the eye to cure ophthalmia [46].

The methanol extracts of the leaves and stem bark *V. africana* were evaluated for *in vitro* anthelmintic activity by determining the effects of the extracts on the paralytic and death time of *P. posthuma* using albendazole as reference. The bark extract (20–50 mg/mL) demonstrated a significant ($p < 0.001$) concentration dependent anthelmintic effect by decreasing the paralysis and death times of worms. At 50 mg/mL, the stem bark extract caused worm paralysis within 7.03 ± 0.491 min and death at 14.77 ± 0.117 min [63].

2.17 *Xylopia aethiopica* (Dunal) A. Rich. (Apocynaceae)

Xylopia aethiopica is popularly known as the African pepper and locally called '*Hwentia*' in the Ghanaian Akan language meaning slender nose, referring to the shape of the fruit. *X. aethiopica* is known for its numerous medicinal properties in African traditional medicine. The bark infusion is used in the treatment of asthma, stomach aches and rheumatism. The bark powder is also applied topically on ulcerous wounds and used locally for the treatment of cancer and stomach ulcers. The root powder is known to relief toothache and pyorrhea [71].

The ethanolic extract of the dried fruits and leaves (300–300 mg/mL) were investigated for anthelmintic activity against earth worms. The anthelmintic activity of the fruit extract was more potent that the leaf extract. Both extracts demonstrated a concentration dependent activity with the fruit extract demonstrating significant paralytic and death times ($p < 0.001$) at 100 and 300 mg/mL [72].

3. Plants with cercaricidal and anti-schistosomal activities identified from Ghana

See **Table 1**.

3.1 *Azadirachta indica* A. Juss (Meliaceae)

[Refer to Section 2.3 for plant description].

The methanol leaf extract of *A. indica* was investigated for cercaricidal activity against freshly shed cercariae of *Schistosoma mansoni*. At a concentration range of 31.2–1000 µg/mL, the leaf extract caused a steady increase in the number of dead cercariae during an observation period of 15 to 180 mins. At 60 mins, 250 µg/mL of extract was found to cause 100% mortality of cercariae. At the end of the observation period (180 mins) the leaf extract recorded an IC_{50} of 27.62 µg/mL which was about four times lower than the effect of the positive control *Balanites aegyptiaca* (IC_{50} of 5.95 µg/mL) [73].

The effect of *A. indica* leaf extract on the viability of adult schistosome worms (i.e. adulticidal effect) was further investigated. At the end of 120 h, the extract at 62.5–1000 µg/mL was found to be lethal to the incopula adult worms. Further in an *in vivo* study, the ability of the leaf extract (500 mg/kg *p.o.*) to reduce the worm recovery and worm burden in *S. mansoni* infected mice was investigated.

Plant	Family	Common name	Part Investigated	Activity Type
Alcornea cordifolia	Euphorbiaceae	Christmas Bush	Leaves	Anthelmintic activity against *Pheretima posthuma* [33]
Alstonia boonei	Apocynaceae	Alstonia	Roots, stem bark	Anthelmintic activity against *Pheretima posthuma, Lubricus terretris* [35, 36]
Azadirachta indica	Meliaceae	Neem	Seeds Leaves	Anthelmintic activity against *Hymenolepsis diminuta*, *Enterbius vermicularis* and hookworm [38] Cercaricidal and adulticidal activity against *Schistosoma mansoni* [73]
Carica papaya	Caricaceae	Pawpaw	Leaves, stem bark, seeds	Anthelmintic activity against *Pheretima posthuma* [41]
Combretum mucronatum	Combretaceae	—	Leaves	Anthelmintic activity against *Caenorhabditis elegans* [40, 43, 45]
Cyperus difformis	Cyperaceae	—	Whole plant	Anthelmintic activity against *Pheretima posthuma* [47]
Dichapetalum crassifolium	Dichapeltaceae	—	Stems, roots	Anti-schistosomal activity against eggs obtained from clinical isolates of *Schistosoma haematobium* [75]
Erythrophloem ivorense	*Euphorbiaceae*	—	Leaves, stem bark Roots	Cercaricidal activity against post-infective larvae (schistosomule) and adult parasite of *Schistosoma mansoni* [77] Cercaricidal activity against freshly shed cercariae from *Schistosoma haematobium* [78]
Garcinia cola	Gutifferae	Bitter kola	Stem bark	Anthelmintic activity against *Pheretima posthuma* [51]
Holarrhena floribunda	Apocynaceae	—	Stem bark	Cercariae from *Schistosoma haematobium*
Morinda lucida	Rubiaceae	—	Stem bark	Anthelmintic activity against *Pheretima posthuma* [51] Cercaricidal activity against *Schistosoma mansoni* cercariae Adulticidal effect against *S. mansoni* adult worms [73]
Moringa oleifera	Moringaceae	Moringa	Foliage	Anthelmintic activity against *round worms in wild edible snails (Achatina achatina)* [55]
Nauclea latifolia	Rubiaceae	African peach	Stem bark	Cercaricidal activity against *Schistosoma mansoni* cercariae Adulticidal effect against *S. mansoni* adult worms [73]
Ocimum basilicum	Lamiaceae	Basil	Fruits	Anthelmintic activity against *Eudrilus eugeniae* [57]

Plant	Family	Common name	Part Investigated	Activity Type
Paullinia pinnata	Euphorbiaceae	—	Roots	Anthelmintic activity against the free-living nematode *Caenorhabditis elegans* and larval stages of the parasitic helminths: *Ancylostoma caninum*, *Haemonchus contortus*, *Toxocara cati* and *Trichuris vulpis* [60]
Plumbago zeylanica	Plumbaginaceae	—	Leaves	Anthelmintic activity against *Pheretima posthuma* [62]
Phyllanthus amarus	Euphorbiaceae	—	Leaves	Cercaricidal activity against *Schistosoma mansoni* cercariae
Rauwolfia vomitoria	Apocynaceae	Snakeroot	Leaves, roots Roots, stem bark	Anthelmintic activity against *Pheretima posthuma* [63] Cercaricidal activity against *Schistosoma mansoni* cercariae Adulticidal effect against *S. mansoni* adult worms [73]
Sclerocarya birrea	Anacardiaceae		Roots	Anthelmintic activity against *Lumbricus terrestris* [66]
Vernonia amygdalina	Asteraceae	Bitter leaf	Leaves, stem bark Leaves	Anthelmintic activity against *Lumbricus terrestris* [36] Cercaricidal activity against *Schistosoma mansoni* cercariae Adulticidal effect against *S. mansoni* adult worms [73]
Voacanga africana	Apocynaceae	—	Leaf, stem bark	Anthelmintic activity against *Pheretima posthuma* [63]
Xylopia aethiopica	Apocynaceae	African black pepper	Fruits, leaves	Anthelmintic activity against *Lumbricus terrestris* [72]

Table 1.
Medicinal plants from Ghana with anthelmintic and anti-schistosomal activity.

After a two-week period of treatment, the mean number of worms recovered from *A. indica*-treated mice was 19.80 ± 8.194 which was significantly lesser than that of the untreated mice (40.20 ± 3.072) [73].

The effect of the extract on the weight of spleen and liver of infected mice were all significantly lesser in the *A. indica*-treated group than that of the untreated group ($p < 0.05$). Organ histology also revealed only few granulomas which were smaller in diameter in the treatment groups whereas those in the untreated were severe ($p < 0.05$). Treated cercariae-infected mice group also had relatively less severe inflammatory cell infiltration compared with untreated group [73].

3.2 *Dichapetalum crassifolium* Chodat (Dichapetalaceae)

Dichapetalum crassifolium is a scandent shrub, about 1.5 m tall usually found growing in the rain forest, shady places, primitive woods and rocky areas of African countries including Ghana, Angola, Benin, Cameroon, Ivory Coast, Liberia, Nigeria, Sierra Leone, Tanzania, Togo and Zambia [74].

Crude extracts (pet-ether, ethyl acetate and methanol) and isolated triterpenoids from the stems and roots of *D. crassifolium* were investigated for anti-schistosomal activity against eggs obtained from clinical isolates of *Schistosoma haematobium* using the 96-well plate-egg hatch assay [75].

For the stem extracts, the ovicidal potency was in the following order petroleum ether (IC_{50} = 443.70) > EtOAc (IC_{50} = 638.00) > MeOH (IC_{50} = 893.70 μg/mL). The IC_{50} values for the root extracts were 248.60, 546.40, and 566.30 μg/mL respectively for the EtOAc, pet-ether and MeOH extracts.

The isolated compounds (Friedelan-3-one, β-Sitosterol/stigmasterol, Dichapetalin M and Dichapetalin A) showed higher ovicidal activity than the extracts though activities for both extracts and compounds were lower compared to the standard drug, praziquantel. The highest ovicidal potency was exhibited by β-sitosterol/stigmasterol mixture with an IC_{50} of 177.90 μg/mL which was about 11 times less potent than praziquantel (15.47 ± 0.06 μg/mL). The next highest was dichapetalin A (151.10 μg/mL) whiles friedelan-3-one showed the least potency with IC_{50} of 378.10 μg/mL. From the root extract, Dichapetalin M showed ovicidal effect with IC_{50} of 191.00 μg/mL [75].

3.3 *Erythrophleum ivorense* Afzel (Euphorbiaceae)

E. ivorense is a large tree which grows to about 40 m tall, with a cylindrical bole, sometimes fluted at the base. It is widely distributed in the evergreen primary and secondary forests of tropical Africa where it is commonly called by names like 'forest ordeal tree', 'red water tree' and 'sasswood tree'. Among the Akan tribe in Ghana, it is known as '*potrodum*'. The stem-bark and roots are usually employed in the treatment of epilepsy, emesis, pain, oedema, constipation and worm infestations [76].

The cercaricidal activity of the leaf and stem bark extracts of *E. ivorense* was investigated against two developmental stages of *Schistosoma mansoni* namely: the post-infective larvae (schistosomule) and the adult parasite. Various solvent fractions were assayed against the schitosomules at a concentration range of 0.31–100 μg/mL and against adult parasites at 1.25 mg/mL. The acetone fractions of both leaf and bark demonstrated the highest anti-schistosomal activity causing severe phenotypic alterations (immobility/inactivity, change in shape, translucence, surface disintegration) and death of schistosomules at all dilutions (except 0.31 μg/mL) at 24 h and 48 h. For adult parasites, severe phenotypic changes specifically damage to the adult parasite's tegument (surface) was observed for the acetone fraction of the stem bark extract. The adult worms were observed to be uncoordinated by 5 h, darkened in color by 24 h and died at 48 h exhibiting tegumental damage [77].

In another study, the *in vitro* cercaricidal activity of solvent fractions and isolated compounds from the root bark of *E. ivorense* was investigated against freshly shed cercariae from *Schistosoma haematobium*. Whereas the cercariae showed normal viability without any morphological changes (tail loss) throughout the entire duration of the experiment in the untreated group, exposure of cercariae to the crude hydro-ethanolic extract, its fractions and compounds caused a concentration and time-dependent decrease in viability of cercariae. Within two hours of incubation, all cercariae died at the various concentrations of test compounds and extracts. Eriodictyol, was the most potent compound with an IC_{50} of 1.23 ± 0.05 μg/mL. All test samples exhibited a much higher cercaricidal activity than the standard drug praziquantel which caused only 40% mortality of cercariae at the highest concentration tested (IC_{50} = 695.50 ± 0.05 μg/mL) [78].

3.4 *Holarrhena floribunda* (G. Don) Dur. & Schinz. (Apocynaceae)

Holarrhena floribunda is native to West Africa and is known in Ghana as '*osese*' among the Akans. The plant is traditionally used in the treatment of malaria, fever and bareness in females. It has antifungal, antibacterial and antidiabetic properties [79, 80].

The hydroethanolic and alkaloidal extracts from the stem bark of *H. floribunda* were tested on cercariae from *Schistosoma haematobium* at concentrations between 15.625 and 500.00 μg/mL. After 180 mins of contact with test samples, the ethanolic extract exhibited the highest cercaricidal potency with an IC_{50} of 20.09 ± 1.11 μg/mL higher than the effect of paraziquantel (IC_{50} = 695.50 ± 1.12). The alkaloidal extract also exhibited cercaricidal potency with an IC_{50} of 53.20 ± 1.33 μg/mL. The isolated compounds: holonamine, holadienine and conessine exhibited cercaricidal potency with IC_{50} values of 53.24 ± 1.28, 470.80 ± 1.00 and 33.28 ± 1.04 respectively. The results confirmed the activity of *Holarrhena floribunda* against *S. heamatobium* ceracriae [81].

3.5 *Morinda lucida* Benth (Rubiaceae)

[Refer to Section 2.8 for plant description].

In a previous study, the cercaricidal activity of the methanol stem bark extract of *M. lucida* was carried out. The extract at a concentration of 500 μg/mL elicited 100% mortality of *S. mansoni* cercariae within 120 mins of exposure giving an IC_{50} value of 262.3 μg/mL, which was however lower than the effect of the positive control *Balanites aegyptiaca* (IC50 of 5.95 μg/mL). Further, the *in vitro* adulticidal effect of the stem bark extract on adult schistosome worms revealed that at a concentration of 125–1000 μg/mL, the extract was found to be lethal to the adult worms within 120 h of exposure [73].

3.6 *Nauclea latifolia* Carl Lin. (Rubiaceae)

Nauclea latifolia, commonly called the African peach, is a deciduous shrub with an open canopy distributed throughout tropical and savanna regions of Africa and Asia. It varies widely in height from around 10–30 m according to soil and moisture conditions. The plant is used against various medical conditions such as diabetes, fever, indigestion and cough [82].

Previous studies on the cercaricidal activity the methanolic extract of stem bark of *N. latifolia* revealed 100% mortality of *S. mansoni* cercariae at a concentration of 250 μg/mL at 120 min (IC_{50} = 195.9 μg/mL). Further the extract was found to exhibit schistomicidal effect being lethal to the adult incopula worms at a concentration range of 500–1000 μg/mL within 120 mins of exposure [73].

3.7 *Phyllanthus amarus* Schum. and Thonn. (Euphorbiaceae)

P. amarus is a small herb bearing ascending herbaceous branches normally found around coastal and muddy areas. The whole plant is used in the treatment of gonorrhea, menorrhagia and other urinary and sexually transmitted infections. It is useful in gastropathy, diarrhea, dysentery, intermittent fevers, ophthalmopathy, scabies, ulcers and wounds [83].

The methanolic extract (250 μg/mL) of *P. amarus* leaves exhibited moderate cercaricidal activity on freshly shed *S. mansoni* cercariae causing 100% mortality of cercariae within 180 mins of exposure (IC_{50} = 250.4 μg/mL). It was further established that at 125–1000 μg/mL, the extract caused a drastic reduction in the viability of adult worms [73].

3.8 *Rauwolfia vomitoria* Afzel. (Apocynaceae)

[Refer to Section 2.13 for plant description].

The root and stem bark of *Rauwolfia vomitoria* were evaluated for schistosomicidal effect on two different parasitic stages of *Schistosoma mansoni* i.e. cercariae and adult worms [84].

The ethanolic extract of the root and stem bark were both found to be active against the cercariae and adult worms. At a concentration range of 62.5–1000 μg/mL the stem bark extract exhibited significant anti-cercarial activity ($p < 0.05$) with an LC_{50} of 207.4 and 61.18 μg/mL after 1 and 2 h of exposure respectively. At the highest concentration (1000 μg/mL), there was 100% mortality of cercariae within 90 min of exposure. The roots were less active than the stem bark showing activity at a higher concentration range of 250–1000 μg/mL. The schistomicidal activity of the stem bark and roots were further determined against adult worms. All adult worms exposed to the concentrations range of 250–1000 μg/mL for both plant parts died within 120 h of incubation [84].

3.9 *Vernonia amygdalina* Del. (Asteraceae)

[Refer to Section 2.15 for plant description].

In a previous study, the evaluation of the cercaricidal and schistosomicidal activities of the methanol extract of the leaves of *V. amygdalina* revealed significant potency response. At 250 μg/mL, the extract exhibited cercaricidal activity with an IC_{50} of 35.84 μg/mL within 180 min of exposure. Further, the extract was found to reduce the viability of adult schistosome worms *in vitro* at 250–1000 μg/mL.

The ability of the leaf extract (500 mg/kg *p.o.*) to reduce the worm recovery and worm burden in *S. mansoni* infected mice was further investigated in an *in vivo* study. After a two-week period of treatment, the mean number of worms recovered from *V. amygdalina*-treated mice was 12.00 ± 1.549, indicating 48.9%, worm burden which was significantly lower than that of the untreated group (40.20 ± 3.072). While there was significant increase in the weight of the liver and spleen of the untreated infected mice with marked formation of granuloma, *V. amygdalina*-treat infected mice showed no increase in liver or spleen size and had few granulomas which were smaller in diameter with relatively less severe inflammatory cell infiltration compared [73].

4. Conclusion

The anthelmintic and anti-schistosomal activities of some medicinal plants employed in Ghanaian traditional medicine have been validated. For most of these plants however, the specific bioactive constituents are not yet identified. It is therefore imperative that further studies to isolate and verify the constituents responsible for the observed activities be performed. Further, the evaluation of safety profiles will add substantial value to the reported bioactivities and make these plants attractive for adaptation to pharmaceutical companies for further development.

Conflict of interest

Authors have no conflict of interest to declare.

Author details

Evelyn Asante-Kwatia[1*], Abraham Yeboah Mensah[1], Lord Gyimah[1] and Arnold Donkor Forkuo[2]

1 Department of Pharmacognosy, Faculty of Pharmacy and Pharmaceutical Sciences, College of Health Sciences, Kwame Nkrumah University of Science and Technology, Kumasi, Ghana

2 Department of Pharmacology, Faculty of Pharmacy and Pharmaceutical Sciences, College of Health Sciences, Kwame Nkrumah University of Science and Technology, Kumasi, Ghana

*Address all correspondence to: emireku@yahoo.com

References

[1] Hotez PJ, Kamath A. Neglected tropical diseases in sub-Saharan Africa: review of their prevalence, distribution, and disease burden. PLoS Neglected Tropical Diseases. 2009; 3(8):e412.

[2] Conteh L, Engels T, Molyneux DH. Socioeconomic aspects of neglected tropical diseases. The Lancet. 2010; 375(9710):239-247.

[3] Mitra AK, Mawson AR. Neglected tropical diseases: epidemiology and global burden. Tropical Medicine and Infectious Disease. 2017; 2(3):36.

[4] Knopp S, Steinmann P, Keiser J, Utzinger J. Nematode infections: soil-transmitted helminths and Trichinella. Infectious Disease Clinics. 2012; 26(2):341-358.

[5] Savioli L, Stansfield S, Bundy DA, Mitchell A, Bathia R, Engels D, et al. Schistosomiasis and soil-transmitted helminth infections: forging control efforts. Transactions of the Royal Society of Tropical Medicine and Hygiene. 2002; 96(6):577-579.

[6] King CH, Dangerfield-Cha M. The unacknowledged impact of chronic schistosomiasis. Chronic Illness. 2008; 4(1):65-79.

[7] Ross AG, Chau TN, Inobaya MT, Olveda RM, Li Y, Harn DA. A new global strategy for the elimination of schistosomiasis. International Journal of Infectious Disease. 2017; 54:130-137.

[8] Fenwick A, Savioli L, Engels D, Bergquist NR, Todd MH. Drugs for the control of parasitic diseases: current status and development in schistosomiasis. Trends in Parasitology. 2003; 19(11):509-515.

[9] Danso-Appiah A, De Vlas S, Bosompem K, Habbema J. Determinants of health-seeking behaviour for schistosomiasis-related symptoms in the context of integrating schistosomiasis control within the regular health services in Ghana. Tropical Medicine & International Health. 2004; 9(7):784-794.

[10] Adenowo AF, Oyinloye BE, Ogunyinka BI, Kappo AP. Impact of human schistosomiasis in sub-Saharan Africa. Brazilian Journal of Infectious Diseases. 2015; 19(2):196-205.

[11] Ukwandu NCD, Nmorsi O. The perception, beliefs and practices toward genitourinary schistosomiasis by inhabitants of selected endemic areas (Edo/Delta States) in south-eastern Nigeria. Revista do Instituto de Medicina Tropical de São Paulo. 2004; 46(4):209-216.

[12] Van Wyk AS, Prinsloo G. Health, safety and quality concerns of plant-based traditional medicines and herbal remedies. South African Journal of Botany. 2020; 133:54-62.

[13] Cock I, Selesho M, Van Vuuren S. A review of the traditional use of southern African medicinal plants for the treatment of selected parasite infections affecting humans. Journal of Ethnopharmacology. 2018; 220:250-264.

[14] Mwangi VI, Mumo RM, Nyachieo A, Onkoba N. Herbal medicine in the treatment of poverty associated parasitic diseases: A case of sub-Saharan Africa. Journal of Herbal Medicine. 2017; 10:1-7.

[15] Danso-Appiah A, Stolk WA, Bosompem KM, Otchere J, Looman CW, Habbema JDF, et al. Health seeking behaviour and utilization of health facilities for schistosomiasis-related symptoms in Ghana. PLoS Neglected Tropical Diseases. 2010; 4(11):e867.

[16] Asase A, Hesse DN, Simmonds MS. Uses of multiple plants prescriptions for

treatment of malaria by some communities in southern Ghana. Journal of Ethnopharmacology. 2012; 144(2):448-452.

[17] Adeniyi A, Asase A, Ekpe PK, Asitoakor BK, Adu-Gyamfi A, Avekor PY. Ethnobotanical study of medicinal plants from Ghana; confirmation of ethnobotanical uses, and review of biological and toxicological studies on medicinal plants used in Apra Hills Sacred Grove. Journal of Herbal Medicine. 2018; 14:76-87.

[18] Bethony J, Brooker S, Albonico M, Geiger SM, Loukas A, Diemert D, et al. Soil-transmitted helminth infections: ascariasis, trichuriasis, and hookworm. The Lancet. 2006; 367(9521):1521-1532.

[19] Montresor A, Trouleau W, Mupfasoni D, Bangert M, Joseph S, Mikhailov A, et al. Preventive chemotherapy to control soil-transmitted helminthiasis averted more than 500 000 DALYs in 2015. Transactions of The Royal Society of Tropical Medicine and Hygiene. 2017; 111(10):457-463.

[20] Crompton DWT, Nesheim MC. Nutritional impact of intestinal helminthiasis during the human life cycle. Annual Review of Nutrition. 2002; 22(1):35-59.

[21] Hotez PJ, Bundy DA, Beegle K, Brooker S, Drake L, de Silva N, et al. Helminth infections: soil-transmitted helminth infections and schistosomiasis. Disease Control Priorities in Developing Countries. 2nd edition. 2006.

[22] Geerts S, Gryseels B. Drug resistance in human helminths: current situation and lessons from livestock. Clinical Microbiology Reviews. 2000; 13(2):207-222.

[23] French MD, Evans D, Fleming FM, Secor WE, Biritwum N-K, Brooker SJ, et al. Schistosomiasis in Africa: improving strategies for long-term and sustainable morbidity control. PLoS Neglected Tropical Diseases. 2018; 12(6):e0006484.

[24] World Health Organization, Investing to overcome the global impact of neglected tropical diseases: third WHO report on neglected tropical diseases. Vol. 3; Geneva: WHO; 2015.

[25] Colley DG, Bustinduy AL, Secor WE, King CH. Human schistosomiasis. The Lancet. 2014; 383(9936):2253-2264.

[26] Sacko M, Magnussen P, Keita AD, Traoré MS, Landouré A, Doucouré A, et al. Impact of *Schistosoma haematobium* infection on urinary tract pathology, nutritional status and anaemia in school-aged children in two different endemic areas of the Niger River Basin, Mali. Acta Tropica. 2011; 120:S142-S150.

[27] Fenwick A, Jourdan P. Schistosomiasis elimination by 2020 or 2030? International Journal for Parasitology. 2016; 46(7):385-388.

[28] Ross AG, Olveda RM, Li Y. An audacious goal: the elimination of schistosomiasis in our lifetime through mass drug administration. Lancet (London, England). 2015; 385(9983):2220-2221.

[29] Olveda DU, McManus DP, Ross AG. Mass drug administration and the global control of schistosomiasis: successes, limitations and clinical outcomes. Current Opinion in Infectious Diseases. 2016; 29(6):595-608.

[30] Doenhoff MJ, Kusel JR, Coles GC, Cioli D. Resistance of *Schistosoma mansoni* to praziquantel: is there a problem? Transactions of the Royal Society of Tropical Medicine and Hygiene. 2002; 96(5):465-469.

[31] Danso-Appiah A, De Vlas SJ. Interpreting low praziquantel cure rates

of *Schistosoma mansoni* infections in Senegal. Trends in Parasitology. 2002; 18(3):125-129.

[32] Boniface PK, Ferreira SB, Kaiser CR. Recent trends in phytochemistry, ethnobotany and pharmacological significance of *Alchornea cordifolia* (Schumach. & Thonn.) Muell. Arg. Journal of Ethnopharmacology. 2016; 15(191):216-244.

[33] Osei Akoto C, Acheampong A, Boakye YD, Akwata D, Okine M. *In vitro* anthelminthic, antimicrobial and antioxidant activities and FTIR analysis of extracts of *Alchornea cordifolia* leaves 2019; 8(4): 2432-2442.

[34] Adotey JPK, Adukpo GE, Opoku Boahen Y, Armah FA. A review of the ethnobotany and pharmacological importance of *Alstonia boonei* De Wild (Apocynaceae). International Scholarly Research Notices. 2012; 2012:1-9.

[35] Klu MW, Apenteng JA, Mintah DN, Addy BS, Nyarko-Danquah I, Afriyie SB. *In vitro* anthelmintic activity of stem and root barks of *Alstonia boonei* De Wild. Journal of Medicinal Plants Research. 2016; 10(13):179-182.

[36] Danquah CA, Koffuor G, Annan K, Ketor E. The anthelmintic activity of *Vernonia amygdalina* (Asteraceae) and *Alstonia Boonei* de wild (Apocynaceae). Journal of Medical and Biomedical Sciences. 2012; 1(1):21-27.

[37] Hashmat I, Azad H, Ahmed A. Neem (*Azadirachta indica* A. Juss)-A nature's drugstore: an overview. International Research Journal of Biological Science. 2012; 1(6):76-79.

[38] Lawson B, Tuani G, Dompreh A. Comparative Performance of Neem (*Azadiracta Indica*) Seed Extract and Mebendazole (Vermox) against Naturally Occuring Helminths in Rats (Rattus Nor Vegicus). Journal of Science and Technology. 2003; 23(2):1-9.

[39] Vij T, Prashar Y. A review on medicinal properties of *Carica papaya* Linn. Asian Pacific Journal of Tropical Disease. 2015; 5(1):1-6.

[40] Agyare C, Spiegler V, Sarkodie H, Asase A, Liebau E, Hensel A. An ethnopharmacological survey and *in vitro* confirmation of the ethnopharmacological use of medicinal plants as anthelmintic remedies in the Ashanti region, in the central part of Ghana. Journal of Ethnopharmacology. 2014; 158:255-263.

[41] Goku PE, Orman E, Quartey ANK, Ansong GT, Asare-Gyan EB. Comparative Evaluation of the *In Vitro* Anthelminthic Effects of the Leaves, Stem, and Seeds of *Carica papaya* (Linn) Using the *Pheretima posthuma* Model. Evidence-Based Complementary and Alternative Medicine. 2020; 2020:1-8.

[42] Boadu AA, Asase A. Documentation of herbal medicines used for the treatment and management of human diseases by some communities in southern Ghana. Evidence-Based Complementary and Alternative Medicine. 2017; 2017:1-12.

[43] Spiegler V, Sendker J, Petereit F, Liebau E, Hensel A. Bioassay-guided fractionation of a leaf extract from *Combretum mucronatum* with anthelmintic activity: oligomeric procyanidins as the active principle. Molecules. 2015; 20(8):14810-14832.

[44] Spiegler V, Hensel A, Seggewiß J, Lubisch M, Liebau E. Transcriptome analysis reveals molecular anthelmintic effects of procyanidins in *C. elegans*. Plos one. 2017; 12(9):e0184656.

[45] Herrmann FC, Spiegler V. *Caenorhabditis elegans* revisited by atomic force microscopy–ultra-structural changes of the cuticle, but not in the intestine after treatment with

Combretum mucronatum extract. Journal of Structural Biology. 2019; 208(2):174-181.

[46] Derakhshan A, Gherekhloo J. Factors affecting *Cyperus difformis* seed germination and seedling emergence. Planta daninha. 2013; 31(4):823-832.

[47] Adu F, Agyare C, Sam GH, Boakye YD, Boamah V. Anthelmintic resistance modifying properties of extracts of *Cyperus difformis* L.(Cyperiaceae). Investigational Medicinal Chemistry and Pharmacology. 2018; 1(1):1-12.

[48] Odebunmi E, Oluwaniyi O, Awolola G, Adediji O. Proximate and nutritional composition of kola nut (*Cola nitida*), bitter cola (*Garcinia cola*) and alligator pepper (*Afromomum melegueta*). African Journal of Biotechnology. 2009; 8(2):308-310.

[49] Irvine FR. Woody plants of Ghana. Woody plants of Ghana. 1961.

[50] Farombi EO, Bitter kola (*Garcinia kola*) seeds and hepatoprotection. In Nuts and seeds in health and disease prevention. Academic Press; 2011. pp. 221-228.

[51] Apenteng JA, Mintah DN, Klu MW, Quartey AK, Oppong AB, Harrison E, et al. *In vitro* anti-infective and antioxidant activities of *Garcinia cola* Heckel and *Morinda lucida* Benth. Journal of Medicinal Plants Research. 2017; 11(32):507-512.

[52] Zimudzi C, Cardon D. *Morinda lucida* Benth. Plant Resources of Tropical Africa / Ressources végétales de l'Afrique tropicale. 2005. Available from: http://www.prota4u.org/search.asp [Accessed: 11 February, 2021].

[53] Bosch CH. *Moringa oleifera* Lam. Plant Resources of Tropical Africa / Ressources végétales de l'Afrique tropicale). 2004. Available from: http://www.prota4u.org/search.asp [Accessed: 11 February, 2021]

[54] Abdull Razis AF, Ibrahim MD, Kntayya SB. Health benefits of *Moringa oleifera*. Asian Pacific Journal of Cancer Prevention. 2014; 15(20):8571-8576.

[55] Aboagye IF, Mensah D, Boadu F. Anthelmintic Effect of *Moringa oleifera* Lam. in Wild-caught *Achatina achatina* Linnaeus, 1758 from the Sefwi Wiawso District, Ghana. West African Journal of Applied Ecology. 2015; 23(2):27-33.

[56] Danso-Boateng E. Effect of drying methods on nutrient quality of Basil (*Ocimum viride*) leaves cultivated in Ghana. International Food Research Journal. 2013; 20(4):1569.

[57] Osei Akoto C, Acheampong A, Boakye YD, Naazo AA, Adomah DH. Anti-inflammatory, antioxidant, and anthelmintic activities of *Ocimum basilicum* (Sweet Basil) fruits. Journal of Chemistry. 2020; 2020:1-7.

[58] Annan K, Gbedema S, Adu F. Antibacterial and radical scavenging activity of fatty acids from *Paullinia pinnata* L. Pharmacognosy Magazine. 2009; 5(19):119.

[59] Spiegler V, Peppler C, Werne S, Heckendorn F, Sendker J, Liebau E, et al. Anthelmintic activity of a traditionally used root extract from *Paullinia pinnata*. Planta Medica. 2016; 82(S 01):P925.

[60] Spiegler V. Anthelmintic A-Type Procyanidins and Further Characterization of the Phenolic Composition of a Root Extract from *Paullinia pinnata*. Molecules. 2020; 25(10):2287.

[61] Datta S, Mishra R. *Plumbago zeylinica* Linn.(Chitrak)-Review as rasayan (rejuvenator/antiaging). International Journal of Research in Pharmaceutical and Biomedical Sciences. 2012; 3(1):250-267.

[62] Apenteng JA, Brookman-Amissah MG, Osei-Asare C, Oppong EE, Ogundeyi M. *In Vitro* Anti-infective and Antioxidant Activity of *Plumbago zeylanica* Linn. International Journal of Current Research in Biosciences and Plant Biology. 2016; 3(8):131-137.

[63] Adu F, Apenteng JA, Akanwariwiak WG, Sam GH, Mintah DN, Bortsie EB. Antioxidant and *in-vitro* anthelminthic potentials of methanol extracts of barks and leaves of *Voacanga africana* and *Rauwolfia vomitoria*. African Journal of Microbiology Research. 2015; 9(35):1984-1988.

[64] Gouwakinnou GN, Lykke AM, Assogbadjo AE, Sinsin B. Local knowledge, pattern and diversity of use of *Sclerocarya birrea*. Journal of Ethnobiology and Ethnomedicine. 2011; 7(1):1-9.

[65] Hall JB. *Sclerocarya birrea* (A.Rich.) Hochst. Plant Resources of Tropical Africa / Ressources végétales de l'Afrique tropicale) 2002. [Accessed 01 February, 2021].

[66] Akoto CO, Acheampong A, Boakye YD, Kokloku BK, Kwarteng G. In vitro anthelmintic, anti-inflammatory, antioxidant activities and FTIR analysis of *Sclerocarya birrea* root. Journal of Pharmacognosy and Phytochemistry. 2020; 9(2):1389-1401

[67] Ucheck Fomum F. *Vernonia Amygdalina* Delile. Plant resources of Tropical Africa/Ressources végétales de l'Afrique tropicale 2004 01-08-2019]. Available from: http://www.prota4u.org/search.asp. [Accessed 01 February, 2021]

[68] Ogidi OI, George DG, Esie NG. Ethnopharmacological properties of *Vernonia amygdalina* (Bitter Leave) medicinal plant. Journal of Medicinal Plants. 2019; 7(2):175-181.

[69] Mintah DN, Quartey AK, Oppong AB, Ayensu I, Apenteng JA. Synergistic *in-vitro* anthelmintic potentials of *Vernonia amygdalina* Delile stem and *Carica papaya* Lin seeds. World Journal of Pharmacy and Pharmaceutical Sciences. 2017; 6(10):103-115.

[70] Koroch AR, Juliani HR, Kulakowski D, Arthur H, Asante-Dartey J, Simon JE. *Voacanga africana*: Chemistry, quality and pharmacological activity. In: Juliani, H, et al. editors. African Natural Plant Products: New Discoveries and Challenges in Chemistry and Quality; ACS Symposium Series. Washington DC: American Chemical Society; 2010.

[71] Woode E, Ameyaw EO, Boakye-Gyasi E, Abotsi WK. Analgesic effects of an ethanol extract of the fruits of *Xylopia aethiopica* (Dunal) A. Rich (Annonaceae) and the major constituent, xylopic acid in murine models. Journal of Pharmacy & Bioallied Sciences. 2012; 4(4):291.

[72] Apenteng JA, Ogundeyi M, Oppong EE, Osei-Asare C, Brookman-Amissah MG. *In vitro* Anti-infective and Antioxidant activity of *Xylopia aethiopica* [Dun.] A. Rich: A comparison of the fruits and leaves extracts. Journal of Medicinal Plants Studies. 2016; 4:24-29.

[73] Acheampong DO, Owusu-Adzorah N, Armah FA, Aninagyei E, Asiamah EA, Thomford AK, et al. Ethnopharmacological evaluation of schistosomicidal and cercaricidal activities of some selected medicinal plants from Ghana. Tropical Medicine and Health. 2020; 48:1-10.

[74] Breteler FJ. The African Dichapetalaceae IX. The Netherlands: Wageningen Agricultural University; 1986. Available from: https://library.wur.nl/WebQuery/wurpubs/

fulltext/286414 [Accessed 01 February, 2021]

[75] Chama MA, Onyame HA, Fleischer C, Osei-Safo D, Waibel R, Otchere J, et al. In vitro activities of crude extracts and triterpenoid constituents of *Dichapetalum crassifolium* Chodat against clinical isolates of *Schistosoma haematobium*. Heliyon. 2020; 6:e04460.

[76] Adu-Amoah L, Agyare C, Kisseih E, Ayande PG, Mensah KB. Toxicity assessment of *Erythrophleum ivorense* and *Parquetina nigrescens*. Toxicology Reports. 2014; 1:411-420.

[77] Kyere-Davies G, Agyare C, Boakye YD, Suzuki BM, Caffrey CR. Effect of phenotypic screening of extracts and fractions of *Erythrophleum ivorense* leaf and stem bark on immature and adult stages of *Schistosoma mansoni*. Journal of Parasitology Research. 2018; 2018:1-7

[78] Armah FA, Amoani B, Henneh IT, Dickson RA, Adokoh CK, Amponsah IK, et al. *In vitro* Cercaricidal Activity of fractions and isolated compounds of *Erythrophleum ivorense* (Fabaceae) root bark against *Schistosoma haematobium*. International Journal of Tropical Disease & Health, 2018; 34(3): 1-9.

[79] Tamboura H, Bayala B, Lompo M, Guissoe I, Sawadogo L. Ecological distribution, morphological characteristics and acute toxicity of aqueous extracts of *Holarrhena floribunda* (g. don) durand & schinz, *Leptadenia hastata* (pers.) decne and *Cassia sieberiana* (dc) used by veterinary healers in Burkina Faso. African Journal of Traditional, Complementary and Alternative Medicines. 2005; 2(1):13-24.

[80] Hoekou YP, Tchacondo T, Karou SD, Yerbanga RS, Achoribo E, Da O, et al. Therapeutic potentials of ethanolic extract of leaves of *Holarrhena floribunda* (G. Don) Dur. & Schinz (Apocynaceae). African Journal of Traditional, Complementary and Alternative Medicines. 2017; 14(2):227-233.

[81] Amponsah IK, Armah FA, Alake J, Harley BK, Ampofo EK, Asante-Kwatia E, Clinton B, Amoani B, Henneh TI. In-vitro Anti-cercarial activity of extracts and steroidal alkaloids from the stem bark of *Holarrhena floribunda* (G. Don) Dur. & Schinz International Journal of Phytomedicine. 2020; 12(3):069-073

[82] Abdel-Rahman NA-G, *Nauclea latifolia* (Karmadoda): Distribution, Composition and Utilization. In: Mariod A. editor. Wild Fruits: Composition, Nutritional Value and Products. Springer Cham; 2019. pp. 435-445.

[83] Patel JR, Tripathi P, Sharma V, Chauhan NS, Dixit VK. *Phyllanthus amarus*: ethnomedicinal uses, phytochemistry and pharmacology: A Review. Journal of Ethnopharmacology. 2011; 138(2):286-313.

[84] Tekwu EM, Bosompem KM, Anyan WK, Appiah-Opong R, Owusu KB-A, Tettey MD, et al. *In vitro* assessment of anthelmintic activities of *Rauwolfia vomitoria* (Apocynaceae) stem bark and roots against parasitic stages of *Schistosoma mansoni* and cytotoxic study. Journal of Parasitology Research. 2017; 2017:1-11.

Chapter 3

Traditional African Medicine

Motamarri V.N.L. Chaitanya, Hailemikael Gebremariam Baye, Heyam Saad Ali and Firehiwot Belayneh Usamo

Abstract

African traditional medicine is defined as one of the holistic health care system comprised of three levels of specializations namely divination, spiritualism, and herbalism. The traditional healer provides healing services based on culture, religious background, knowledge, attitudes, and beliefs that are prevalent in his community. Hence the current chapter focuses on the different types of african healing system, traditional healers, traditional practices and modern herbalism and also describes the phytochemical and pharmacological evidences of the traditional african herbs like Acanthus montanus (Acanthaceae), *Amaranthus spinosus* (Amaranthaceae), Bridelia ferruginea (Euphorbiaceae) etc.

Keywords: traditional African medicine, traditional healers, divination, spiritualism, and herbalism

1. Introduction

As per World Health Organization (2002), The "Traditional medicine" may be defined as health practices, approaches, knowledge and beliefs incorporating plant, animal and mineral based medicines, spiritual therapies, manual techniques and exercises, applied singularly or in combination to treat, diagnose and prevent illnesses or maintain well-being [1].

Africa is one of the heritage continent also known as cradle of human being and the concept of traditional medicine in africa is existed long back with out documentation as a hidden evidence less practices for human beings who have been struggling with various unknown diseases. African people have their own ancestral practices to heal using different methods [2].

According to World Health Organization report more than 80% of the people in Africa depend on traditional medicine for their health care needs (WHO, 2003). The African people have been depending on various plants and animals source for their drug to treat various physical and mental illness. Nearly, 4000 medicinal plants have been documented towards their various Pharmacological activities [3].

Any Traditional medicine consists of medical treatment with an ancient roots that has been passed over generations to maintain health, as well as to prevent, diagnose, improve or treat illnesses. Various cultural and historic conditions have been influenced in the development of traditional medicine. The common basis for any Traditional medicine concept is a holistic healing to maintain life equilibrium between the body, mind and soul with external environment. Even the Traditional African Medicine is not an exception from this universal holistic approach.

Some of the Traditional healing systems and concepts have been supported by huge volume of literature and are in transition towards evidence healing concepts. However

Traditional African Medicine is still has not been documented and under process of documentation as from generations to generations, this was hidden as secret concept of healing. Still to date, in most parts of the africa, the major population have been continue to rely on traditional medicine to meet their primary health care needs [4].

In Africa, Traditional medicine is a healing belief system having its own health and disease concept. This is considered as a hidden treasure or knowledge that will pass from father to his only one beloved son of that family. The various healing concepts in Traditional African medicine (TAM) includes herbalism, surgery, bone setting, spinal manipulation, psychotheraphy, hydrotherapy, occultism, hydrotheraphy etc. However, lack of indepth scientific validation of these african traditional medicine and their documentation is a greatest lacuna and very much attention is required by the modern herbalists to safegaurd this healing concept. In the herbalism, vegetable, animal, and mineral substances have been used. In the metaphysical healing concept, Spirtualism concepts like prayers, invocations, or incantations have been offered to some mysterious and powerful forces in the various belief concept system like exorcism, divination, libation etc., were also been practicing to heal several diseases, how ever, scientific validation and documentation is still challenging [5, 6]. Some of plants used in herbalism by traditional african healers are *Foeniculum vulgare* Mill [7], *Rauvolfia serpentina* Linn [8], *Cinchona pubescens* Vahl [9], *Digitalis purpurea* Linn [10], etc.

1.1 Historical development of African traditional medicine

As per the Traditional African Medicine, Illness is considered as disorder that having both natural and supernatural causes. This must be treated by both physical and spiritual means, using various procedures like divination, incantations, animal sacrifice, exorcism, and herbs. It is a type of holistic health care system based on three levels of basic principles known as divination, spiritualism, and herbalism. The health care services provided by the healers are based on culture, religious background, knowledge attitudes, and their community beliefs.

The historical Development of African Traditional Medicine consists of two periods/Eras:

a. Colonial Era

b. Modern Period

1.1.1 Colonial era

In this era, the traditional medicinal practices have been considered as primitive and backward. Under colonial rule, traditional healers and their practices were not recognized by the scientific world as they were wrongly predicted as witchcraft and black magic. These practices in this era was considered by many nations as an illegal by the colonial authorities. Even in this period, attempts have been made to control the sale of herbal medicines. With the spread of colonaialism and christianity, christian missionaries built private ones and allopathy system of medicine is on widespread to treat various diseases [11].

1.1.2 Modern period

However, in this era, the modern world showed more interest in some of the medicinal plants that have been using in the african traditional healing systems as a bioprospecting tools towards discovery of drug leads and drugs. Still there are many hidden mysteries that need to be learnt from traditional african practices by the modern world. In the writings of Dr. T. Adeoze Lambo, a Nigerian psychiatrist, stated in 1979, always there is a mystery in the african traditional healing concepts especially in the treatment of neurosis [12].

1.2 Principles, methods and areas in African traditional medicine

The pricinples involved in African traditional medicine is organized into three levels namely divination, spiritualism, and herbalism. It is a holistic approach that considers illness may be due to both physical and spirtual means that can be healed by using the concepts of divination, incantations, animal sacrifice, exorcism, and herbs. Traditional healers in the africa have been occupied prime position in the living community that uses herbs, minerals, animals and other spirtual and cultural belief to cure various diseases.

1.2.1 Divination

It is the spirtual healing process which is an act to contact between spirtual world and the mundane world for getting the guidance to heal [13]. The traditional healers were known as diviners. As per the belief systems, divination is a part of witchcraft and is a sign of metaphysical curses to block ones living energy. The local tribes in Africa have strong belief in metaphysical healing systems incompare to western medicine with respect to some ailments which are not clearly understood by the allopathy system of medicine. These type of ailments were considered as spiritual illnesses. The healing protocols in divination includes following of ancestor instructions and sacrifices to the spirtual world. Sometimes the diagnosis in divination protocol includes dream interpretations like apperance of ancestors, nightmares, omens, owls in diseased patients. The traditional healers through the secret knowledge of divination and interprets the dreams through the communication to spirtual world and the healing process will be done as per the spirtual communication by the healer which includes secret recipes of herbal bath, herbal decoctions, sacrifices, incantations and wearing of herbal parts as a protective medicine in form of talisman [14, 15]. However clear scientific validation over divination healing is still unclear and lack of scientific evidences. The traditional healers never expose their herbo-magical remedies to the other people. The different procedures in the divination includes tarot card and readings, celetic ogham, Norse Runes, tasseography or tasseomancy, Pendulum readings, osteomancy, lithomancy, fullmoon water scrying, pshycic automatic writing. To understand divination, it is always better to watch a documented video (Video 1; Available from (can be viewed at) https://www.youtube.com/watch?v=yZpqdICyq UM.).

1.2.2 Spirtualism

The spiritualism or spirtual healing is an important healing practices in traditional african medicine and includes the following healing procedures.

a. Spirtual protection

Africans believe that some unknown illness are may be due to an attack by the evil spirits. In this case the spirtual healer prescribes talisman, charm, amulets, specially designed body marks, and a spiritual bath to drive evil spirits away. These rituals are helpul in driving off evil and dangerous dark spirtual forces or elements to ward off the evils or dangers that may have befallen a individual or family or community [16].

b. Sacrifices

Sacrifices are also part of spirtual healing process, sometimes offered at the request of spirits, gods, and ancestors which includes the sacrifice and burrial of various animals like dogs, cats, buffaloes etc. which are burried alive at midnight to save the individual from the evil attacks. Even these sacrifices includes some secret herbs, in which the healers believe that without these herbs, the process of sacrifices is incomplete. In some parts of Africa like Southafrica, this sacrifices also includes human sacrifices which is known as muti or ritual murder, which includes in identification of young child and sacrifices by spirtual healers for various beleifs [17].

c. Spirtual cleanising

This is also a part of spirtual healing also known as spirtual bath. This remedy is prescribed to the disease person with procedure and how many times per day. This includes secret herbal bath, holy water bath, and animal blood poured from head to toe and these practices are common in the african countries like Ghana.

After this bath, the diseased individual is required to offer certain items for sacrifice or libation like dove, dog, cat, got, fowl etc., along with local gin, cola nut, eggs, and plain white, red, or black cloth. It is belief that these items after sacrifice will be taken by the Gods.

The Gods guide the traditional healer for specific bath and specific item for sacrificing. All the specified things will be tied in cloth and thrown into flowing river after sacrifice and left to degrade. Sometimes, these things will left at the cross roads at the outskirts of community depending on the nature and severity of the case.

In somecases these spirtual cleansing techniques also known as a Ritual sacrifice baths. The Hausa-Fulani women of Zaria, Nigeria during cold or respiratory illness, they practices these type of ritual bath with hot water splashes along with the twigs of tamarind or neem tree [18, 19].

d. Exorcism

This is also a part of spirtual healing practising in several parts of the Africa. This includes expelling demons or evil spirits from the person with illness. Africans believes that certain diseases are may be due to possession of evil spirits. The diseases like mental illness is considered due to these evil possessions. The skill of exorcism is only be performed by the traditional leader or preist in that community. This includes various rituals like rosary chanting, dances, drums, music, songs, bibilical verses, touches the ill person with animal tails, and other objects [20].

The other healing techniques in the spirtualism includes libation, which is defined as the pouring of some liquids like gin, aromas on to the ground as an offering to get release from the illness due to karmic events. This can be practised through various techniques like invocation, supplication, and conclusion. The process of libation is to win over the evil wishes and curses by enemies. This includes pouring or offering various liquids to the ground along with prayers and chants [21]. It is always better to watch a documented video (Video 2; available from (can be viewd at) https://youtu.be/lKZeDWpcoEw) to understand more about the spirtualism practices.

1.2.3 Herbalism

Even use of herbal remedies is a part of Traditional African Medicine since ancient times. However, In Africa, the herbalism is in transition towards modernisation. It is the study, standardization, quality control and use of different herbs with evidences or non evidences. The herbalism existed since ancient times. The modern medicine is also evolved from herbalism for eg: Aspirin (from willow bark), Quinine (From cinchona bark).

The herbal practitioners will take an extensive questionarrie consists of case histories and examines the patient physically. The patient medical history and symptoms will also be given an attention by the traditional healers. This includes the examination of every day physiological process like appetite, digestion, urination, defacation, and sleep.

The prescription includes individualized herbs or combinational herbs, usually in form of tinctures, extracts, fractions, decoctions, distillates, snuffs, gruels, teas, syrups, pills, ointments, polutices, etc. [22].

Various parts will be used in herbalism like roots, bulbs, and rhizomes from various healing herbs like *Acacia senegal* Linn (Leguminosae: Mimosoideae), *Aloe ferox* Mill. (Xanthorrhoeaceae), *Artemisia herba-alba* Asso (Med)(Asteraceae), *Aspalathus linearis* (Brum.f.) R. Dahlg. (Fabaceae), *Centella asiatica* (L.) Urb. (Apiaceae), *Catharanthus roseus* (L.) G. Don (Apocynaceae), *Cyclopia genistoides* (L.) Vent. (Fabaceae), *Harpagophytum procumbens* (Burch.) DC. (Pedaliaceae), *Momordica charantia* Linn. (Cucurbitaceae), *Pelargonium sidoides* DC. (Geraniaceae) are some of the important herbs in the traditional African herbalism [23]. To understand the modern and traditional herbalism in africa it is always best from video material (video 3, video 4) (Video 3 available from (can be viewed at) https://youtu.be/7IvP3SSU2nM and Video 4 available from (can be viewed at) https://youtu.be/DV69JKi29Mk).

2. African traditional medicine in different regions

2.1 Traditional Ethiopian medicine

The first recorded epidemic that occurred in Ethiopia dates back to 849 during the rule of Abba Yohannes as the head of the Ethiopian church. The disease and famine in those days was perceived as God's punishment for Yohannes' misdeeds. In the documented letter to Abba Yohannes, from the Ethiopian emperor and it was wrote that "great tribulations have come upon our land, and all our men are dying of the plague and all of our beasts and cattle have perished." [24]. However prediction of pinpoint of the birth of medicine in Ethiopia is impossible due to lack of clear document evidences. However the Ethiopian traditional medicine (ETM) is poorly documented. In Ethiopia, the traditional healers usually follows herbalism, spirtual healing, bone setting and minor surgical procedures in treating disease. The ETM is highly complex and diverse because the principals of healing will vary from one ethnic group to others. The Ethiopian traditional medicine beleives that disease is mystical and natural imbalance concept and follows holistic approach in the treatment [25].

Despite Western medicine becoming more widespread in Ethiopia, Still today Ethiopians trust and highly depend on traditional healing principles because of the easy and cheap access in urban parts. The first traditional elixir in Etthiopia is holy water and holy oil (Tesbel in Ahmaric) and moslems called as zemzem. Ethiopians beleives that holy water heals every ailment, when it is drunk or had bathed in. Even the orthodox christians in Ethiopia believes in the healing actions of holy annoited oil for treating minor ailments, this gave the concept of herbalism and spirtuality in Ethiopian traditional medicine [26, 27].

The traditional medicine in Ethiopia is a combined concept of spiritualism and herbalism. The ETM IS a perceptive and own generational understanding of healing concept but scientific explaination and validation is yet to be explored. Traditional practitioners are wide with different concepts and they includes bonesetters, birth attendants, tooth extractors, (called 'Wogesha' and yelimd awalaj' respectively in amharic) herbalists, as well as 'debtera', 'tenquay' (witch doctors), and spiritual healers such as 'weqaby' and 'kalicha.' Like other african medicine, Ethiopian medicine also believes that some diseases like mental illness is due to evil curses and healing concepts like exorcism also present in TEM.

Traditional Healing in Ethiopian traditional medicine is not only concerned with curing of diseases but also consists of protection and promotion of human physical, spiritual, social, mental and material wellbeing. The concept of Ethiopian medicine is kept hidden and considered as a hidden treasure, which will be passed orally from father to his favorite child.

The various traditional practices includes herbal medications, medications for pshyco social conditions like Exorcisms for Zar, Aganint, Buda, Ayene Tilla etc., Fumigation (inhalation), and Holy water or blessed water. The traditional practices includes Bone setting, Surgery, Cauterisation, Counter-irritation, Bleeding, Cupping, Steam Bath, Vapor Bath (woushba) and Moxibustion.

The Ethiopian traditional medicine includes Orthodox Christian literate healer (debetera), Orthodox Christian astrologer (Metsehaf Gelach), Mystique spiritual healer (Bale Zar), Divine healer (Psychi, Tenquay), Bone-setter (Woggesha), Kitab ketabi (Amult maker), Islamic Literature healer (Kabir), Islamic medical teacher (Sheki), and Cushitic healer (Qaalluu, Qaallicha, Argessa). the practices in ETM consists of preventive, curative, and surgical care. Traditional Ethiopian medicine includes several elements or disease prevention.

The best example of preventive care in ETM is the prevention of diseases like small pox. It was prevented by the traditional healers in following the social distance protocol and people were vaccinated through innoculating by taking pus from sick person during special rituals. Incase of preventive practices, the following protocols like Sweeping or covering floors with particular plants is another traditionally practiced disease preventive measure. The other methods of prevention include kitabs, which are also used for the purpose of protecting an individual against evil eye, as well as snake and scorpion bites. As healers beleive that contagious disease is an evil act or causing by evil eye. The traditional healers also suggest Amulets, arm rings, hair style and eye make-up (antimony or kool) are also supposed to protect from the evil eye.

There are some secret herbs that are useful as charms against an enemy. In addition, cultural rituals and sacrifices are commonly involved in preventive care. In curative practices of ETM, the diseases like gastrointestinal disturbances, respiratory disorders, sexually transmitted infections, tuberculosis, impotency, hemorrhoids, rabies, intestinal parasites, skin problems, liver diseases, mental disorders, hypertension, diabetes, gynecological conditions rheumatism, malaria etc. will be treated using knowledge of traditional herbalism.

One of the well recognized groups of these healers are the secular medhanit awakis (kitel betashs) herbalists using plants as their primary means of providing treatment. All these awakis have their own traditional pharmacopeias.

The surgical practices includes Traditional practices like bone-setting, uvulectomy, circumcisions, bleeding and cupping, cautery, scarification and tooth extraction. The setting of bones is regarded as an important surgical procedure which requires a certain degree of skill and experience on the part of the healer. In most places, the healer involved in bone-setting is the local wogesha. How ever these practices are crude and aseptic with or without the application of medicines.

The documentation of Ethiopian traditional medicine is resitricted only to the scientific literature and the knowledge of traditional pharmacopeia is not published and it was secret/hidden with the healers and practitioners. The ethiopian ethnophamacological information of medicinal plants is fast disappearing and this is more happening in industrialized countries, the erosion of popular information on plants is much faster than in developing ones.

In view of the rapid loss of such knowledge, its documentation as well as a better understanding of its botanico-historical and holistic roots has become an essential task to preserve and restore ethno-allied disciplines [28].

Some of the famous examples in the herbalism of Ethiopian Traditional medicine are as follows

i. *Ocimum basalicum* Linn (Lamiaceae) which is locally known as besobila in amharic, commonly used in ethiopia as culinary spice or herb, traditionally beleived to heal various ailments like malaria, head ache, a common herb in treatment of various microbial infestations. It is also a common insect repellent in Ethiopia [29].

ii. *Brassica nigra* Linn (Brassicaceae) which is locally known as senafitch in amharic, commonly used herb in Ethiopian Traditional Medicine to heal various ailments like stomach disorders, wound healing, and abortifacient [30].

iii. *Nigella sativa* Linn (Ranunculaceae) which is locally known as Tiqur azmud in amharic, commonly used herb in Ethiopian traditional medicine to heal various ailments like bronchitis, rheumatism and related inflammatory diseases [31].

iv. *Capsicum annuum* Linn (Solanaceae) which is locally known as kara or berebere or mita mita in ahmaric,commonly used herb in Ethiopian traditional medicine to heal various ailments like culinary herb, dysentry and vomiting [32].

v. *Cinnamomum zeylanicum* Breyn (Lauraceae) which is locally known as quarafa in amharic, commonly used herb in Ethiopian traditional medicine to heal various ailments like microbial infestations, aphrodisiac [33].

vi. *Coriandrum sativum* Linn (Apiaceae) which is locally known as dimbelal in amharic, commonly used herb in Ethiopian traditional medicine to heal various ailments like stomach and gastro intestinal disorders [34].

vii. *Cuminum cyminum* Linn (Apiaceae) which is locally known as ensilal in amharic, commonly used herb in Ethiopian traditional medicine to heal various ailments like bronchopulmonary disorders, dyspeptic head ache, and stomach bloating [35].

viii. *Linum usitatissimum* (Linn.) (Linaceae) which is locally known as telba in amharic, commonly used herb in Ethiopian traditional medicine to heal various ailments like dietary fiber, purgative, immunomodulatory, anti hyperlipidemic, and wound healing [36].

ix. *Catha edulis* Forsk (Celastraceae) which is locally known as chat in amharic, commonly used herb in Ethiopian traditional medicine to heal various ailments like stimulant, mental illness, gonorrhea and common cold [37].

x. *Ruta chalepensis* Linn (Rutaceae) which is locally known as tenadam in amharic, commonly used herb in Ethiopian traditional medicine to heal various ailments like stomach ache, diarrohea and influenza [38].

2.2 Ghana traditional medicine

Ghanians have developed an unique indigenous healing traditional knowledge which was adapted and defined from their culture, beliefs, and environment, which satisfied their health needs over centuries.

In Ghana, people depends on the traditional knowledge as primary source to heal from various ailments especially in rural areas due to lack of sphisticated medical facilities and being western medicine is an expensive task. The traditional healers and patients ratio in Ghana is 1: 200 approximately. Hence Traditional medicine plays an important role in Ghanian health care. The traditional knowledge is in the hands of spiritual healers but every family have some sort of traditional healing knowledge which was inherited from many generations as folklore medicine.

The healing in Ghanian traditional system includes physical and spirtual aspects. The traditional healers are also known as herbal spiritualist collectively known as "bokomowo", who having occult knowledge towards divination, exorcism and spirtual herbalism. There are various local names for the Ghanian traditional healers like "gbedela" (Ewe), "kpeima" (Dagomba), "odunsini" (Akan), and "isofatse" (Ga). These healers have their own different concepts and principles of understanding the concept of disease. Today The modern herbalism is a major part of Ghanian health care and having their own traditional medicine directorate eastablished by the Ministry of health to provide validated traditional medicine which is quality, safe, and efficacious.

The Ghanian excellence in traditional and alternative medicine have reached to the level of a standardized herbal medicine which is an essential part of modern herbalism. The ministry of health has been established various research centers to validate herbal medicine and incorprated herbalism as a part of university curriculum. There are various degree - awarding medical schools and training many students to graduate them as certified traditional medical doctors. The Ghanian traditional herbal pharmacopeia is also an important acheivement in the journey of Ghanian herbalism [39, 40]. Some of the important herbs in Ghanian herbalism are

i. *Allophylus africanus* P. Beauv. (Sapindaceae) locally known as Odwendwena, the folk uses the bark decoction in the treatment of hemorrhoids and as lactogenic [41].

ii. *Ananas comosus* L. Merr. (Bromeliaceae) locally known as aborobe, the folk uses the fruit and root of this plant to cure jaundice [41].

iii. *Bidens pilosa* Linn. var. Radiata (Asteraceae) locally known as Gyinantwi, the folk uses the whole plant decoction to treat jaundice and hypertension [42].

iv. *Citrus aurantifolia* Linn. (Rutaceae) locally known as anka, the folk uses the fruit juice in treating urinary tract infections [43].

v. *Dialium guineense* Willd. (Fabaceae) locally known as Osenafo, the folk uses the fruit as an anti-diarrheal [44].

vi. *Ocimum canum* Sims. (Lamiaceae) locally known as mme, the folk uses the leaf decoction in treatment of poisoning and malaria [45].

vii. *Oncoba spinosa* Forssk. (Salicaceae) locally known as Astrotoa, the folk uses the leaf and root decoction or powder in the treatment of cough and wounds [46].

viii. *Paullinia pinnata* Linn. (Sapindaceae) locally known as twentini, the folk uses the root and leaf decoction in the treatment of rheumatism, Sexual weakness, and stroke [47].

ix. *Phyllanthus fraternus* G.L. Webster. (Phyllanthaceae) locally known as Bomagueakire, the folk uses the leaf decoction in the treatment of Tuberculosis [48].

x. *Xylopia aethiopica* (Dunal) A. Rich. (Annonaceae) locally known as Hwenetia, the folk uses the seed decoction in the treatment of Chicken pox, stomachache, bladder trouble [49].

2.3 Traditional medicine in South Africa

Traditional Medicine is a part of everyone life in South Africa. Wide variety of concepts and beleifs existed in south african traditional healing system. The herbalism plays an important role in the traditional healing, The combined herbs use in traditional medicine is known as muti and the market where these herbs will sold is known as muti-market. Herbs, animals, mineral drugs will be available in this market. Eleven herbs are very important in muti market and plays a crucial role in isicakathi (Herbalism) and they are as follows

i. *Commelina africana Linn. (Commelinaceae)*: locally known as geeleendagsblom, the folks uses the root concoction in treatment of microbial infestations, venereal diseases, and to treat women with menstrual cramps [50].

ii. *Agapanthus africanus* Linn.(Amaryllidaceae): locally known as kleinbloulelei, the folks uses this herb in pregnancy care and the traditional healers uses to treat ailments related to pregnancy. They use as leaf or root decoction Orally or rectally, to facilitate easy delivery and a healthy child [51].

iii. *Chlorophytum comosum* Linn.(Asparagaceae): locally known as hen-en-kuikens, the folks uses this herb as a protective charm or as an aumlet during pregnancy time against evil spirits for both mother and child [52].

iv. *Ledebouria socialis* (Baker) Jessop. *(Hyacinthaceae)*: commonly known as South African scilla or scilla in the local market. The folks uses this herb in treatment of pregnancy, diarrhea, influenza, backaches, skin irritations and wounds [53].

v. *Ranunculus multifidus* Forssk. (Ranunculaceae): commonly known as isijojokazane in Zulu. The folks uses this herb in treatment of cures for headaches, urinary complaints, throat ulcers and coughs [54].

vi. *Thunbergia atriplicifolia* E.Mey. ex Nees.(Acanthaceae): commonly known as swartoognooi. The local folks uses the herbal infusion as an antiseptic wash in the treatment of sores and swellings [55].

vii. *Kohautia amatymbica* Eckl. & Zeyh.(Rubiaceae): commonly known as Ikhubalo Elimnyama. The local folks uses the root infusion as an emetic [56].

viii. *Plantago afra* Linn. (*Plantaginaceae*): commonly known as plantain. The local folks uses the mucilage in the treatment of wounds, inflammations, and eye irritations [57].

ix. *Gazania linearis* Linn. (Asteraceae): commonly known as african daisy. The local folks used to prevent miscarriage and tooth ache [58].

x. *Helichrysum pedunculatum* Hilliard & B.L.Burtt. (Asteraceae): commonly known as impepho. The local folks used as an antiseptic and wound healing [58].

xi. *Senecio coronatus* (Thunb.) Harv. Aka. (Asteraceae): commonly known as *Indlebe Yebokwe*. The local folks used to get rid of pubic lice [58, 59].

2.4 Other parts of Africa

Eritrea is one of African country and has been known for its traditional medicine practice. They use different plants for different diseases. In Eritrea plant called *Kalanchoe marmorata* baker and belongs to the family of Crassulaceae. In local name it is known as Saniaco which has been using for Cold, Intestinal parasites & Burns. some of the important herbs that have been using in Eriterea are mentioned in the [60].

The folks of Uganda as other Africa countries have rich in traditional health care knowledge for addressing various health problems and 80% people uses medicinal herbs atleast once in daily life. For instances in northern sector of kibale national park, the folks have been using *Vernonia amygdalina* Del. (Asteraceae) and Albizia coriaria. Welw. ex Oliv (Fabaceae) in the treatment of malaria and cough respectively. Some other common medicinal plants that have been using *Coffea arabica* Linn (Rubiaceae), *Coffea canephora* Pierre ex A.Froehner (Rubiaceae) for treatingdiarrhoea, *Crassocephalum vitellinum* S.Moore (Rubiaceae) in treating wound, Turraea africana (Welw.) Cheek (Meliaceae) for worm infestations [60]. In Zambia, 75% of people depend on traditional medicinal knowledge to treat various ailments like infertility, wound healing. The famous plant used in zambian societies are *Aloe vera Linn (Aloeaceae)* as an antiseptic, wound healing, antitussive, skin irritant. Various herbs like *Terminalia sericea* Burch. ex DC. (Combretaceae), Strychnos cocculoides Baker. (Strychnaceae), *Ximenia caffra* Sond.(Olacaceae), *Cassia abbreviata* oliv. (Fabaceae), *Combretum hereroense* Schinz. (Combretaceae), *Combretum imberbe* Wawra. (Combretaceae), *Dichrostachys cinerea* Linn. (Fabaceae), *Boscia foetida* Schinz. (Capparaceae), *Momordica balsamina* Linn. (Cucurbitaceae) and *Peltophorum africanum* Sond (Fabaceae) [61].

3. Conclusion

In Africa, Traditional medicine is one of the important health care system till today. The traditional african medicine is now evolved as an evidence based healing system and serving as a good prime element in reverse pharmacology and drug discovery. Hence Traditional african medicine is a break through in the drug discovery process.

Even now a days, serious attention has been made on the quality, Safety, and efficacy from the evolved african countries like South Africa, Ethiopia, Ghana, and kenya and there is a paradigm shift from non- validated traditional herbalism to validated modern herbalism which is even accepting by the modern doctors about the magical healing effects of these herbs in diseases like cancer, Covid-19, HIV, and tuberculosis. The current demand for African traditional healing concepts like spiritualism, humanisim, and herbalism towards the disease preventive care and African herbalism has been became an important bioprospecting tool in drug discovery of new drug molecules.

S.no	Scientific name	Family name	Local name	African location	Parts of used	Traditional uses/ Folk remedies
1.	*Agapanthus africanus* Linn	Amaryllidaceae	kleinbloulelei	South Africa	Leaf, Root	To treat various ailments related to pregnancy. They use as leaf or root decoction Orally or rectally, to facilitate easy delivery and a healthy child
2.	*Allium sativum Linn*	Alliaceae	Shigueti tseada	Eritrea	bulb	Hypertension,Malaria and Asthma
3.	*Allophylus africanus* P. Beauv.	Sapindaceae	Odwendwena	Ghana	bark	hemorrhoids and as lactogenic
4.	*Aloe macrocarp* Todaro.	Aloeaceae	Tsebir	Eritrea	Latex	Impotency, Malaria & easing labor
5.	*Ananas comosus L. Merr.*	Bromeliaceae	aborobe	Ghana	Fruit and Root	jaundice
6.	*Brassica nigra* Linn.	Brassicaceae	Senafitch	Ethiopia	Leaves, seeds	Stomach disorders, wound healing, and abortifacient
7.	*Bidens pilosa Linn.* var. Radiata.	Asteraceae	Gyinantwi	Ghana	Whole plant	Decoction to treat jaundice and hypertension
8.	*Capsicum annuum* Linn.	Solanaceae	kara or berebere	Ethiopia	Fruits	Culinary herb, dysentry and vomiting
9.	*Carica papaya* Linn.	Caricaceae	papayo	Eritrea	seed	Diabetes, Amoeba and Typhoid fever
10.	*Catha edulis Forsk.*	Celastraceae	Chat	Ethiopia	Leaf	Stimulant, mental illness, gonorrhea and common cold
11.	*Cinnamomum zeylanicum Breyn.*	Lauraceae	Quarafa	Ethiopia	Bark	Microbial infestations, aphrodisiac
12.	*Citrus aurantifolia* Linn.	Rutaceae	Anka	Ghana	Fruit juice	Treating urinary tract infection
13.	*Chlorophytum comosum* Linn.	Asparagaceae	Hen-en-kuikens	South Africa	Root/ stem	Protective charm or as an aumlet during pregnancy time against evil spirits for both mother and child
14.	*Commelina africana Linn.*	Commelinaceae	geeleendagsblom	South Africa	Root	Microbial infestations, venereal diseases, and to treat women with menstrual cramp
15.	*Dialium guineense* Willd.	Fabaceae	Osenafo	Ghana	Fruit	Antidiarroheal
16.	*Dodonaea viscosa* Linn.	Sapindaceae	Taheses	Eritrea	leaves	Worm infestation and dandruff

S.no	Scientific name	Family name	Local name	African location	Parts of used	Traditional uses/ Folk remedies
17.	*Gazania linearis* Linn.	Asteraceae	African daisy	South Africa	Aerial parts	Local folks used to prevent miscarriage and tooth ache
18.	*Helichrysum pedunculatum* Hilliard & B.L.Burtt.	Asteraceae	Impepho	South Africa	Aerial parts	Antiseptic and wound healing
19.	*Kohautia amatymbica* Eckl. & Zeyh.	Rubiaceae	Ikhubalo Elimnyama	South Africa	Root	Root infusion as an emetic
20.	*Ledebouria socialis (Baker)* Jessop.	Hyacinthaceae	South African scilla	South Africa	Aerial parts	Treatment of pregnancy, diarrhea, influenza, backaches, skin irritations and wounds.
21.	*Linum usitatissimum* (Linn.)	Linaceae	Telba	Ethiopia	Seeds, husk	Dietary fiber, purgative, immunomodulatory, anti hyperlipidemic, and wound healing.
22.	*Nigella sativa* Linn.	Ranunculaceae	Tiqur azmud	Ethiopia	Leaves, seeds	Bronchitis, rheumatism and related inflammatory diseases.
23.	*Ocimum basalicum* Linn.	Lamiaceae	Besobila	Ethiopia	Whole plant / Leaves	Malaria, head ache, a common herb in treatment of various microbial infestations and mosquito repellant.
24.	*Ocimum canum Sims.*	Lamiaceae	mme	Ghana	Leaf decoction	Treatment of poisoning and malaria.
25.	*Oncoba spinosa* Forssk.	Salicaceae	Astrotoa	Ghana	Leaf and Root	Treatment of cough and wounds.
26.	*otostegia fruticosa ssp.* Schimperi.	Lamiaceae	Fashadima	Eritrea	Leaves	Tonsillitis arthritis and Endo parasite.
27.	*Paullinia pinnata* Linn.	Sapindaceae	twentini	Ghana	Root and leaf	Rheumatism, Sexual weakness, and stroke.
28.	*Phyllanthus fraternus* G.L. Webster.	Phyllanthaceae	Bomagueakire	Ghana	Leaf	Tuberculosis
29.	*Plantago afra* Linn.	Plantaginaceae	Plantain	South Africa	Aerial Parts	Mucilage in the treatment of wounds, inflammations, and eye irritations
30.	*Pollichia campestris* Aiton	Carryophllaceae	Hareg bait	Eritrea	Root	Snake bite, tonsillitis, eye disease

S.no	Scientific name	Family name	Local name	African location	Parts of used	Traditional uses/ Folk remedies
31.	*Prosopis juliflora (SW) D.C.*	Fabaceae	Temer musa	Eritrea	Pods	Lactation, digestion disturbance.
32.	*Ranunculus multifidus Forssk.*	Ranunculaceae	Isijojokazane	South Africa	Aerial parts	Folks uses this herb in treatment of cures for headaches, urinary complaints, throat ulcers and coughs
33.	*Senecio coronatus (Thunb.) Harv. Aka.*	Asteraceae	Indlebe Yebokwe.	South Africa	Seeds	To get rid of public lice.
34.	*Thunbergia atriplicifolia* E.Mey. ex Nees.	Acanthaceae	Swartoognooi	South Africa	Leaves	Folks uses the herbal infusion as an antiseptic wash in the treatment of sores and swellings
35	*Xylopia aethiopica* (Dunal) A. Rich.	Annonaceae	Hwenetia	Ghana	Seed	Chicken pox, stomachache,bladder trouble

Table 1.
Some common medicinal plant used in different parts of Africa.

Figure 1.
Traditional African medicine.

The main lacuna of this traditional medicine in Africa are documentation evidences regulation, Quality control, and standardization. This is possible only by integrating the traditional medicine with modern medicine. Hence the current chapter is focused on the importance of Traditional african medicine in terms of herbalism and belief based healing system. To make the chapter for easy understanding, video materials, **Table 1** and **Figure 1** were incorporated.

Acknowledgements

We would like to thank Department of pharmacy, Dilla university, Dilla, Ethiopia for providing necessary resources and encouraging, we would like to thank Mrs. Pavithra Motamarri for her support in preparing the photo materials with traditional african healer Mr. Lulama at pretoria.

Conflict of interest

The authors declare no conflict of interest.

DOI: http://dx.doi.org/10.5772/intechopen.96576

Author details

Motamarri V.N.L. Chaitanya[1*], Hailemikael Gebremariam Baye[1], Heyam Saad Ali[2] and Firehiwot Belayneh Usamo[1]

1 Department of Pharmacy (Pharmacognosy), College of Medicine and Health Sciences, Dilla University, Ethiopia

2 Pharmaceutics Department, University of Khartoum, Khartoum, Sudan

*Address all correspondence to: chaitanya.phyto@gmail.com

References

[1] WHO. Legal Status of Traditional Medicine and Complementary/ Alternative Medicine: World Wide Review. Geneva [Internet]. 2001. Available from : https://apps.who.int/ iris/handle/10665/42452.

[2] Dawit A, Ahadu A. Medicinal plants and enigmatic health practices of northern Ethiopia [Internet]. 1993. Available from: https:// agris.fao.org/agris-search/search. do?recordID=XF2015013726.

[3] WHO. Traditional Medicine [Internet]. 2003. Available from: https:// apps.who.int/gb/archive/pdf_files/ WHA56/ea5618.pdf

[4] WHO. Traditional Medicine [Internet]. 2020. Available from: http:// www.emro.who.int/health-topics/ traditional-medicine/introduction.html.

[5] Busia K. Medical provision in Africa -- past and present. Phytother Res.2005; 19: 919-23. DOI: 10.1002/ptr.1775.

[6] James PB, Wardle J, Steel A, Adams J. Traditional, complementary and alternative medicine use in Sub-Saharan Africa: a systematic review. BMJ Global Health. 2018;3: e000895-13. DOI:10.1136/bmjgh-2018-000895.

[7] Badgujar SB, Patel VV, Bandivdekar AH. Foeniculum vulgare Mill: a review of its botany, phytochemistry, pharmacology, contemporary application, and toxicology. Biomed Res Int. 2014;2014:842674-06. DOI:10.1155/2014/842674.

[8] Dhuguru J, Skouta R. Role of Indole Scaffolds as Pharmacophores in the Development of Anti-Lung Cancer Agents. Molecules. 2020;25:1615-50. DOI:10.3390/molecules25071615.

[9] Oreagba IA, Oshikoya KA, Amachree M. Herbal medicine use among urban residents in Lagos, Nigeria. BMC Complement Altern. Med. 2011 ;11:117-25. DOI: 10.1186/1472-6882-11-117.

[10] Ezekwesili-Ofili JO, Okaka ANC. Herbal Medicines in African Traditional Medicine, Philip F editor. Herbal Medicine. Ist ed. Builders: IntechOpen; 2019. DOI: 10.5772/intechopen.80348. Availablefrom: https://www.intechopen. com/books/herbal-medicine/herbal-medicines-in-african-traditional-medicine.

[11] Burke-Gaffney HJ. The history of medicine in the African countries. Med Hist. 1968;12:31-41. doi:10.1017/ s0025727300012746.

[12] Abdullahi AA. Trends and challenges of traditional medicine in Africa. Afr J Tradit Complement Altern Med. 2011;8:115-23. doi:10.4314/ajtcam.v8i5S.5

[13] Larson P, Leeming D.A, editors. Encyclopedia of Psychology and Religion. Ist ed. Springer, Boston, MA, 2014. 174 P. DOI: https://doi. org/10.1007/978-1-4614-6086-2_174.

[14] Omonzejele PF, Maduka C. Metaphysical and value underpinnings of traditional medicine in West Africa. Chin J Integr Med. 2011; 17:99-04. doi: 10.1007/s11655-011-0649-y.

[15] Wigington P. Methods of Divination [Internet]. 2019. Available from: https://www.learnreligions.com/ methods-of-divination-2561764.

[16] Johnson KS, Elbert-Avila KI, Tulsky JA. The influence of spiritual beliefs and practices on the treatment preferences of African Americans: a review of the literature. J Am Geriatr Soc. 2005 ;53:711-9. doi: 10.1111/j.1532-5415.2005.53224.x.

[17] Scholtz HJ, Phillips VM, Knobel GJ. Muti or ritual murder. Forensic Sci Int. 1997; 87:117-23. doi: 10.1016/ s0379-0738(97)02132-4.

[18] Mabogunje OA. Ritual hot baths (wankan-jego) in Zaria, Nigeria. Newsl Inter Afr Comm Tradit Pract Affect Health Women Child. 1990;9:10-20. PMID: 12157981.

[19] Rabiu A, Garba I, Abubakar IS. Ritual hot bath (wankan jego) in Kano: Are they still practicing? What are the implications?Sahel Med J. 2016;19:215-9. doi: 10.4103/1118-8561.196368.

[20] Béliveau VG, Fernández NS. We are body, soul and spirit: Person, disease and processes of healing and exorcism in contemporary Catholicism in Argentina. Salud Colect. 2018; 14:161-77. doi: 10.18294/sc.2018.1504.

[21] Nehusi K. Libation: A Ritual of Heritage in African Life [Internet].2013. Available from: https://africanholocaust.net/african-libation/.

[22] Vickers A, Zollman C, Lee R. Herbal medicine. West J Med. 2001;175s:125-8. doi:10.1136/ewjm.175.2.125.

[23] Fawzi Mahomoodally M. Traditional Medicines in Africa: An Appraisal of Ten Potent African Medicinal Plants. Evid Based Complement Alternat Med. 2013; 1: 1-15.

[24] Alevtina G, Zerihun S. Ethiopian Traditional Medications and their Interactions with Conventional Drugs [Internet]. 2009. Available from: https://ethnomed.org/resource/ethiopian-traditional-and-herbal-medications-and-their-interactions-with-conventional-drugs/

[25] Bishaw M. Promoting traditional medicine in Ethiopia: a brief historical review of government policy. Soc Sci Med. 1991;33:193-200. doi: 10.1016/0277-9536(91)90180-k.

[26] Grossman E. [Complementary medicine--the facts]. Harefuah. 2011. 150:657-8,

[27] Romeo N, Gallo O, Tagarelli G. From Disease to Holiness: Religious-based health remedies of Italian folk medicine (XIX-XX century). J Ethnobiol Ethnomed. 2015. 6;11-50. doi: 10.1186/s13002-015-0037-z.

[28] Leonti M, Casu L. Traditional medicines and globalization: current and future perspectives in ethno-pharmacology. Front Pharmacol. 2013.4; 92-05. doi:10.3389/fphar.2013.00092.

[29] Bekalo TH, Woodmatas SD, Woldemariam ZA. An ethnobotanical study of medicinal plants used by local people in the lowlands of Konta Special Woreda, southern nations, nationalities and peoples regional state, Ethiopia. J Ethnobiol Ethnomed. 2009.5; 26-55. doi:10.1186/1746-4269-5-26.

[30] Muluye AB, Melese E, Adinew GM. Antimalarial activity of 80% methanolic extract of Brassica nigra (L.) Koch. (Brassicaceae) seeds against Plasmodium berghei infection in mice. BMC Complement Altern Med. 2015.15; 367-75. doi: 10.1186/s12906-015-0893-z.

[31] Edeget M. Identification of Nigella sativa for Access and Benefit sharing purpose [Internet]. 2016. Available from: http://et.chm-cbd.net/news/identification-of-nigella-sativa-for-access-and-benefit-sharing-purpose#:~:text=Traditional%20uses%20of%20Nigella%20sativa and text=Black%20seeds%20and%20their%20oil, rheumatism%20and%20related%20inflammatory%20diseases.

[32] Mesfin F, Demissew S, Teklehaymanot T. An ethnobotanical study of medicinal plants in Wonago Woreda, SNNPR, Ethiopia. J Ethnobiology Ethnomedicine.2009.5; 28-46. doi: 10.1186/1746-4269-5-28.

[33] Nasir M, Tafess K, Abate D. Antimicrobial potential of the Ethiopian Thymus schimperi essential oil in comparison with others against certain fungal and bacterial species. BMC Complement Altern Med. 2015.15; 260-78. doi: 10.1186/s12906-015-0784-3.

[34] Eguale T, Tilahun G, Debella A, Feleke A, Makonnen E. In vitro and in vivo anthelmintic activity of crude extracts of *Coriandrum sativum* against Haemonchus contortus. J Ethnopharmacol. 2007.4; 428-33. doi: 10.1016/j.jep.2006.10.003.

[35] Johri RK. Cuminum cyminum and *Carum carvi*: An update. Pharmacogn Rev. 2011; 5: 63-72. doi:10.4103/0973-7847.79101.

[36] Ramin A, Mohammad MZ, Amir HD. A Review on Pharmacological and Clinical Aspects of Linum usitatissimum L. Curr. Drug Discov. Technol. 2019; 16: 148-56. doi: 10.2174/1570163815666180521101136.

[37] Wabe NT. Chemistry, pharmacology, and toxicology of khat (catha edulis forsk): a review. Addict Health. 2011;3: 137-49.

[38] Seble WY, Zemede A, Kelbessa E. Ethnobotanical study of medicinal plants used by local people in Menz Gera Midir District, North Shewa Zone, Amhara Regional State, Ethiopia.J. Med. Plant Res.2018;12: 296-14.

[39] Sawadogo WR, Schumacher M, Teiten MH, Dicato M, Diederich M. Traditional West African pharmacopeia, plants and derived compounds for cancer therapy. Biochem Pharmacol. 2012; 15:1225-40. doi: 10.1016/j.bcp.2012.07.021.

[40] Boadu AA, Asase A. Documentation of Herbal Medicines Used for the Treatment and Management of Human Diseases by Some Communities in Southern Ghana. Evid Based Complement Alternat Med. 2017; 2017: 3043061-74. doi: 10.1155/2017/3043061

[41] Agbovie T, Amponsah K, Crentsil OR, Dennis F, Odamtten GT, Ofusohene-Djan W. Conservation and sustainable use of medicinal plants in Ghana: Ethnobotanical survey [Internet]. 2002. Available from : http://www.unepwcmc.org/species/plants/ghana

[42] Addo-Fordjour P, Anning A K, Belford E J D, Akonnor D. Diversity and conservation of medicinal plants in the Bomaa community of the Brong Ahafo region, Ghana. J. Med. Plant Res. 2008; 2: 226-33.

[43] Alexandra P, Cox P. An Ethnobotanical Survey of the Uses for *Citrus aurantium* (Rutaceae) in Haiti. Econ. Bot.1995; 49: 249-56.

[44] Martial SG, Samy I, Maxime R, Jean-Michel B, Edwige D, Philippe P. Dialium guineense Willd. *Parkia biglobosa* (Jacq.) R. Br. Ex Benth. and *Tamarindus indica* L.: Review of known and synergetic bioactive compounds. J. Med. Plants. Stud. 2018; 6:103-11. orcid: ffhal-01998395f.

[45] Akono NP, Baldovini N, Mouray E, Mambu L, Belong P, Grellier P. Activity of *Ocimum basilicum*, Ocimum canum, and *Cymbopogon citratus* essential oils against *Plasmodium falciparum* and mature-stage larvae of Anopheles funestus. Parasite. 2014;21:33-41. doi:10.1051/parasite/2014033

[46] Djouossi MG, Tamokou JD, Ngnokam D, Kuiate JR, Tapondjou LA, Harakat D, Voutquenne-Nazabadioko L. Antimicrobial and antioxidant flavonoids from the leaves of Oncoba spinosa Forssk. (Salicaceae). BMC Complement Altern Med. 2015; 28:134-41. doi: 10.1186/s12906-015-0660-1.

[47] ZambleA, CarpentierM, KandoussiA, Sahpaz S, Petrault O, Ouk T, Hennuyer N, Fruchart JC, Staels B, Bordet R, Duriez P, Bailleul F, Martin-Nizard F. Paullinia pinnata extracts rich in polyphenols promote vascular relaxation via endothelium-dependent mechanisms. J Cardiovasc Pharmacol. 2006;47:599-08. doi: 10.1097/01.fjc.0000211734.53798.1d.

[48] Zamble A, Carpentier M, Kandoussi A, Sahpaz S, Petrault O, Ouk T, Hennuyer N, Fruchart JC, Staels B, Bordet R, Duriez P, Bailleul F, Martin-Nizard F. Paullinia pinnata extracts rich in polyphenols promote vascular relaxation via endothelium-dependent mechanisms. J Cardiovasc Pharmacol. 2006;47:599-08. doi: 10.1097/01.fjc.0000211734.53798.1d.

[49] Yin X, Chávez León MASC, Osae R, Linus LO, Qi LW, Alolga RN. Xylopia aethiopica Seeds from Two Countries in West Africa Exhibit Differences in Their Proteomes, Mineral Content and Bioactive Phytochemical Composition. Molecules. 2019;24:1979-92. doi:10.3390/molecules24101979

[50] Kudumela RG, Masoko P. In Vitro Assessment of Selected Medicinal Plants Used by the Bapedi Community in South Africa for Treatment of Bacterial Infections. J Evid Based Integr Med. 2018;23:2515690X18762736-46. doi: 10.1177/2515690X18762736.

[51] Kaido TL, Veale DJ, Havlik I, Rama DB. Preliminary screening of plants used in South Africa as traditional herbal remedies during pregnancy and labour. J Ethnopharmacol. 1997; 55:185-91. doi: 10.1016/s0378-8741(96)01499-7.

[52] Dlisani PB, Bhat RB. Traditional Health Practices in Transkei with Special Emphasis on Maternal and Child Health . Pharm. Biol. 1999; 37:32-6. doi:10.1076/phbi.37.1.32.6316.

[53] PBS (Pacific Bulb Society). Ledebouria socialis. [Internet]. 2018. available from: https://www.pacificbulbsociety.org/pbswiki/index.php/Ledebouria_socialis.

[54] Naidoo D, van Vuuren SF, van Zyl RL, de Wet H. Plants traditionally used individually and in combination to treat sexually transmitted infections in northern Maputaland, South Africa: antimicrobial activity and cytotoxicity. J Ethnopharmacol.2013;149:656-67. doi: 10.1016/j.jep.2013.07.018.

[55] Sultana KW, Chatterjee S, Roy A, Chandra I. An Overview on Ethnopharmacological and Phytochemical properties of Thunbergiasp. Med Aromat Plants. 2015; 4:5-8. doi: 10.4172/2167-0412.1000217.

[56] Steenkamp V. Traditional herbal remedies used by South African women for gynaecological complaints. J Ethnopharmacol. 2003; 86: 97-08. doi:10.1016/s0378-8741(03)00053-9.

[57] Adom MB, Taher M, Mutalabisin MF, Amri MS, Abdul Kudos MB, Wan Sulaiman MWA, Sengupta P, Susanti D. Chemical constituents and medical benefits of *Plantago major*. Biomed Pharmacother. 2017;96:348-60. doi: 10.1016/j.biopha.2017.09.152.

[58] Bolofo RN, Johnson CT. "The identification of 'Isicakathi' and its medicinal use in Transkei". Bothalia. 1988; 18: 125-30. doi:10.4102/abc.v18i1.995.

[59] Yemane B, Medhanie G, Reddy KS. Survey of Some Common Medicinal Plants Used In Eritrean Folk Medicine. International Journal of Medicinal Plants. Photon.2018; 112: 865-76.

[60] Omara T, Kiprop AK, Ramkat RC, et al. Medicinal Plants Used in Traditional Management of Cancer in Uganda: A Review of Ethnobotanical Surveys, Phytochemistry, and Anticancer Studies. Evid Based Complement Alternat Med. 2020;2020:3529081-07. doi:10.1155/2020/3529081.

[61] Chinsembu KC. Ethnobotanical Study of Plants Used in the Management of HIV/AIDS-Related Diseases in Livingstone, Southern Province, Zambia. Evid Based Complement Alternat Med. 2016;2016:4238625-39. doi:10.1155/2016/4238625.

Chapter 4

Ethnomedicine Study on Medicinal Plants Used by Communities in West Sumatera, Indonesia

Skunda Diliarosta, Monica Prima Sari, Rehani Ramadhani and Annisa Efendi

Abstract

Currently, the development of conventional medicine is getting more advanced, it cannot be denied that medicinal plants still occupy their main role as medicine for various human diseases, especially in developing countries. This is rooted in the knowledge of the local community about plants that can be used as medicine for various diseases. Ethnomedicine is a field of study that raises local knowledge of the community to maintain their health. From numerous studies on the field, 33 species of plants have been found which are believed by the natives to West Sumatra as medicine. Ethnomedicinal data were analyzed using Index of Cultural Significance (ICS) value. The results of the analysis showed that the species of plants that is voted most important for the community were soursop (*Annona muricata*) and red betel (*Piper sp.*). In general, the part of plant that is most often used as medicine is the leaf, and the way to consume it is by boiling it so that you can get the herbs from the plant extract.

Keywords: ethnomedicine, quantitative analysis, medicinal plants, local community, West Sumatera

1. Introduction

Indonesia has around 25,000-30,000 species of plants and is inhabited by around 300- 700 ethnicities. These ethnic groups use it for various purposes, one of which is for medicinal purposes. The use of plants as medicinal substances is mostly passed down orally, so they are prone to degradation. Ethnomedicine study is a method that can be used to document the use of plants by ethnic groups with scientifically acceptable research methods. This paper aims to explain the study of ethnomedicine especially in West Sumatra and its research methods.

One of the local wisdoms possessed by Indonesians is to utilize the natural biological resources in the vicinity. Every local community uses their vegetable resources to fulfill their daily needs, one of which is to maintain their health which is known as medicinal plants. Knowledge on the use of medicinal plants is generally passed down orally, so that knowledge is limited to certain groups of people and is susceptible to degradation due to cultural acculturation and modernization.

The use of plants to maintain health has long been carried out in Indonesia in line with the development of civilization. Indonesia has been formulating and using medicinal plants (traditional medicine) since the era of Hindu-Javanese kingdom. In West Sumatra itself, medicinal plants have been an alternative treatment since the time of our ancestors. One of the plants that is widely used in the experiment is the soursop plant (Latin name), betel plant (Latin name) and castor plant (Latin name). The part of the plant that is used as medicine is the leaves.

2. Ethnomedicine study on medicinal plants used by local communities in West Sumatera, Indonesia

2.1 Ethnomedicine

Ethnomedicine is a branch of medical anthropology which deals with the origin of disease, causes, and treatment according to certain groups of people. The ethnomedicine aspect is an aspect that appears along with the development of human culture. In the field of medical anthropology, ethnomedicine gives rise to various therminologies. This branch is often called folk medicine, primitive medicine, however ethnomedicine is considered more appropriate [1].

Ethnomedicine is a field of ethnobotany studies that reveals local knowledge of various ethnicities in maintaining their health. Empirically, it can be seen that traditional medicine uses both plants and animals. However, in terms of the number and frequency of use, plants are more widely used than animals. Eventually, this resulted in traditional medicine being identical to medicinal plants.

Currently, ethnomedicine research is aimed at finding new chemical compounds that are useful in the manufacture of modern drugs for dangerous diseases, such as cancer drugs. Up until now, most of the drugs used for cancer treatment are still extracted directly from plants because synthetic compounds cannot be made or their production costs are much more expensive than direct extraction from plants. In addition, treatment for diseases which are currently developing, the new purpose of ethnomedicine research is to find new compounds with fewer side effects, the emergence of resistant effects from existing drugs, and also to anticipate the emergence of new diseases. This has resulted in ethnomedicine research continuing to develop, especially in countries rich in biodiversity such as Indonesia.

The use of plants as herbs and medicine traditionally or often referred to as empirical is often associated with uses that have no scientific basis at all. Even though research is so advanced, it is very possible that in the past the use of traditional medicine was only based on lineage and undocumented experience and there was no scientific data. Now, there are numerous recent studies that support the practice of using plants for the treatment of various diseases. Exploration of Local Knowledge of Ethnomedicine and Community-Based Medicinal Plants in Indonesia, hereinafter referred to as Research on Medicinal Plants and Herbs (RISTOJA) has succeeded in collecting data related to the use of plants for medicinal purposes in almost every ethnicity in the territory of Indonesia (34 Provinces) [2].

2.2 West Sumatera

West Sumatra is one of the provinces in Indonesia located on the west coast in the central part of Sumatra island which consists of lowlands on the west coast and volcanic plateaus formed by the Bukit Barisan on the eastern side. This province has a land area of 42,297.30 km^2 which is equivalent to 2.17% of Indonesia's area. More than 45.17% of this area is still covered by protected forest. The coastline of this province is entirely in contact with the Indian Ocean along 2,420,357 km with a sea area of 186,580 km^2. Mentawai Islands, which are located in the Indian Ocean, are included in this province (**Figure 1**).

Astronomically, West Sumatra is located between 00.54 'North Latitude and 30.30' South Latitude and between 98.36' – 101.53 'East Longitude and is traversed by the equator or the equator. Based on its geographical position, West Sumatra Province has the following boundaries: North - North Sumatra and Riau Provinces; South - Indian Ocean; West - Indian Ocean; East - Jambi and Bengkulu Provinces. Located on the west coast of the central part of the island of Sumatra with an area of approximately 42.2 thousand square kilometers.

Like other regions in Indonesia, the climate of West Sumatra is generally tropical with temperatures quite high, between 22.6 ° C to 31.5 ° C. This province is also traversed by the equator, precisely in Bonjol city, Pasaman district. There is a number of large rivers flow from this province into the east coast of Sumatra, such as Batang Hari, Siak, Inderagiri (referred to as Batang Kuantan in the upper part), and Kampar. Meanwhile, the rivers that flow into the west coast are Batang Anai, Batang Arau, and Batang Tarusan.

2.3 Study area

This research was conducted to identify plants used by the people of West Sumatra as medicinal plants. This research was conducted in several areas in West Sumatra, namely Padang city, Padang Pariaman district, Pariaman city, Padang Panjang city, Bukittinggi city, and Payakumbuh city.

2.4 Data collection

Field observations were carried out in January-April 2020. Using the purposive sampling method, ethnomedicine data were collected through semi-structural interviews and discussions from 18 informants. Information regarding local medicinal plants, parts used, method of application and preparation is recorded. Data on the gender, age and educational status of informants were also collected. Plant specimens were also collected to help identify the medicinal plant species obtained.

Figure 1.
Map of West Sumatra. Source: perpustakaan.menlhk.go.id.

2.5 Demographic data of informants

In this study, 18 informants were involved, consisting of 12 men and 6 women. 4 informants came from Padang city, 3 from Padang Pariaman district, 3 from Pariaman city, 5 from Padang Panjang city, 2 from Bukittinggi city and 1 from Payakumbuh city. The highest number of informants is over 50 years old (**Table 1**).

From this research, it can be seen that older informants have more knowledge about medicinal plants than younger informants. This may be due to the lifestyle of the younger group which is more modern, they are not too interested in natural medicinal plants and prefer modern medicines obtained from doctors [3, 4].

Knowledge about the use of medicinal plants is largely derived orally (75%) from their ancestors, this type of inheritance method is common for traditional healers [5]. Thus seldom is there documentation for their practice and therefore there is an urgent need to document all information about the traditional practice of using medicinal plants especially for the treatment of growth determinants.

2.6 Plant inventory

There are 33 medicinal plants (**Figure 2**) that are believed by the local community as medicinal plants that can cure various diseases, a complete list of plants is presented in the **Table 2**. Local names, taxonomic names, parts used and method of preparation are also given.

2.7 Plant parts being used and preparation

From the PPV (**Figure 3**) it was revealed that the leaves were the most widely used part, namely 58% followed by fructus 11%. The findings of this study are similar to those of other ethnomedicine studies [6–8], and most traditional healers use the leaves perhaps because it is relatively abundant and also to preserve and preserve medicinal plant species. Herb, direct eating (raw parts or juice) and direct use (crushed plant parts topically) are various methods used in traditional healing

Category	Group	n
Gender	Men	12
	Girl	8
Age	20-40 years	5
	41-50 years	6
	51-60 years	7
	More than 60 years	2
District/City	Padang	4
	PadangPariaman	3
	Pariaman	3
	Padang Panjang	5
	Bukittinggi	3
	Payakumbuh	2

Table 1.
Informant demographic data (n =).

Figure 2.
Plant specimens.

Determinants	Local name	Taxonomic name	Method of preparation	The part used
Cancer	Sirsak	*Annona muricata*	Dec	Jui
Natural betadine	Parancih Betadin	*Jatropha multifida Linn*	Du	Jui
Cancer	Jengkol	*Archidendron pauciflorum*	De	Fru
Cancer, Antioxidants	Dalimo	*Punica granatum*	De	Fru
Maintain body temperature stability	Petai	*Parkia speciosa Hassk*	De	Fru
Cancer	Benalu Kopi	*Scurrula ferruginea (Jack) dancer*	Dec	Jui
Diabetes	Sambung Nyawa	*Gynura procumbens*	Dec	Jui
Internal Medicine	Sitawa	*Andrographis paniculata*	Du	Rhi
Cancer	Bawang Dayak	*Eleutherine palmifolia*	Dec	Tub
Hemorrhoids, fever	Sidingin	*Kalanchoe pinnata*	Du	Jui
Antimicrobial	Cikarau	*Enhydra fluctuans Lour*	Dec	Jui
Lowering blood pressure and blood sugar levels	Sarai Harum	*Cymbopogon nardus (L.) Rendl.*	Dec	Cau
Colds and diarrhea	Kayu Manis	*Cinnamomum sp*	Dec	Cau
Cancer	Sirih Merah	*Piper sp*	Dec	Jui
Indigestion	Daun Sikaduduak	*Melastoma candidum*	Dec	Jui,Rad, Fru, Sem
Diarrhea, dysentery, colds	Jeriangau	*Acorus calamus*	Dec	Rhi
Antioxidants	Bunga Kembang Sepatu	*Hibiscus rosasinensis*	Dec	Jui
Antioxidants	Lagundi	*Vitex trifolia*	Dec	Jui
Reducing fever	Binahong	*Anifere cordifolia*	Dec	Jui
Relieve rheumatism, tuberculosis	Seringan-seringan	*Flemingia strobilifera*	Dec, Du	Jui
Diabetes, rheumatism, cancer	Pandan cina	*Pandanus odorus*	Dec	Jui
Lower uric acid levels	Sijanggi	*Cosmos caudatus*	Dec, Du	Jui
Diabetes mellitus, cholesterol	Teh Afrika	*Nernobia amygdaliris*	Dec	Jui
Itchy	Pugaran	-	Du	Jui
Kidney Damage	Benalu Jeruk	*Dendrophthoe Glabresseris*	Dec	Jui
Cancer	Cemara Sumatera	*Taxus sumaterana*	Dec	Jui, Cor
Diabetes	Keji Beling	*Strobilanthes crispa*	Dec	Jui
Ulcer, low blood pressure, cancer	Benalu Coklat	*D. Pentandra*	Dec	Jui
Skin inflammation, indigestion	Kunyit Putih	*Curcuma Zedoaria*	Dec	Rhi

Determinants	Local name	Taxonomic name	Method of preparation	The part used
Cough, asthma, bronchitis	Akar Wangi	*Polygala paniculata*	Due	Jui
Antimicrobial	Akar kuning	*Arcangelisia flava Merr.*	Due	Cau
Hypertension	Pohon Mindi	*Melia azedarach*	Due	Rad
Blood circulation	Pohon Mahoni	*Swietenia mahagoni*	De	Fru

*Abbreviation: **ICS**: Index of Cultural Significance; **PPV**: Plant Part Value; **Fru**: Fructus (Fruit); **Sem**: Semen (Seed); **Tub**: Tuber; **Rhi**: Rhizome; **Cor**: Cortex; **Jui**: Juice; **Rad**: Radix (Root); **Cau**: Caulis; **Dec**: Decoction; **De**: Direct eat (raw part or the juice); **Du**: Direct use (topical).*

Table 2.
Medicinal plants.

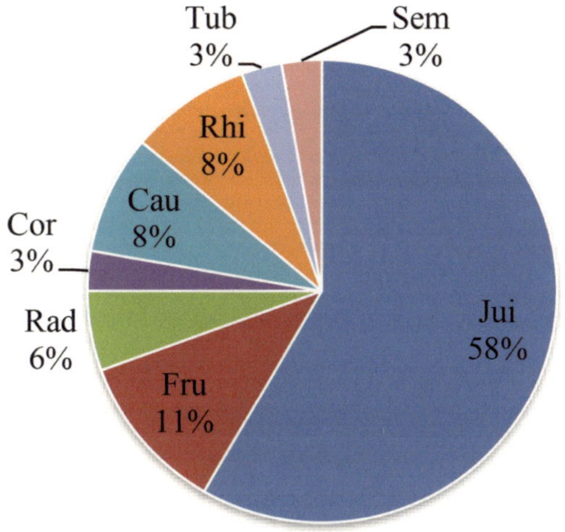

Figure 3.
The percentage value of the plant part.

for medicinal plants. Decoction of plant parts is the most commonly used method followed by direct consumption of plants (raw or juice parts).

This study aims to document and quantify medicinal plants used by communities in West Sumatra, Indonesia. The plant species collected from this study are mostly common medicinal plants used by traditional healers and are frequently documented in ethnobotany studies in Indonesia [9–13].

2.8 Quantitative analysis of ethnomedicine data

2.8.1 Cultural significance index

The significance value of plant species in the study area is calculated using the following formula:

$$ICS \sum_{i}^{n} = 1(q \times i \times e)ui$$

Where q is the quality value for each species (1-5), I refer to the intensity value and e is the exclusivity value. The results obtained from the ICS calculations are then categorized based on the value proposed by Turner [14] shown in **Table 3**.

ICS value	Category
100 and more	Significance is very high
55-99	High significance
20-49	Medium significance
5-19	Low significance
1-4	Significance is very low
0	Can be ignored

Table 3.
The ICS value was proposed by Turner [14].

No.	Local name	Taxonomy Name	ICS	ICS category
1	Sirsak	*Annona muricata*	30	Medium significance
2	Parancih Betadin	*Jatropha multifida Linn*	9	Low significance
3	Jengkol	*Archidendron pauciflorum*	12	Low significance
4	Dalimo	*Punica granatum*	9	Low significance
5	Petai	*Parkia speciosa Hassk*	12	Low significance
6	Benalu Kopi	*Scurrula ferruginea (Jack) dancer*	12	Low significance
7	Sambung Nyawa	*Gynura procumbens*	9	Low significance
8	Sitawa	*Andrographis paniculata*	12	Low significance
9	Bawang Dayak	*Eleutherine palmifolia*	9	Low significance
10	Sidingin	*Kalanchoe pinnata*	12	Low significance
11	Cikarau	*Enhydra fluctuans Lour*	9	Low significance
12	Sarai Harum	*Cymbopogon nardus (L.) Rendl.*	9	Low significance
13	Kayu Manis	*Cinnamomum sp*	9	Low significance
14	Sirih Merah	*Piper sp*	30	Medium significance
15	Daun Sikaduduak	*Melastoma candidum*	9	Low significance
16	Jeriangau	*Acorus calamus*	9	Low significance
17	Bunga Kembang Sepatu	*Hibiscus rosasinensis*	9	Low significance
18	Lagundi	*Vitex trifolia*	9	Low significance
19	Binahong	*Anifere cordifolia*	9	Low significance
20	Seringan-seringan	*Flemingia strobilifera*	9	Low significance
21	Pandan cina	*Pandanus odorus*	9	Low significance
22	Sijanggi	*Cosmos caudatus*	9	Low significance
23	Teh Afrika	*Nernobia amygdaliris*	9	Low significance
24	Pugaran	-	9	Low significance
25	Benalu Jeruk	*Dendrophthoe glabresseris*	12	Low significance
26	Cemara Sumatera	*Taxus sumaterana*	12	Low significance
27	Keji Beling	*Strobilanthes crispa*	9	Low significance
28	Benalu Coklat	*D. Pentandra*	9	Low significance
29	Kunyit Putih	*Curcuma zedoaria*	9	Low significance
30	Akar Wangi	*Polygala paniculata*	9	Low significance

No.	Local name	Taxonomy Name	ICS	ICS category
31	Akar kuning	*Arcangelisia flava Merr.*	12	Low significance
32	Pohon Mindi	*Melia azedarach*	9	Low significance
33	Pohon Mahoni	*Swietenia mahagoni*	9	Low significance

Table 4.
ICS for each plant.

2.8.2 Index of cultural significance (ICS)

The index of cultural significance (ICS) is a reference used to calculate and predict the level of importance of a plant species in a certain area [14, 15], its value can be seen in **Table 4**. ICS analysis is usually carried out to calculate the usefulness of complete plants (food, medicine, rituals, construction, etc.) [14, 16], but because this study focuses more on medicinal plants that are trusted by the local community or known as ethnomedicine so for this study the only use calculated for medicinal purposes only.

Turner calculates the ICS value using the researcher's subjective allocation approach. Turner only uses three variables to calculate the ICS value, namely quality of use, intensity of use, and exclusivity of use [14]. The ICS value is obtained from the result of the multiplication of three variables when the calculation can also occur the addition of the product, this is done if a plant species has more than one use. Turner allocated 5 weight scales for the variables of use quality and intensity of use, namely 5, 4, 3, 2, 1 and allocated 3 scales for the use exclusivity variable, namely 0.5, 1, and 2.

From the research conducted and based on the mathematical calculation of the ICS value, it was seen that only two plant species had moderate significance (ICS = 30). The two plants that have ICS of moderate significance are red betel and soursop. These two plants are trusted by the public to prevent and treat cancer, so these plants are very popular and are considered important by the local community. As for the other plants, it is categorized as low significance. From the ICS calculation data, a plant for the local community can be used as raw material for medicine or herbal plants [17]. Although Turner's ICS ranks plant species used for food, especially staple foods, as the type of quality that has the highest score in determining cultural importance (CS).

3. Conclusion

From the ethnomedicine research conducted, it is known that there are 33 plants that are used as medicinal plants by people in West Sumatra, Indonesia. Mostly, the plant leaves are used as raw material for medicine through direct consumption (juice of the plant parts) or by boiling, which is the most common way of preparation. Quantitative ethnomedicine data can be analyzed using the Index of Cultural Significance (ICS). To determine ICS, three variables are needed, namelyquality of use, intensity of use, and exclusivity of use.

Acknowledgements

The author would like to thank the Indonesian Ministry of Research and Technology for providing the budget through the Primary Research Scheme for

Higher Education 2020-2022 and also to informants who have shared knowledge about medicinal plants.

Conflict of interest statement

The author states that there is no conflict of interest regarding the publication of this paper.

Author details

Skunda Diliarosta, Monica Prima Sari, Rehani Ramadhani* and Annisa Efendi
Universitas Negeri Padang, Padang, Indonesia

*Address all correspondence to: annisa.efendi31@gmail.com

References

[1] Sudarmin. 2014. *Pendidikan karakter, etnosains dan kearifan lokal (Konsep dan penerapannya dalam penelitian dan pembelajaran sains)*. Semarang: Fakultas matematika dan ilmu pengetahuan alam universitas negeri semarang.

[2] Mujahid, Rohmat, dkk. 2019. *Studi Etnomedicine Pengobatan Luka Terbuka dan Sakit Kulit pada Beberapa Etnis di Provinsi Kalimantan Timur*. Kartika: Jurnal Ilmiah Farmasi Vol 7 No 1.

[3] Alves RRN, Rosa IML. *Biodiversity, traditional medicine and public health: Where do they meet?*. J Ethnobiol Ethnomed. 2007; 3: 14.

[4] Mahato G, Hansda B, Banerjee N. *Ethnobotanicals used for the Treatment of Skin Diseases with Special Emphasis on Carbuncle Disease from Purulia District of West Bengal in India*. Pharmacogn J. 2019; 11(4): 745-753.

[5] Hong L, Guo Z, Huang K, Wei S, Liu B, Meng S, et al. *Ethnobotanical study on medicinal plants used by Maonan people in China*. J Ethnobiol Ethnomed. 2015; 11(1).

[6] Gomez-Beloz A. *Plant use knowledge of the Winikina Warao: The case for questionnaires in Ethnobotany*. Econ Bot. 2002; 56(3): 231-241.

[7] Senouci F, Ababou A, Chouieb M. *Ethnobotanical Survey of the Medicinal Plants used in the Southern Mediterranean*. Case study: The region of Bissa (northeastern Dahra Mountains, Algeria). Pharmacogn J. 2019; 11(4): 647-659.

[8] Boadu AA, Asase A. *Documentation of herbal medicines used for the treatment and management of human diseases by some communities in southern Ghana*. Evid based Complement Altern Med. 2017; 2017:3043061.

[9] Roosita K, Kusharto CM, Sekiyama M, Fachrurozi Y, Ohtsuka R. Medicinal plants used by the villagers of a Sundanese community in West Java, Indonesia. J Ethnopharmacol. 2008; 115(1): 72-81.

[10] Taek MM, Banilodu L, Neonbasu G, Watu YV, E.W. BP, Agil M. *Ethnomedicine of Tetun ethnic people in West Timor Indonesia: philosophy and practice in the treatment of malaria*. Integr Med Res. 2019; 8(3): 139-144.

[11] Smd R, Rostiana O, Hernani ERP, Penelitian B, Obat T, Ekosistem C. *Penggalian Iptek Etnomedisin Di Gunung Gede Pangrango. Bul Penelit Tanam Rempah dan Obat*. 2015;18(1):13-28.

[12] Novaryatiin S, Indah I. *The Medicinal Plants Used in Anjir Pulang Pisau, Central Kalimantan-Indonesia*. Pharmacogn J . 2019;11(6).

[13] Fajarini U, Zaharah, Sina I. *Traditional knowledge on malaria of gayo people in central aceh, Indonesia.* Stud Ethno-Medicine. 2016;10(4):498-502.

[14] Turner NJ. "*The Importance of a Rose": Evaluating the Cultural Significance of Plants in Thompson and Lillooet Interior Salish*. Am Anthropol. 1988; 90(2): 272-290.

[15] Phillips O, Gentry AH, Reynel C, Wilkin P, B. CG-D. *Quantitative Ethnobotany and Amazonian Conservation*. Conserv Biol. 1994; 8(1): 225-248.

[16] Hoffman B, Gallaher T. *Importance indices in ethnobotany*. Ethnobot Res Appl. 2007; 5: 201-218.

[17] Musa MS, Abdelrasool FE, Elsheikh EA, Ahmed LAMN, Mahmoud ALE, Yagi SM. *Ethnobotanical study of medicinal plants in the Blue Nile State, South-eastern Sudan*. J Med Plants Res. 2011; 5(17): 4287-4297.

Chapter 5

Advanced Pharmacological Uses of Marine Algae as an Anti-Diabetic Therapy

Thilina Gunathilaka, Lakshika Rangee Keertihirathna and Dinithi Peiris

Abstract

Marine seaweeds are a promising source of bioactive secondary metabolites that can be utilized in drug development and nutraceuticals. Diabetes mellitus is a leading non-communicable disease, and it is the third leading cause of death worldwide. Among the types of diabetes, type 2 became the major health problem as it is associated with severe health complications. Since available oral hypoglycemic drugs cause several adverse effects, it is worth searching for a natural cure with fewer or no side effects that may benefit patients with type 2 diabetes. Among the marine seaweeds, brown and red seaweeds are extensively studied for the anti-diabetic activity compared to the green seaweeds. Bioactive compounds present in marine seaweeds possess anti-diabetic potential through diverse mechanisms, mainly by reducing postprandial hyperglycemia and associated complication. Most of the studies emphasized that the marine seaweeds control the hyperglycemic condition by inhibiting carbohydrate hydrolyzing α-amylase,α glucosidase enzymes, and the inhibitory effect of dipeptide peptidase-4 that are involved in the degradation of incretins. Similarly, bioactive compounds in marine seaweeds can reduce diabetes complications by inhibiting angiotensin-converting enzymes, aldose reductase, protein tyrosine phosphatase 1B enzyme. This chapter focuses on the anti-diabetic potential of marine brown, green, and red seaweeds through different mechanisms.

Keywords: Marine seaweeds, microalgae, bioactive compound, diabetes, drug discovery, mechanisms of action

1. Introduction

The prevalence of diabetes has increased rapidly over the past few years, mainly in low to middle-income countries, and became one of the major causes of premature death worldwide. According to the WHO statistics, 422 million people were estimated as diabetes in 2014, and 1.6 million deaths were reported [1]. The International Diabetes Federation estimated that the world's diabetic population has increased to 592 million by 2035. The largest number of diabetes cases was reported in the Western Pacific region (132 million), while 71.4 million diabetes cases were reported in the South Asian area [2].

Diabetes mellitus is a chronic metabolic disease characterized by hyperglycemia due to defects in insulin secretion, insulin action, or both. It is mainly classified as insulin-dependent diabetes mellitus (Type 1 DM) and non-insulin-dependent diabetes mellitus (type 2 DM). Type 1 DM is associated with deficiency of insulin, which occurs due to the destruction of pancreatic ß-cells via an autoimmune process. In contrast, type 2 DM is linked with insulin resistance, which reduces insulin utilization by peripheral tissues and results in hyperglycemia and obesity [3]. Type 2 DM became a major health problem worldwide associated with microvascular and macrovascular health complications. Microvascular and microvascular complications include diabetic retinopathy, neuropathy, nephropathy, and cerebrovascular diseases, peripheral arterial diseases, respectively [4]. Therefore, natural therapeutic approaches [5] should be developed to maintain the blood glucose level and long-term complications in patients with type 2 DM.

As currently available treatment regimens for type 2 DM have adverse side effects, it is necessary to search for an effective drug that helps maintain the blood glucose level and complications in patients with type 2 DM. Even though most of the researchers focused on herbal medicine, none have a full beneficial effect on curing patients with type 2 DM [6]. Hence, it is worth emphasizing marine seaweeds as they have been identified as a rich source of promising bioactive compounds synthesized from their biochemical and physiological mechanisms. Besides, most marine seaweeds are survived in extremely harsh environments, which provide enormous potential to produce complex bioactive compounds to withstand extreme conditions. As a result, the composition of the bioactive compounds in marine seaweeds can vary depending on the geographic area and seasonal changes [7]. As most marine seaweeds are a potential source of bioactive compounds with various therapeutic effects, this chapter mainly emphasizes the pharmacological uses of marine algae as an anti-diabetic therapy.

2. Therapeutic targets for type 2 diabetes mellitus

As type 2 DM is a progressive disorder, the search for effective treatments is essential to maintain hyperglycemia and its associated diabetic complication. Insulin resistance and impaired beta-cell function lead to hyperglycemia due to alteration in glucose homeostasis, which in turn cause loss of postprandial glucose control. Therefore, postprandial blood glucose maintenance is essential to manage the hyperglycemic condition and associated complications in type 2 diabetes patients [8]. Postprandial hyperglycemia in type 2 DM patients can be controlled by inhibiting metabolic enzymes such as α-amylase, α-glucosidase, dipeptide peptidase-IV, gut-derived peptide hormones (incretins), and glucagon-like peptide-1 hormone. The glucose-dependent insulinotropic peptide, aldose reductase, angiotensin-converting enzyme, and protein tyrosine phosphatase 1B are involved with diabetic complications [9].

Alpha-amylase and alpha-glucosidase are exo-acting glycoside hydrolase enzymes involved in carbohydrate digestion. Alpha-amylase is involved in the digestion of long-chain carbohydrates, while alpha-glucosidase catalyzes the end step hydrolysis of starch or disaccharides into simple glucose units. Therefore, inhibitors of these enzymes delay glucose absorption, reducing the postprandial blood glucose level [10].

Dipeptide peptidase-IV is a protease enzyme involved in the degradation of incretins, a group of metabolic hormones that stimulate ß cells of Langerhans' islet to release insulin. Incretins are released after nutrient intake, and they delayed gastric emptying and decrease glucagon secretion in addition to stimulation of insulin secretion [11]. Contrarily, the incretin effect on insulin secretion gradually decreases once the patient becomes euglycaemic [12]. Hence, inhibitors of dipeptide peptidase IV are efficient therapeutic means to reduce the degradation of incretins, which help maintain hyperglycemic conditions in type 2 DM.

Similarly, aldose reductase is a rate-limiting enzyme involved in the polyol pathway, which catalyzes glucose reduction into sorbitol in an NADPH-dependent pathway. As the aldose reductase has broad substrate specificity, it binds with glucose and converts it into sorbitol once the hexokinase is saturated and the blood glucose level is high. As a result, produced sorbitol is accumulated within the cells and creates an osmotic effect, leading to cataracts and diabetic neuropathy [13, 14]. Thus, aldose reductase inhibitors prevent secondary diabetes complications.

Similarly, the angiotensin-converting enzyme plays a vital role in the renvital angiotensin-aldosterone system, a hormone system responsible for maintaining the blood pressure and fluid balance in the body. Angiotensin-converting enzymes convert angiotensin I into angiotensin II, a potent vasoconstrictor that mainly acts on arterioles that stimulate the release of aldosterone from the renal cortex and improve sodium reabsorption from the kidney. Therefore, activation of the renin-angiotensin-aldosterone system leads to increased blood pressure, resulting in microvascular and macro-vascular complications in patients with type 2 DM. Thus, inhibitors of the angiotensin-converting enzyme reduce the long-term microvascular and macrovascular complications by lowering the arterial and venous blood pressure [15]. A study reported by Ustundag et al. [16] confirms that angiotensin-converting enzyme activity is increased in diabetic patients compared to normal individuals.

Correspondingly, protein tyrosine phosphatase IB (PTP IB) is a negative regulator of the insulin signaling pathway that dephosphorylate tyrosine residues in insulin receptor and insulin receptor substrate-1. Which in turn reduces insulin sensitivity [17]. Hence, inhibition of the PTP IB enzyme leads to lower blood glucose levels by enhancing insulin sensitivity. The stable hyperglycemic condition in type 2DM patients leads to the accumulation of advanced glycated end products in various tissues resulting in diabetic complications such as neuropathy, nephropathy, retinopathy, and other chronic diseases [18]. Therefore, the natural compounds, which inhibit the formation of advanced glycation end products, would be a promising therapeutic target to suppress the diabetic complications associated with glycated products.

3. Bioactive compounds present in marine seaweeds

Marine seaweeds are categorized into three algal classes; red (*Rhodophyceae*), green (*Chlorophyceae*), and brown algae (*Phaeophyceae*) based on the presence of natural pigments. *Phaeophyceae* contains brown color fucoxanthin pigment, whereas *Rhodophyceae* possess red color pigments phycoerythrin and phycocyanin, and *Chlorophyceae* is rich in green color pigment chlorophyll. Among the three varieties, most marine algae are referred to as "edible" that can be used for human consumption. Asians and South Africans mainly consume these edible seaweeds as a promising complementary and alternative medicine [19].

Recently, marine seaweeds have been identified as a rich source of bioactive secondary metabolites with human health benefits. In particular, polyphenols, sterols, alkaloids, flavonoids, tannins, proteins, peptides, essential fatty acids, enzymes, vitamins, and pigments are extensively synthesized by marine seaweeds. These compounds exhibit significant chemical and biological properties such as anti-diabetic, antioxidant, cytotoxic, anti-fungal, anti-bacterial, anti-coagulant, anti-inflammatory, and antiproliferative activities, etc., [19, 20]. The marine seaweeds are a rich source of sulfated polysaccharides (**Figure 1**), which have been reported to possess beneficial human effects. Fucoidan, alginates, and laminarans

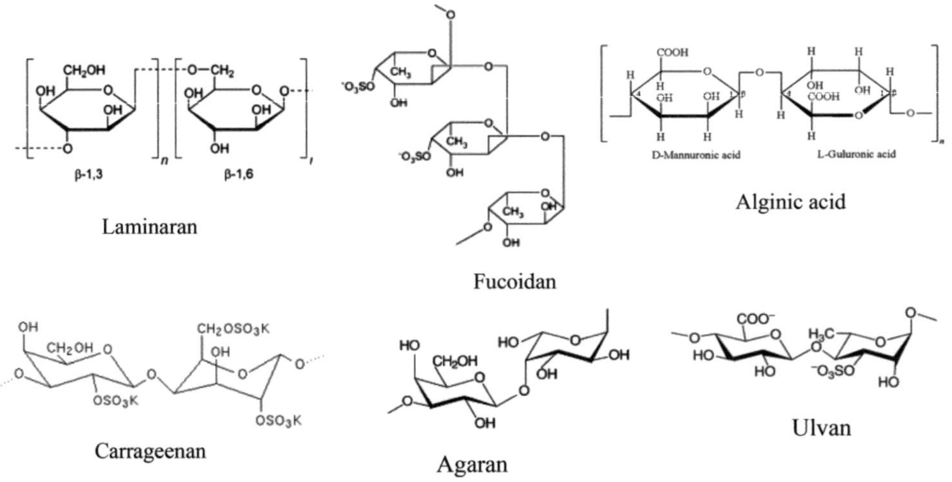

Figure 1.
Chemical structures of sulfated polysaccharides present in marine seaweeds.

are sulfated polysaccharide found in brown seaweeds and reported to exhibit anti-diabetic, antioxidant, and anti-inflammatory activities [21]. Carrageenans and agarans are sulfated polysaccharides found in red seaweeds. Similarly, ulvan is the sulfated polysaccharide found in green seaweeds [22]. The sulfated polysaccharides are known to possess anti-viral, anti-tumor, and anti-coagulant activities [23].

Marine seaweeds are rich in polyphenolic compounds, including flavonoids, bromophenols, phlorotannins, mycosprine-like amino acids, and phenolic terpenoids (**Figure 2**). The mycosprine-like amino acid is a small molecule with hydroxylated aromatic rings. Phlorotannins are polyphenolic metabolites found in brown seaweeds. They can be classified into six subgroups; fuhalols, phlorethols, fucophlorethols, and fucols, eckols, and carmalols based on their linkage between phloroglucinol units and hydroxyl groups. Flavonoids, bromophenol, phenolic terpenoids, phenolic acids, and mycosporine-like compounds are reported to possess antioxidant, anti-diabetic, anti-inflammatory, anti-allergic, and anticancer properties [24–26].

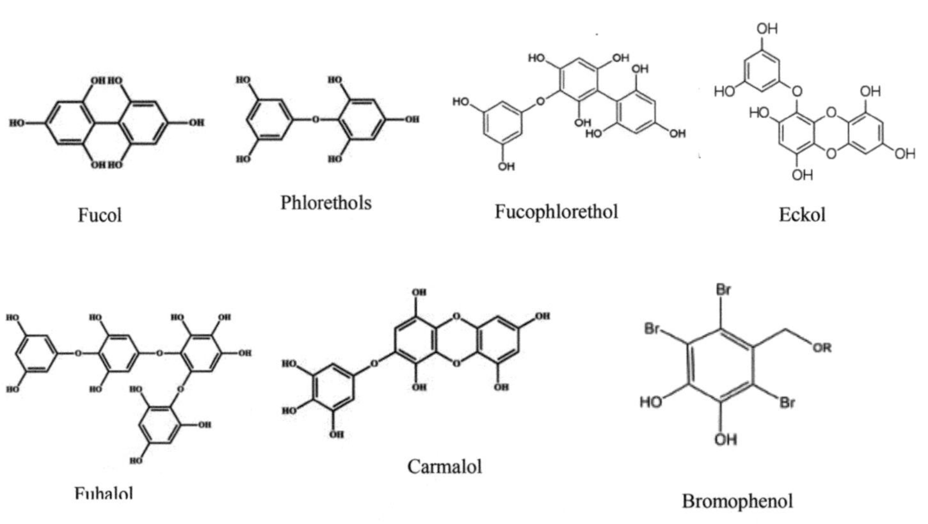

Figure 2.
Organic structures of phlorotannins and bromophenols.

Among the bioactive compounds present in marine seaweeds, marine algae-derived accessory pigments are important as they possess beneficial biological activities [27]. Fucoxanthin is the most abundant accessory pigment found in brown seaweeds and reported to have potent biological activities such as anticancer, antioxidant, anti-diabetic activities due to the presence of unusual allenic bond and a 5, 6-monoepoxide in its structure [21]. Phycobiliproteins, a water-soluble accessory pigment found in red seaweeds, can be divided into three main categories; phycocyanins, allophycocyanins, and phycoerythrin. Phycoerythrins are abundantly found in red seaweeds and reported to possess immuno-modulating and anticancer activities. Similarly, chlorophylls are found in green seaweeds and are said to have antioxidant activity [27].

Similarly, marine algae-derived peptides have been identified to possess a wide range of biological activities such as antioxidant, anti-diabetic, anti-microbial, antihypertensive properties, etc. Hence, most algal-derived proteins have been widely used in food and pharmaceutical industries [28]. The protein content of the marine seaweeds differs depending on the seasonal period and type of species. The brown seaweeds usually contain low protein content compared to the red and green seaweeds. Despite this, some brown algal species such as *Choonospora minima*, *Padina gymnospora*, *Dictyota menstrualis*, and *Sargassum vulgare* possess high protein content up to 10–15%. According to the reported studies, green seaweed contains an average protein content level, ranging between 10–26%. In contrast, the highest protein content was reported in red seaweeds such as *Phorphyra tenure* and *Palmaria palmata*, which was around 47% [29].

4. Anti-diabetic potentials of marine algae

Marine seaweeds have been widely studied for their anti-diabetic potential through different mechanisms due to bioactive secondary metabolites. Several *in-vitro* and *in-vivo* studies have been conducted so far to confirm the hypoglycemic effect of marine algae in addition to its ability to suppress diabetic complications. This section emphasizes the anti-diabetic potential of marine brown, red, and green seaweeds through diverse mechanisms.

4.1 Inhibitory activity of carbohydrate hydrolyzing α-amylase and α-glucosidase enzymes

a. Brown seaweeds

Among the brown seaweeds, "Ecklonia" and "Eisenia" genera have been reported to exert hypoglycemic effects through α-amylase and α-glucosidase inhibitory activities [30]. The observed hypoglycemic activity can be attributed to the presence of phlorotannins; eckol, dieckol, 6,6′-bieckol, phlorofucofuroeckol-A, and phloroglucinol, and 7-phloroeckol [31]. According to the reported studies, methanol extract of *Ecklonia cava* exercises its hypoglycemic effects through the inhibitory activity of α-glucosidase enzymes (IC_{50}–10.7 μM), compared to the standard acarbose used. Similar results were reported with phlorotannins isolated from *Ecklonia stolonifera* against the α-glucosidase enzyme. Dieckol (IC_{50}:1.61 μM) and phlorofucofuroeckol-A (IC_{50}:1.37 μM) isolated from *Ecklonia stolonifera* reported to exhibit the potent inhibitory activity of α-glucosidase enzymes compared to the standard drug (IC_{50}: 51.65 μM). Similarly, eckol (IC_{50}: 11.16 μM) isolated from *Ecklonia maxima* demonstrated strong α-glucosidase inhibitory activity comparable to the isolated phloroglucinol (IC_{50}: 1991 μM). Besides, *Eisenia bicyclis* from genus

Eisenia reported possessing 87% of inhibitory effect on α-amylase at 1 mM concentration in addition to the inhibitory effect on α-glucosidase and advanced glycation end products. Moreover, isolated eckol (IC_{50}: 22.78 μM), dioxinodehydroeckol (IC_{50}: 34.60 μM) and phloroglucinol (IC_{50}: 141.18 μM) from *Eisenia bicyclis* exhibited potent α-glucosidase inhibitory activity [32].

A brown seaweed *Sargassum hystrix* reported to exhibit inhibitory effect on α-amylase IC_{50}: 0:58 ± 0:01 mg/ml; IC_{50} acarbose: 0:53 ± 0:00 mg/ml) and α-glucosidase (IC_{50}: 0:59 ± 0:02 mg/ml; IC_{50} acarbose: 0:61 ± 0:01 mg/ml) enzymes compared to the standard acarbose [33]. This was further confirmed by an *in-vivo* study using streptozotocin-induced rats and observed that the deduction of pre-prandial (186.4 mg/ml) and postprandial (186.9 mg/ml) blood glucose levels at 300 mg/kg concentration comparable to the standard drug glibenclamide (5 mg/kg) (Pre-prandial;195.6 mg/ml; postprandial:104.8 mg/ml) without any adverse effects. Correspondingly, ethanol (150 mg/kg) and aqueous (300 mg/kg) extracts of *Sargassum polycystum* reported to reduce hyperglycaemic condition in diabetic rats [34]. Further studies has reported that a brown seaweed *Ascophyllum nodosum* effectively inhibited α-amylase (IC_{50}: 0.1 μg/ml) and α-glucosidase enzymes ((IC_{50}: 19 μg/ml) due to the presence of phlorotannins [35].

b. Green seaweeds

Green seaweeds belong to the genus "Ulva." They have been reported to possess hypoglycemic activity, and they have been used for various food dishes in Asians due to the presence of high soluble fiber content. The aqueous extract of green seaweeds *Ulva lactuca* (Inhibition-α-amylase: 83.4%; α-glucosidase: 61.81%) and *Ulva reticulate* (Inhibition-α-amylase: 89.1%; α-glucosidase: 76.02%) were effective against α-amylase and α-glucosidase enzymes at a concentration of 100 μg/ml after 8 hours of extraction period at 37 °C in a water bath as it gets more time to release the phytochemicals and colloids to the extract [36]. Similarly, the crude extract of *Ulva ohnoi* exhibited α-amylase inhibition by 41.7% and complete α-glucosidase inhibition at 10 mg/mL [37].

The methanol extract of a green seaweed *Chlorodesmis* inhibited α-amylase enzyme by 72% at 500 μg/ml with IC50 of 408.9 μg/ml without any effect on α-glucosidase enzymes. Similarly, chloroform extract of *Chaetomorpha aerea exhibited a potent inhibitory effect on α-amylase enzyme with* IC_{50} of 147.6 μg/ml. Besides, methanol extract of green seaweeds *Enteromorpha intestinalis (59%) and Cladophora rupestris (14%) exhibited a moderate and lower effect on the α-amylase inhibitory activity at a concentration of* 500 μg/ml [38]. Moreover, crude extracts of green seaweeds *Derbesia tenuissima and Oedogonium intermedium* were reported to exhibit lower α-amylase (53.6% and 49.2%) and potent α-glucosidase (73.98% and 69.5%) inhibitory effect at a concentration of 10 mg/ml [39]. Further studies reported that the green seaweed *Chlorella pyrenoidosa* could suppress the hyperglycaemic condition by inhibiting α-amylase and α-glucosidase enzymes. Besides, a green seaweed *Cladophora rupestris* has been reported to exhibit a hypoglycemic effect through α-amylase and α-glucosidase inhibitory mechanisms [40].

c. Red seaweeds

Among the marine red seaweeds, the genus "Gracillaria" was reported to possess the hypoglycemic effect through the inhibitory effect on α-amylase and α-glucosidase enzymes. Gunathilaka *et al*, [41] reported that the ethyl acetate fraction of red seaweed *Gracillaria edulis* exhibited potent α-amylase (IC_{50}: 279.48 μg/ml) and α-glucosidase (IC_{50}: 87.92 μg/ml) inhibitory activity compared to the standard acarbose ($IC_{50amylase}$: 87.43 μg/ml; $IC_{50glucosidase}$: 0.38 μg/ml) due to the presence of reported anti-diabetic compound 1H-Indole-2-carboxylic acid,6-(4-ethoxy phenyl)-3-methyl-4-oxo- 4,5,6,7-tetrahydro-isopropyl ester. Further studies reported that the aqueous extract of *Gracillaria edulis* inhibited the α-amylase and α-glucosidase enzyme by 87.86% and 79.55% at a concentration of 100 μg/ml. Similarly, *Gracilaria corticata* and *Acanthophora spicifera* had an inhibitory effect on α-amylase (84.66%; 54.73%) and α-glucosidase (73.53%; 46.86%) enzyme at a concentration of 100 μg/ml [36].

4.2 Inhibitory activity of dipeptidyl peptidase-IV (DPP-IV)

Dipeptidyl peptidase-IV (DPP-IV) is an enzyme involved in the degradation of incretin hormones, maintaining postprandial blood glucose levels. Among three types of seaweeds, brown seaweeds have been extensively reported to possess a dipeptidyl peptidase-IV (DPP-IV) inhibitory effect compared to red and green seaweeds [42].

a. Brown seaweeds

The brown seaweeds *Padina sulcata*, *Sargassum binderi*, and *Turbinaria conoides* have been reported to exhibit a potent inhibitory effect on the DPP-4 enzyme in a dose-dependent manner. The maximum inhibitory effect of *Padina sulcata*, *Sargassum binderi*, and *Turbinaria conoides* were observed as 83.09%, 81.75%, and 76.20%, at a concentration of 10 mg/ml, respectively. Further, crude water extracts of the above three brown seaweeds could secrete glucagon-like peptide-1 (GLP-1) to a greater extent than prevent hyperglycaemic conditions [42]. Similarly, ethyl acetate: methanol fraction (IC50: 0.013 mg/ml) of *Sargassum wightii* has been reported to exhibit an inhibitory effect on DPP-4 enzymes compared to the standard drug diprotein-A (IC50: 0.007 mg/ml) [43]. The methanol extract of *Turbinaria ornate* exhibited a strong inhibitory effect on the DPP-4 enzyme by 55.4% at 80 μg/ml than the standard drug diprotin A (65%) might attribute to the presence of fucoids and sulfated polysaccharides in *T. ornata* [44].

b. Green seaweeds

The previous study conducted by Chin *et al.* [42] reported the inhibitory activity of green seaweed *Halimeda macroloba* on the DPP-4 enzyme. *Halimeda macroloba* inhibited the DPP-4 enzyme by 60.53% at a 10 mg/ml concentration compared to the positive control Berberine (75.92% at 1 mg/mL). Moreover, crude water extract of *H. macroloba* was able to stimulate glucagon-like peptide-1 (GLP-1) secretion.

c. Red seaweeds

The sulfated polygalactans isolated from red seaweeds *Kappaphycus alvarezii* and *Gracilaria opuntia* have been reported to possess the inhibitory effect on the DPP-4 enzyme. According to the results, sulfated galactans isolated from *Gracilaria opuntia* (IC50 0.09 mg/mL) significantly inhibited the DPP-4 enzyme than the sulfated galactans of *Kappaphycus alvarezii* (IC50 0.12 mg/mL) compared to the standard diprotin A (IC50 1.54 mg/mL). The observed activity might be due to the reaction between functional groups of sulfated polygalactan with DPP-4 by H-bonding and hydrophilic interactions [45]. Similarly, aqueous, alkaline, and a mixture of aqueous/alkaline fractions of a red seaweeds *Palmaria palmate* have exhibited a potent inhibitory effect on DPP-4 enzyme with IC_{50} of 2.52 ± 0.05 mg/ml, 4.60 ± 0.09 mg/ml, and 4.24 ± 0.02 mg/ml respectively [46]. Further studies reported that the red seaweed *Palmaria palmate's protein hydrolysate* had a potential inhibitory effect on the DPP-4 enzyme [40]. These results confirmed the possible inhibitory effect on DPP-4 enzymes of red seaweed extracts.

4.3 Inhibitory activity of aldose reductase (AR)

a. Brown seaweeds

The ethyl acetate fraction of brown seaweed, *Ecklonia stolonifera* has been reported to possess a strong inhibitory effect on aldose reductase enzymes due to the presence of phlorotannins such as 7-phloroeckol and 2-phloroeckol in ethyl acetate fraction [47]. Similarly, phlorofucofuroeckol-A isolated from *Eisenia bicyclis* exhibited a potent inhibitory effect of aldose reductase enzyme (IC50: 6.22 μM). They also confirmed the inhibitory effect of fucosterol in the rat lens. Carotenoids isolated from *Ecklonia stolonifera* exhibited potent inhibitory activity on aldose reductase enzyme (IC50: 18.94 μM) compared to the standard positive control quercetin (IC50: 1.34 μM). The presence of porphyrin derivatives (pheophytin-A and pheophorbide-A) in the dichloromethane fraction of *Saccharina japonica* caused excellent inhibitory effects on aldose reductase (AR) in rat lens [48]. Moreover, fucoxanthin isolated from *Undaria pinnatifida* and *Eisenia bicyclis* reported acting as a competitive inhibitor on the aldose reductase enzyme [49].

b. Green seaweeds

The chloroform and ethanol fractions of green seaweed, *Capsosiphon fulvescens* showed a potent inhibitory action on the AR enzyme [50]. The authors further carried out isolation of compounds, and the isolated compounds (capsofulvesin A, B, and chalinasterol) demonstrated high inhibitory action on AR enzyme with IC_{50} values of 52.53, 101.92, and 345.27 μM, respectively.

c. Red seaweeds

Regarding red seaweeds, the bromophenol compounds present in red seaweeds have been identified as effective therapeutic agents. The bromophenols such as bis (2,3,6-tribromo-4,5 -dihydroxy phenyl) methane, 2,2′,3,6,6′-pentabromo-3′,4,4′,5-tetrahydroxydibenzyl ether, and 2,2′,3,5′,6-pentabromo- 3′,4,4′,5-tetrahydroxydiphenylmethane isolated from red seaweed, *Symphyocladia latiuscula* are well known for their inhibitory effects on aldose reductase. This enzyme is responsible for the fructose formation in the polyol pathway [25].

4.4 Inhibitory activity of protein tyrosine phosphatase 1B (PTP 1B)

a. Brown seaweed

The brown seaweeds belonged to the genus "Sargassum" as reported to exhibit the potent inhibitory activity of PTP 1B enzyme due to the presence of secondary bioactive compounds. Ali *et al.* [51] reported that the hexane fraction (IC_{50}: 1.83 μg/ml) of *Sargassum serratifolium* strongly inhibited the PTP 1B enzyme than the standard ursolic acid (IC_{50}: 1.12 μg/ml). During the compound isolation, three plastoquinones (sargachromenol, sargahydroquinoic acid, and sargaquinoic acid) were identified, and among them, sargahydroquinoic acid exhibited a potent PTP 1B inhibitory effect (IC50:5.14 μg/ml). Similarly, the chloroform extract of *Sargassum yezoense* (54.4%), *Sargassum fluvellum* (36.1%), *Sargassum horneri* (46.2%), Sargassum *sagmianum (21.4%),* Sargassum *hemiphyllum* (44.1%), and *Sargassum siliquastrum* (14.8%) could inhibit PTP 1B enzymes at 15 μg/ml of concentration [52]. Further, the phlorotannins such as eckol, 7-phloroeckol, and phlorofucofuroeckol-A isolated from *Ecklonia stolonifera, Ecklonia cava*, and *Eisenia bicyclis* could act as non-competitive inhibitors on PTP 1B enzyme [53]. Moon *et al.* [32] further confirmed the inhibitory effect of phlorofucofuroeckol-A (IC_{50}: 0.56 μM), 7-phloroeckol (IC_{50}: 2.09 μM), and eckol (IC_{50}: 2.64 μM) isolated from *Ecklonia stolonifera* and *Eisenia bicyclis. Moreover, fucosterol isolated from Eisenia* bicyclis and *Ecklonia* stolonifera also showed PTP 1B inhibitory effect [54].

b. Green seaweeds

Several studies have been reported to elucidate the anti-diabetic potential of green seaweeds by enhancing insulin sensitivity through the mechanism of PTP 1B inhibition. Among the marine green seaweeds, Crude chloroform and methanol extract of a green seaweed *Derbesia marina* has been reported to exhibit an inhibitory effect on PTP 1B enzyme by 61.7% and 80.65 respectively at a concentration of 15 μg/ml. Further, the crude chloroform and methanol extract of edible green sea lettuce "*Ulva pertusa*" has exhibited potent PTP 1B inhibition at 15 μg/ml by 25.8% and 48.1%, respectively. Similarly, the crude chloroform and methanol extract of *Enteromorpha linza* (42.1%:35.4%) and *Codium adhaerens* (51.5%:71.2%) increased insulin sensitivity by inhibiting PTP 1B enzyme at 15 μg/ml concentration [52]. *Further, the compounds isolated from the green seaweeds belonged to the genus "Caulerpa" had a potent anti-diabetic effect by the mechanism of PTP 1B inhibition.* Racemosin C, Caulerpin, Caulerpic acid isolated from *Caulerpa racemosa,* and Caulersin isolated from *Caulerpa serrulata* have reported significant PTP1B inhibitor [55].

c. Red seaweeds

Most of the red seaweeds belonged to the genus "Chondus" exhibited anti-diabetic activity via PTP 1B enzyme inhibition. According to the recorded studies, chloroform extract of *chondus ocellanthus* and *chondus crispus* inhibited PTP 1B enzymes by 41.5% and 27.6% at a concentration of 15 μg/ml. Similarly, red seaweeds belonged to the genus "Laurencia" had a potential inhibitory effect on PTP 1B enzyme. The methanol extract of *Laurencia okamurae* (33.1%) and chloroform extract of *Laurencia intermedia* (43.3%) could inhibit PTP 1B enzyme at 15 μg/ml of concentration. In addition to that 15 μg/ml concentration of chloroform extract of *Corallina pilulifera*, *Gymnogongrus flabelliformis*, and *Gracillaria textori* inhibited PTP 1B enzyme by 58.3%, 38.6%, and 24.9%, respectively [46]. Further, the compounds bromophenol and 3, 4-dibromo-5-(2-Bromo-3, 4-dihydroxy-6-(ethoxymethyl)benzyl)benzene-1,2-diol isolated from red seaweed, *Rhodomela confervoides* could increase insulin sensitivity via inhibition of PTP 1B enzyme [40].

4.5 Inhibitory activity of angiotensin-converting enzymes (ACE)

a. Brown seaweeds

Phlorotannins eckol, phlorofucofuroeckol-A, and dieckol isolated from brown seaweed, *Ecklonia stolonifera* could inhibit angiotensin-converting enzyme with IC50 values of 70:82 μM, 12:74 μM, and 34:25 μM, respectively. Among the isolated phlorotannins, dieckol acted as a non-competitive inhibitor of ACE [56]. Similarly, the phloroglucinol isolated from the ethyl acetate fraction of Sargassum (56.96 μg/ml) wightii significantly inhibited the ACE compared to the positive control captopril (51.79 μg/ml) [57]. An amino acid sequence isolated from edible brown seaweed, *Undaria pinnatifida,* could significantly inhibit angiotensin-converting enzymes [57]. The protein-derived hydrolysate of *Undaria pinnatifida* exhibited a potent antihypertensive effect via inhibiting ACE [39]. Further, the enzymatic hydrolysate of *Ecklonia cava* has been reported to show a potent inhibitory effect on ACE with IC50 values from 2.33 up to 3.56 μg/mL [58].

b. Green seaweeds

Among the green seaweeds, few studies have been reported regarding the inhibitory effect on the angiotensin-converting enzyme. Crude and saponified extracts of *Ulva ohnoi*, *Derbesia tenuissima,* and *Oedogonium intermedium* exhibited an inhibitory effect on the angiotensin-converting enzyme. The crude extract of *Ulva ohnoi*, *Derbesia tenuissima,* and *Oedogonium intermedium* had a less potent inhibitory effect at 10 mg/ml. In contrast, the saponified extract of *Ulva ohnoi*, *Derbesia tenuissima,* and *Oedogonium intermedium* inhibited 1.9%, 1.47%, and 7.37% compared to the positive control captopri*l* (6.15% inhibition at 200 μg/ml). However, carotenoids; siphonaxanthin, neoxanthin, 9′-*cis*-neoxanthin, loroxanthin, violaxanthin, lutein, siphonein, α-carotene, and β-carotene present in green seaweeds are found to be poor inhibitors of ACE [37]. Further, a protein-derived hydrolysate of an edible green seaweed *Enteromorpha clathrata* had a potent inhibitory effect on ACE [58].

c. Red seaweeds

The red seaweeds have been widely studied to elucidate the inhibitory effect on angiotthe ensin-converting enzyme, as it plays a crucial role in regulating blood pressure. According to the recorded studies, the aqueous extract at 20 °C of red seaweeds *Lomentaria catenata, Lithophyllum okamurae, Ahnfeltiopsis flabelliformis, and Gracilaria textorii* significantly inhibited the angiotensin-converting enzyme by 98.92%, 89.23%, 73.45%, and 65.40% at a lower concentration of 200 μg/ml. Similarly, the methanol extract at 700C of red seaweeds, *Ahnfeltiopsis flabelliformis, and Laurencia okamurae* has been reported to exhibit a strong inhibitory effect angiotensin-converting enzyme by 97.59% and 78.01% at a concentration of 200 μg/ml. Further, the methanol extract at 70 °C of red seaweeds *Grateloupia filicina, Sinkoraena lancifolia, Grateloupia elliptica, Grateloupia lanceolata,* and *Laurencia okamurae* exhibited an inhibitory effect on ACE by 83.14%, 80.86%,68.13%, 89.04%, and 69.80% at 200 μg/ml of concentration [59]. This study revealed the presence of ACE like inhibitors in red seaweeds. Protein-derived hydrolysate in *Palmaria palmate* (red seaweed) showed marked antihypertensive activity. The antihypertensive activity was exerted via inhibition of angiotensin-converting enzymes [46]. Further, an enzymatic hydrolysate of a red seaweed *Pyropia columbina* exhibited an inhibitory effect on the angiotensin-converting enzyme with an IC50 value of 1.2 mg/ml [58].

4.6 Inhibitory activity of the formation of advanced glycation end products (AGEs)

a. Brown seaweeds

Among the brown seaweeds, *Ecklonia cava* has been extensively studied for its anti-diabetic activity. The phlorotannins isolated from *Ecklonia cava* such as eckol

(IC_{50}: 1:6 × 103 μM), phlorofucofuroeckol-A (IC_{50}: 2:4 × 103 μM), fucofuroeckol A (IC_{50}: 7:4 × 102 μM), and dieckol (IC_{50}: 7:4 × 102 μM) could inhibit the formation of advanced glycation end products comparable to the standard drug aminoguanidine hydrochloride (IC_{50}: 8:1 × 103 μM) [60]. Similarly, the phlorotannins isolated from methanol extract of brown seaweeds *Sargassum polycystum* (IC_{50}: 35:245 μg/ml), Turbinaria Ornate (IC_{50}: 22:7 μg/ml), and Padina pavonica IC_{50}: 15:16 μg/ml) had the ability to suppress the formation of advanced glycation end-products [61]. Further, phlorotannins extracted from the ethyl acetate fraction of *Fucus vesiculosus* (IC_{50}: 0.045 mg/ml) significantly inhibited the AGEs formation compared to the phloroglucinol (IC_{50}: 0.068 mg/ml) [62].

b. Green seaweeds

So far, minimal studies have been reported to demonstrate the inhibitory effect of green seaweeds on the formation of advanced glycation end products. The chloroform, ethanol, and butanol fractions of a green seaweed *Capsosiphon fulvescens* have been reported to exhibit an inhibitory effect on the formation of advanced glycation end-products [50].

c. Red seaweeds

Regarding the red seaweeds, the ethyl acetate fraction (IC_{50}: 586.54 μg/ml) of *Gracillaria edulis* has been reported to exhibit the inhibitory effect on the formation of advanced glycation end products compared to the standard drug rutin (IC_{50}: 11.55 μg/ml) [34]. Similarly, carrageenan extract from red algae could inhibit progressive glycation end product uptake by macrophage-like RAW 264.7 cells [63].

5. Conclusions

Recently, marine seaweeds have been extensively studied for their therapeutic effects due to promising bioactive compounds. Among the non-communicable diseases, diabetes mellitus is the third leading cause of death associated with vascular complications. As it is a progressive disorder, it is necessary to search for an adequate drug for natural resources with minimum side effects. Therefore, this chapter illustrates the different anti-diabetic mechanisms of marine seaweed extracts and their bioactive compounds.

Acknowledgements

The University of Sri Jayewardenepura, Sri Lanka (ASP/01/RE/SCI/2017/50).

Conflict of interest

The authors declare no conflict of interest.

Author details

Thilina Gunathilaka, Lakshika Rangee Keertihirathna and Dinithi Peiris*
Department of Zoology/Genetics and Molecular Biology Unit (Center for Biotechnology), Faculty of Applied Sciences, University of Sri Jayewardenepura, Nugegoda, Sri Lanka

*Address all correspondence to: dinithi@sjp.ac.lk

References

[1] Zaccardi, F.; Webb, D.R.; Yates, T.; Davies, M.J. Pathophysiology of type 1 and type 2 diabetes mellitus: a 90-year perspective. Postgrad. Med. J. **2016**, *92*, 63-69.

[2] Panunti, B.; Jawa, A.A.; Fonseca, V.A. Mechanisms and therapeutic targets in type 2 diabetes mellitus. Drug Discov. Today Dis. Mech. **2004**, *1*, 151-157.

[3] Heendeniya, S.N.; Keerthirathna, L.R.; Manawadu, C.K.; Dissanayake, I.H.; Ali, R.; Mashhour, A.; Alzahrani, H.; Godakumbura, P.; Boudjelal, M.; Peiris, D.C. Therapeutic Efficacy of Nyctanthes arbor-tristis Flowers to Inhibit Proliferation of Acute and Chronic Primary Human Leukemia Cells, with Adipocyte Differentiation and in Silico Analysis of Interactions between Survivin Protein and Selected Secondary Meta. Biomolecules **2020**, *10*, 165.

[4] Vadivelu, R.; Vijayvergiya, R. Panvascular risk factor - Diabetes. Cor Vasa **2018**, *60*, e18–e29.

[5] Dissanayake, D.M.I.H.; Perera, D.D.B.D.; Keerthirathna, L.R.; Heendeniya, S.; Anderson, R.J.; Williams, D.E.; Peiris, L.D.C. Antimicrobial activity of Plumbago indica and ligand screening of plumbagin against methicillin-resistant Staphylococcus aureus. J. Biomol. Struct. Dyn. **2020**, 1-12.

[6] Sudasinghe, H.P.; Peiris, D.C. Hypoglycemic and hypolipidemic activity of aqueous leaf extract of Passiflora suberosa L. PeerJ **2018**, 6, e4389.

[7] Di Donato, P.; Buono, A.; Poli, A.; Finore, I.; Abbamondi, G.; Nicolaus, B.; Lama, L. Exploring Marine Environments for the Identification of Extremophiles and Their Enzymes for Sustainable and Green Bioprocesses. Sustainability **2018**, *11*, 149.

[8] Hinnen, D.; Strong, J. iGlarLixi: A New Once-Daily Fixed-Ratio Combination of Basal Insulin Glargine and Lixisenatide for the Management of Type 2 Diabetes. Diabetes Spectr. **2018**, *31*, 145-154.

[9] Qaid, M.M.; Abdelrahman, M.M. Role of insulin and other related hormones in energy metabolism—A review. Cogent Food Agric. **2016**, *2*, 1267691.

[10] Hinnen, D.A. Therapeutic Options for the Management of Postprandial Glucose in Patients With Type 2 Diabetes on Basal Insulin: FIGURE 1. Clin. Diabetes **2015**, *33*, 175-180.

[11] Gerich, J. Pathogenesis and management of postprandial hyperglycemia: role of incretin-based therapies. Int. J. Gen. Med. **2013**, *6*, 877-895.

[12] Makrilakis, K. The Role of DPP-4 Inhibitors in the Treatment Algorithm of Type 2 Diabetes Mellitus: When to Select, What to Expect. *Int. J. Environ. Res. Public Health* **2019**, *16*, 2720.

[13] Tang, W.H.; Martin, K.A.; Hwa, J. Aldose Reductase, Oxidative Stress, and Diabetic Mellitus. Front. Pharmacol. **2012**, *3*.

[14] Thomas, V.; Spiro, T.; Moomaw, W.R. Emissions and Exposure to Metals: Cadmium and Lead. In *Industrial Ecology and Global Change*; Socolow, R., Andrews, C., Berkhout, F., Thomas, V., Eds.; Cambridge University Press: Cambridge, 2010; pp. 297-318.

[15] Fountain, J.H.; Lappin, S.L. *Physiology, Renin Angiotensin System*; StatPearls Publishing: Treasure Island, Florida, 2020;

[16] Ustündağ, B.; Canatan, H.; Cinkilinç, N.; Halifeoğlu, I.; Bahçecioğlu, I.H. Angiotensin converting enzyme (ACE) activity levels in insulin-independent diabetes mellitus and effect of ACE levels on diabetic patients with nephropathy. Cell Biochem. Funct. **2000**, *18*, 23-28.

[17] Vieira, M.N.N.; Lyra e Silva, N.M.; Ferreira, S.T.; De Felice, F.G. Protein Tyrosine Phosphatase 1B (PTP1B): A Potential Target for Alzheimer's Therapy? *Front. Aging Neurosci.* **2017**, *9*, 1-9.

[18] Singh, V.P.; Bali, A.; Singh, N.; Jaggi, A.S. Advanced glycation end products and diabetic complications. Korean J. Physiol. Pharmacol. **2014**, *18*, 1-14.

[19] Khalid, S.; Abbas, M.; Saeed, F.; Bader-Ul-Ain, H.; Ansar Rasul Suleria, H. Therapeutic Potential of Seaweed Bioactive Compounds. In *Seaweed Biomaterials*; IntechOpen, 2018; pp. 2-14.

[20] Rengasamy, K.R.; Mahomoodally, M.F.; Aumeeruddy, M.Z.; Zengin, G.; Xiao, J.; Kim, D.H. Bioactive compounds in seaweeds: An overview of their biological properties and safety. Food Chem. Toxicol. **2020**, *135*, 111013.

[21] Gunathilaka, T.L.; Samarakoon, K.; Ranasinghe, P.; Peiris, L.D.C. Antidiabetic Potential of Marine Brown Algae—a Mini Review. J. Diabetes Res. **2020**, *2020*, 1-13.

[22] Wijesekara, I.; Pangestuti, R.; Kim, S.-K. Biological activities and potential health benefits of sulfated polysaccharides derived from marine algae. Carbohydr. Polym. **2011**, *84*, 14-21.

[23] Udayangani, R.M.A.C.; Somasiri, G.D.P.; Wickramasinghe, I.; Kim, S. Potential Health Benefits of Sulfated Polysaccharides from Marine Algae. In *Encyclopedia of Marine Biotechnology*; Wiley, 2020; pp. 629-635.

[24] Ganesan, A.R.; Tiwari, U.; Rajauria, G. Seaweed nutraceuticals and their therapeutic role in disease prevention. Food Sci. Hum. Wellness **2019**, *8*, 252-263.

[25] Cotas, J.; Leandro, A.; Monteiro, P.; Pacheco, D.; Figueirinha, A.; Gonçalves, A.M.M.; da Silva, G.J.; Pereira, L. Seaweed Phenolics: From Extraction to Applications. Mar. Drugs **2020**, *18*, 384.

[26] Chojnacka, K. Biologically Active Compounds in Seaweed Extracts - the Prospects for the Application. Open Conf. Proc. J. **2012**, *3*, 20-28.

[27] Pangestuti, R.; Kim, S.-K. Biological activities and health benefit effects of natural pigments derived from marine algae. J. Funct. Foods **2011**, *3*, 255-266.

[28] Rabiei, S. Marine-Derived Bioactive Peptides with Pharmacological Activities- A Review. J. Clin. DIAGNOSTIC Res. **2017**, *11*, 1-6.

[29] Lafarga, T.; Acién-Fernández, F.G.; Garcia-Vaquero, M. Bioactive peptides and carbohydrates from seaweed for food applications: Natural occurrence, isolation, purification, and identification. Algal Res. **2020**, *48*, 101909.

[30] Gabbia, D.; De Martin, S. Brown Seaweeds for the Management of Metabolic Syndrome and Associated Diseases. Molecules **2020**, *25*, 4182.

[31] Lee, S.-H.; Jeon, Y.-J. Anti-diabetic effects of brown algae derived phlorotannins, marine polyphenols through diverse mechanisms. Fitoterapia **2013**, *86*, 129-136.

[32] MOON, H.E.; ISLAM, M.N.; AHN, B.R.; CHOWDHURY, S.S.; SOHN, H.S.; JUNG, H.A.; CHOI, J.S. Protein Tyrosine Phosphatase 1B and α-Glucosidase Inhibitory Phlorotannins from Edible Brown Algae, Ecklonia stolonifera and Eisenia bicyclis.

Biosci. Biotechnol. Biochem. **2011**, *75*, 1472-1480.

[33] Husni, A.; Pratiwi, T.; Samudra, A.G.; Nugroho, A.E. In vitro anti-diabetic activity of Sargassum hystrix and Eucheuma denticulatum from Yogyakarta Beach of Indonesia. *Proc. Pakistan Acad. Sci. Life Environ.* Sci. **2018**, *55*, 1-8.

[34] Gotama, T.L.; Husni, A. Antidiabetic Activity of Sargassum hystrix Extracts in Streptozotocin-Induced Diabetic Rats. Prev. Nutr. Food Sci. **2018**, *23*, 189-195.

[35] Nwosu, F.; Morris, J.; Lund, V.A.; Stewart, D.; Ross, H.A.; McDougall, G.J. Anti-proliferative and potential anti-diabetic effects of phenolic-rich extracts from edible marine algae. Food Chem. **2011**, *126*, 1006-1012.

[36] Reka, P.; A., T.B.; Seethalakshmi, *M. alpha* AMYLASE AND ALPHA GLUCOSIDASE INHIBITION ACTIVITY OF SELECTED EDIBLE SEAWEEDS FROM SOUTH COAST AREA OF INDIA. *Int. J. Pharm. Pharm. Sci.* **2017**, *9*, 64.

[37] Wang, N. Activities of Tropical Green Algae from Australia School of Chemical Engineering, University of New South Wales, 2016.

[38] Unnikrishnan, P.; Suthindhiran, K.; Jayasri, M. Alpha-amylase inhibition and antioxidant activity of marine green algae and its possible role in diabetes management. Pharmacogn. Mag. **2015**, *11*, 511.

[39] Unnikrishnan, P.S.; Suthindhiran, K.; Jayasri, M.A. Antidiabetic potential of marine algae by inhibiting key metabolic enzymes. Front. Life Sci. **2015**, *8*, 148-159.

[40] Lauritano, C.; Ianora, A. Marine Organisms with Anti-Diabetes Properties. Mar. Drugs **2016**, *14*, 220.

[41] Gunathilaka, T.L.; Samarakoon, K.W.; Ranasinghe, P.; Peiris, L.D.C. In-Vitro Antioxidant, Hypoglycemic Activity, and Identification of Bioactive Compounds in Phenol-Rich Extract from the Marine Red Algae Gracilaria edulis (Gmelin) Silva. Molecules **2019**, *24*, 3708.

[42] Chin, Y.X.; Lim, P.E.; Maggs, C.A.; Phang, S.M.; Sharifuddin, Y.; Green, B.D. Anti-diabetic potential of selected Malaysian seaweeds. J. Appl. Phycol. **2015**, *27*, 2137-2148.

[43] Maneesh, A.; Chakraborty, K.; Makkar, F. Pharmacological activities of brown seaweed Sargassum wightii (Family Sargassaceae) using different in vitro models. Int. J. Food Prop. **2017**, *20*, 931-945.

[44] Unnikrishnan, P.S.; Suthindhiran, K.; Jayasri, M.A. Inhibitory Potential of Turbinaria ornata against Key Metabolic Enzymes Linked to Diabetes. Biomed Res. Int. **2014**, *2014*, 1-10.

[45] Makkar, F.; Chakraborty, K. Antidiabetic and anti-inflammatory potential of sulphated polygalactans from red seaweeds Kappaphycus alvarezii and Gracilaria opuntia. Int. J. Food Prop. **2017**, *20*, 1326-1337.

[46] Admassu, H.; Gasmalla, M.A.A.; Yang, R.; Zhao, W. Bioactive Peptides Derived from Seaweed Protein and Their Health Benefits: Antihypertensive, Antioxidant, and Antidiabetic Properties. J. Food Sci. **2018**, *83*, 6-16.

[47] JUNG, H.A.; YOON, N.Y.; WOO, M.-H.; CHOI, J.S. Inhibitory activities of extracts from several kinds of seaweeds and phlorotannins from the brown alga Ecklonia stolonifera on glucose-mediated protein damage and rat lens aldose reductase. Fish. Sci. **2008**, *74*, 1363-1365.

[48] Jung, H.A.; Islam, M.N.; Lee, C.M.; Oh, S.H.; Lee, S.; Jung, J.H.; Choi, J.S.

Kinetics and molecular docking studies of an anti-diabetic complication inhibitor fucosterol from edible brown algae Eisenia bicyclis and Ecklonia stolonifera. Chem. Biol. Interact. **2013**, *206*, 55-62.

[49] Peng, J.; Yuan, J.-P.; Wu, C.-F.; Wang, J.-H. Fucoxanthin, a Marine Carotenoid Present in Brown Seaweeds and Diatoms: Metabolism and Bioactivities Relevant to Human Health. Mar. Drugs **2011**, *9*, 1806-1828.

[50] Islam, M.N.; Choi, S.H.; Moon, H.E.; Park, J.J.; Jung, H.A.; Woo, M.H.; Woo, H.C.; Choi, J.S. The inhibitory activities of the edible green alga Capsosiphon fulvescens on rat lens aldose reductase and advanced glycation end products formation. Eur. J. Nutr. **2014**, *53*, 233-242.

[51] Hussein, H.A.; Abdullah, M.A. Anticancer Compounds Derived fromMarine Diatoms. Mar. Drugs **2020**, *18*.

[52] Lee, H.J.; Kim, Y.A.; Ahn, J.-W.; Na, H.-J.; Kim, H.-M.; Seo, Y. Screening of Korean marine plants for their inhibitory effect on histamine release from RPMCin vitro. Biotechnol. Bioprocess Eng. **2006**, *11*, 80-83.

[53] Lopes, G.; Andrade, P.; Valentão, P. Phlorotannins: Towards New Pharmacological Interventions for Diabetes Mellitus Type 2. Molecules **2016**, *22*, 56.

[54] Ezzat, S.; Bishbishy, M.; Habtemariam, S.; Salehi, B.; Sharifi-Rad, M.; Martins, N.; Sharifi-Rad, J. Looking at Marine-Derived Bioactive Molecules as Upcoming Anti-Diabetic Agents: A Special Emphasis on PTP1B Inhibitors. Molecules **2018**, *23*, 3334.

[55] Shah, S.A.A.; Hassan, S.S. ul; Bungau, S.; Si, Y.; Xu, H.; Rahman, M.H.; Behl, T.; Gitea, D.; Pavel, F.-M.; Corb Aron, R.A.; et al. Chemically Diverse and Biologically Active Secondary Metabolites from Marine Phylum chlorophyta. *Mar.* Drugs **2020**, *18*, 493.

[56] Thomas, N.V.; Kim, S.-K. Potential pharmacological applications of polyphenolic derivatives from marine brown algae. Environ. Toxicol. Pharmacol. **2011**, *32*, 325-335.

[57] Vijayan, R.; Chitra, L.; Penislusshiyan, S.; Palvannan, T. Exploring bioactive fraction of Sargassum wightii: In vitro elucidation of angiotensin-I-converting enzyme inhibition and antioxidant potential. Int. J. Food Prop. **2018**, *21*, 674-684.

[58] Seca, A.; Pinto, D. Overview on the Antihypertensive and Anti-Obesity Effects of Secondary Metabolites from Seaweeds. Mar. Drugs **2018**, *16*, 237.

[59] Cha, S.H.; Lee, K.W.; Jeon, U.J. Screening of Extracts from Marine Green and Brown Algae in Jeju for Potential Marine Angiotensin-I Converting Enzyme (ACE) Inhibitory Activity. J. Korean Soc. Food Sci. Nutr. **2006**, *35*, 307-314.

[60] Sugiura, S.; Minami, Y.; Taniguchi, R.; Tanaka, R.; Miyake, H.; Mori, T.; Ueda, M.; Shibata, T. Evaluation of Anti-glycation Activities of Phlorotannins in Human and Bovine Serum Albumin-methylglyoxal Models. *Nat. Prod. Commun.* **2017**, *12*, 1934578X1701201.

[61] Shakambari, G.; Ashokkumar, B.; Varalakshmi, P. Phlorotannins from Brown Algae: inhibition of advanced glycation end products formation in high glucose induced Caenorhabditis elegans. Indian J. Exp. Biol. **2015**, *53*, 371-379.

[62] Liu, H.; Gu, L. Phlorotannins from Brown Algae (Fucus vesiculosus) Inhibited the Formation of Advanced

Glycation Endproducts by Scavenging Reactive Carbonyls. J. Agric. Food Chem. **2012**, *60*, 1326-1334.

[63] Nishinaka, T.; Mori, S.; Yamazaki, Y.; Niwa, A.; Wake, H.; Yoshino, T.; Nishibori, M.; Takahashi, H. A comparative study of sulphated polysaccharide effects on advanced glycation end-product uptake and scavenger receptor class A level in macrophages. Diabetes Vasc. Dis. Res. **2020**, *17*, 147916411989697.

Chapter 6

Safety Review of Herbs and Supplements in Heart Disease, Diabetes, and COVID-19

Paula Vieira-Brock

Abstract

Usage of supplements has increased dramatically this last decade. From herbs to vitamins and mineral, consumers are interested in improving health, self-treatment and preventing diseases. Often using information from the internet to self-prescribe, many consumers believe that natural products are safe, while many others avoid using these products because of the lack of an approval process by health officials in many countries. Herbs and other supplements including proteins, vitamins and minerals provide significant benefits to health. The lack of guidance from health professionals however can be problematic. When combined with drugs and disease, herbs can interact and cause side effects. Some of the steps to evaluate the safe use of supplements is to know their mechanism of action, clinical effect, and consumers' medical history. For example, an herb that induces liver enzymes will reduce the effect of a drug that is metabolized by these same enzymes. This can be life threating if the patient depends on this drug for normal function. Based on drug-herb interaction experience and literature review, this book chapter provides insights into safe use of echinacea, licorice, turmeric, and black seed in patients with heart disease, diabetes, and COVID-19.

Keywords: herbs, supplements, drug-herb interaction, safety, COVID-19, heart disease, diabetes

1. Introduction

Dietary supplements are defined in the United States as products that contain one or more dietary ingredient such as vitamins, minerals, herbs, botanicals, and amino acids and are intended to supplement the diet [1]. In other countries dietary supplements are named differently including natural health products, complementary medicines, food supplements, and others [2]. Nonetheless, "dietary supplements" is a general term for products that mostly contain herbs, botanicals, proteins, and/or vitamins and minerals that are used with the intention to promote health. Despite the legal framework, dietary ingredients are often used and recommended for treating or preventing diseases. In this chapter, "dietary supplements" will be used as a general term to encompass several dietary ingredients.

Usage of dietary supplements has increased this last two decades [2]. From herbs, proteins, to vitamins and minerals, consumers are interested in self-treatment and preventing diseases [3]. Often using information from the internet to self-prescribe, many consumers believe that natural products are safe, while many others avoid using these products because of the lack of an approval process by health officials in many countries. Many dietary supplements provide significant benefits to health [4]. However, the lack of guidance from health professionals can be problematic.

Dietary supplements are likely safe when used as prescribed [4, 5]. But, when combined with drugs and disease, these products can interact and cause side effects [6, 7]. Some of the steps to evaluate the safe use of dietary ingredients is to know their mechanism of action, clinical effect, and consumers' medical history. For example, an ingredient that induces liver enzymes will reduce the effect of a drug that is metabolized by these same enzymes. This can be life threating if the patient depends on this drug for normal function.

Due to the benefits that several of these dietary ingredients provide, it is important to evaluate their safety for wide spread recommendation. Particularly due to times of pandemic such as the coronavirus disease 2019 (COVID-19) [8], ways to prevent disease severity and to be used as adjunct treatments are needed. Several dietary ingredients have been reported to be effective against COVID-19 in review articles. For this book chapter, 30 review articles and meta-analysis were evaluated for the selection of the dietary ingredients herein discussed. The selection criterium was based on the number of articles that cited the ingredients as being effective as well as the commonality and accessibility of the ingredients across the globe. Vitamins and minerals were excluded due to their safety being extensively researched. Because COVID-19 severity is worse among patients with diabetes and cardiovascular disease, the safety use of these ingredients in the context of these comorbidities are presented here.

2. Comorbidities and their drug treatments

2.1 COVID-19

COVID-19 is a respiratory infection caused by the virus named "severe acute respiratory syndrome coronavirus 2" (SARS-CoV-2) [8]. COVID-19 is a novel disease officially declared as a pandemic on March 11th, 2020 [9, 10]. SARS-CoV-2 has infected 98.2 million people worldwide and caused 2.1 million deaths as of January 24th, 2021 [11]. COVID-19 is characterized by dry cough, fever, and fatigue symptoms in adults while in children rhinorrhea, abdominal pain, and diarrhea are also present [10]. SARS-CoV-2 binds directly to angiotensin converting enzyme 2 (ACE2) for subsequent entry into cells [10, 12]. Infected cells respond to the virus by generating pro-inflammatory cytokines and chemokines that sometimes lead to a cytokine storm which aggravates the disease [10, 12, 13]. Those with certain underlying health conditions such as respiratory disease, cardiovascular disease, and diabetes as well as older individuals seem to be at a higher risk for developing severe complications from the infection [14, 15]. Because SARS-CoV-2 has approximately 80% genomic homology with SARS-CoV-1, the virus that caused the 2002–2003 epidemic, many research studies have proposed the use of treatments that were effective against SARS-CoV-1 [9]. Current treatments for COVID-19 used in the clinics are ACE2 inhibitors, corticosteroids, chloroquine, anti-inflammatory tocilizumab, comostat, protease inhibitors (lopivavir and ritonavir), and RNA polymerase inhibitors (remdesivir, favipiravir) [16]. Some of the established protocols are: no treatment for mild cases besides acetaminophen for fever; hydroxychloroquine + azithromycin for moderate cases; tocilizumab or sarilumab for worsening respiratory function; and remdesivir, convalescent plasma, corticosteroids for respiratory failure. NSAIDs such as ibuprofen are not recommended due to potential increase in ACE2 expression [17]. Lastly, it has been suggested that reduction in cholesterol decreases viral mRNA [18]. Thus, treatments that reduce cholesterol in addition to antivirals, anti-inflammatories, and respiratory support should be beneficial in managing COVID-19.

2.2 Heart disease and diabetes

As noted above, patients with heart disease and diabetes are more likely to develop severe COVID-19. Thus, many of these patients will be given medications for COVID-19 on top of the current heart/diabetes medications they take. For example, patients continue to take ACE inhibitors or angiotensin II receptor blockers (ARBs) during COVID-19 infection [17]. Furthermore, these are the patients more likely to benefit from dietary ingredients that assist in preventing or treating COVID-19. Due to multiple treatments at once, the likelihood of drug–drug and drug-herb

Drug categories	Liver metabolism	Renal excretion	References
ACE inhibitors: 1. Enalapril 2. Lisinopril	1. Metabolized to active metabolites 2. No liver metabolism	1. ~75% excretion in the urine 2. Excreted unchanged in urine	[21, 22]
ARBs: 1. Losartan 2. Irbesartan	1. CYP3A4 and CYP2C9 2. Glucuronide conjugation and CYP2C9 oxidation	1. Minimal renal excretion with oral administration 2. ~ 20% excretion in the urine	[22]
Hydroxychloroquine	Partially metabolized	Slowly excreted by the kidneys	[21]
Protease inhibitor, ritonavir	CYP3A4 and 2D6 inhibitor	Minimal renal excretion	[16, 23]
RNA polymerase inhibitor, remdesivir	CES1 to form active metabolite	~50% renal excretion	[22]
Corticosteroids and anti-inflammatories 1. Prednisone 2. Tocilizumab	1. Metabolized by CYP3A4 2. Metabolized by CYP3A4	1. Minimal renal excretion	[21, 24]
Antiplatelet, clopidogrel	CYP 2C19 forms active metabolite	~50% excreted in urine	[22]
Beta blocker, atenolol	Minimal metabolism	Major renal excretion	[21]
Cholesterol lowering: 1. Atorvastatin 2. Fluvastatin	1. Metabolized by CYP3A4 2. Metabolized by CYP2C9	1. Minimal renal excretion	[22, 25]
Diuretic, spironolactone	Extensive metabolized	Major renal excretion	[22]
Sulfonylureas 1. Glyburide 2. Glipizide	1. Extensive metabolized 2. Metabolized by CYP2C9	1. ~ 50% renal excretion 2. Minimal renal excretion	[21, 22, 25]
Meglitinides, repaglinide	Metabolized by CYP3A4	Minimal renal excretion	[21]
Metformin	Minimal metabolism	~90% renal excretion	[22]
Glitazones, pioglitazone	Metabolized by CYP2C8	~15–30% renal excretion	[21]

Table 1.
Metabolism and excretion of some common medications used in COVID-19, heart disease and diabetes.

interaction in these patients is high. Drug treatments for heart disease include several types: anticoagulants, antiplatelets, ACE inhibitors, ARBs, beta blockers, calcium channel blockers, cholesterol lowering, diuretics, and vasodilators [19]. For diabetes main medication classes include sulfonylureas, meglitinides, metformin, and glitazones [20]. The metabolism of some commonly prescribed of these medications are listed in **Table 1**. As noted, the most common cytochrome P450 enzyme involved in the metabolism of these drugs are CYP3A4, followed by CYP2C9, 2D6, and 2C8 [21–25]. Approximately half of them are primarily excreted via the kidneys.

3. *Echinacea* spp. (echinacea) – Antiviral and immune support

3.1 Echinacea in COVID-19

Echinacea has antiviral and immunomodulatory effects that seems to be promising against COVID-19 [13, 26]. Several studies have investigated the benefits of echinacea in treating and preventing respiratory tract infections such as the common cold, but not for other health purposes [27]. No studies have yet been completed on echinacea and COVID-19 [28]. A meta-analysis including 17 clinical trials found that echinacea is safe and effective in preventing or treating viral infections. In a separate analysis including 12 clinical trials, echinacea showed to decrease or not change pro-inflammatory cytokines associated with cytokine storm (IL-6, IL-1β, and TNF-α) and increase or not change anti-inflammatory or immune-stimulatory cytokines (IL-10, IL-2, IL-8, IL-3, and IFN-γ). These effects are beneficial during infections since immune stimulatory and anti-inflammatory effects are needed but pro-inflammatory cytokines can aggravate the disease. Adverse events were mild with the most common reported being insomnia, gastrointestinal, and anxiety. One case of serious erythema was reported. Most studies included healthy participants and echinacea dose and method of extraction were quite variable making it difficult to evaluate safety in patients with comorbidities [28].

3.2 Echinacea in diabetes

Not many studies have investigated the effects of echinacea in diabetes. In Wistar rats, 33 days of echinacea root extract showed hypoglycemic activity similar to glibenclamide. No safety parameters were investigated [29]. *Echinacea purpurea* flower extract and caffeic acid derivatives inhibited α-amylase, α-glucosidase, and ACE activities in a concentration-dependent manner [30].

3.3 Echinacea in heart disease

Almost no studies have evaluated the effects of echinacea on heart conditions such as hypertension and hypercholesterolemia. In one study with 374 elderly, 349 reported to use over-the-counter drugs and 43 reported to use herbal medicine. Echinacea was the most common herbal therapy used while aspirin, acetaminophen, laxatives, antiacids, and vitamins were the most common over-the-counter drugs [31]. This single study suggests the potential for interaction of echinacea with drugs.

3.4 Echinacea toxicity

In a review article, echinacea was considered to have a high or medium evidence for efficacy and safety [32]. Debatable concern of hepatotoxicity with echinacea when used for more than 8 weeks has been raised [6]. On the other hand, echinacea has shown hepatic and renal protection against toxins in rats with no effect by itself on liver and kidney parameters including AST, ALT, ALP, blood urea nitrogen and creatinine [33]. No toxicity was found in rats and mice after oral or intravenous injection of *Echinacea purpurea* at high doses. No evidence of mutagenicity in vitro and in mice or carcinogenicity in hamster embryo cells [34]. Echinacea is contraindicated in patients with autoimmune disease. Little evidence to evaluate the effect of echinacea in renal impairment [35].

3.5 Echinacea pharmacokinetics

In vivo pharmacokinetics in 12 healthy men and women, *Echinacea purpurea* root inhibited CYP1A2 and induced CYP3A enzymes [36]. In another similar study with 13 healthy adults, *Echinacea purpurea* induced CYP3A enzymes but did not change the pharmacokinetics of ritonavir. Ritonavir is an inhibitor of CYP3A thus likely counteracted the induction caused by echinacea. No effect on p-glycoprotein was found [37]. In an in vitro study, *Echinacea purpurea* showed mixed effects on CYP3A4 and moderate inhibition of CYP2C9 [38]. A review article found echinacea to have high likelihood of drug-herb interaction [39].

3.6 Echinacea safety summary

Echinacea is likely safe when taken short-term, up to 8 weeks, in healthy adults. Unknown safety in patients with diabetes or heart disease. Caution should be taken when combining with medications metabolized by CYP3A, 1A2, and 2C9 enzymes.

4. *Glycyrrhiza sp.* root (licorice) – Antiviral and respiratory support

4.1 Licorice in COVID-19

Licorice root is used as a flavoring agent in food in many countries. In the United States, anise oil is often used for this purpose. Licorice is promoted as a dietary supplement for digestion, cough, infections, and others [40]. Frequently recommended by herbalists, licorice has recently shown to be the herb most frequently used for COVID-19 treatment [41, 42]. Several review articles have discussed the potential effectiveness of licorice in treating COVID-19 for its antiviral, anti-inflammatory, spasmolytic, and expectorant effects [9, 10, 12–14]. Some in vitro studies showed that the active component glycyrrhizin inhibits the replication of SARS-coronavirus (SARS-CoV) [43, 44]. Other in vitro studies showed that glycyrrhizin may prevent SARS-CoV-2 entry by binding to ACE2 receptors and other protein targets [45, 46]. Clinical trials of licorice use during COVID-19 are ongoing. Daily doses range from 250 mg 25% extract (62.5 mg glycyrrhizin) for 10 days to 2.28 g 3% extract (70 mg glycyrrhizin) for 7 days [47, 48].

4.2 Licorice in diabetes

Not many studies have investigated the effects of licorice in diabetes. In a clinical trial with 58 overweight and obese but otherwise healthy volunteers, 1.5 g licorice extract (<0.01% glycyrrhizin) for 8 weeks decreased insulin and HOMA-IR without side effects [49]. In cell cultures, de-glycyrrhizinated or regular licorice showed to be a potential therapeutic target in diabetic nephropathy [50]. In diabetic mice, licorice hydrophobic flavonoids demonstrated abdominal fat-lowering and hypoglycemic effects [51].

4.3 Licorice in heart disease

Cases of hypokalemia and hypertension have been reported after daily ingestion of licorice tea or after short-term high dose [52–54]. In one case patient was combining licorice with the glucocorticoid medication fludrocortisone [55]. The active components of licorice, glycyrrhizinates, inhibit the enzyme responsible for inactivating cortisol and bind to mineralocorticoid receptors resulting in reversible hyper-mineralocorticoid effects [56]. A meta-analysis with 18 clinical trials found that chronic daily intake of 100 mg glycyrrhizin increases systolic and diastolic blood pressure [57]. In another meta-analysis including 26 clinical trials and 985 subjects, mainly healthy and overweight but some with hypercholesterolemia, found licorice to reduce body weight and BMI but increase diastolic blood pressure. Licorice was given as licorice flavonoid oil with a dose range of 300 mg to 1.8 g/day for 2–16 weeks [58]. In a dose–response relationship investigation in healthy men and women, licorice root with 108 or 217 mg of glycyrrhizin per day for 4 weeks caused no adverse events. However, licorice with 380 and 814 mg glycyrrhizin caused headache, arterial hypertension, hyperkalemia, and peripheral edema. One individual had a family history of hypertension [59]. Lastly, a similar study compared adverse events in patients with hypertension versus normotensive individuals during 100 g licorice containing 150 mg glycyrrhetinic acid per day for 4 weeks. Systolic and diastolic blood pressure were slightly increased in normotensive (3.5 and 3.6 mmHg) but significantly greater increase in hypertensive patients (15.3 and 9.3 mmHg). Increase in urinary cortisol correlated with the rise in blood pressure [60]. These data suggest that glycyrrhizin at dose >200 mg/day short-term and > 100 mg/day long-term in patients or healthy individuals can cause reversible hyperkalemia and hypertension.

4.4 Licorice toxicity

Despite licorice being a substance generally recognized as safe (GRAS) in the United States [61] and regarded as having a high safety profile because it is consumed as food [32], licorice can cause hypertension and hypokalemia in a dose-dependent manner [57]. However, safe dose will vary depending on licorice's composition and the underlying medical conditions. Those with hypertension, heart or kidney disease are more sensitive to licorice toxicity [40]. In a study involving 360 subjects, no clinically significant change in renal function (potassium, blood urea nitrogen, and creatinine levels) were found in 98.3% of the subjects after ~19 days of ~8 g licorice per day taken as dietary supplements that contained other ingredients. The remaining 1.7% of subjects developed hyperkalemia [62]. In a safety and toxicity study with 39 healthy female and male volunteers aged 19–40 years old, glycyrrhizic acid was administered at 1, 2, and 4 mg/kg body weight daily for 8 weeks. A no-effect level of 2 mg/kg was found and applying a 100-safety factor, the acceptable daily intake of 0.2 mg/kg body weight was proposed. This is equivalent to 12 mg glycyrrhizic acid/day for a 60-kg person [63]. Similarly, based on review of in vivo and clinical evidence, an acceptable daily intake has been proposed to be 0.015–0.229 mg glycyrrhizin/kg body weight [64]. The acceptable daily intake without a safety factor is equivalent to 120 mg glycyrrhizic acid. This dose could be considered safe if used short-term in a situation of high benefit versus risk.

4.5 Licorice pharmacokinetics

Glycyrrhiza glabra has shown weak inhibition of CYP3A4 and moderate inhibition of CYP2B6, 2C8, 2C9, and 2C19. *Glycyrrhiza uralensis* showed strong inhibition of CYP2B6, moderate inhibition of CYP2C8, 2C9, and 2C19, and no inhibition of CYP3A4 and 2D6. *Glycyrrhiza inflata* strongly inhibited CYP2C enzymes and moderately inhibited of CYP3A4, 1A2, 2B6, and 2D6. None of the three species inhibited CYP2E1 and 2A6. Glycyrrhizin content was highest in *G. uralensis* suggesting that glycyrrhizin is a weak inhibitor of the major enzymes CYP3A4 and 2D6 [65]. Weak inhibition of CYP3A4 and 2D6 by glycyrrhizin and *G. glabra* were also found in a different study [66].

4.6 Licorice safety summary

A safe daily dose for short-term use consists of licorice with less than 100 mg glycyrrhizin. For daily long-term use a dose of 12 mg glycyrrhizin has been proposed. COVID-19 studies are using short-term doses of <100 mg glycyrrhizin per day. Caution should be taken when combining licorice with medications. Licorice inhibits several cytochrome P450 enzymes including CYP1A2, 2B6, 2C8, 2C9, and 2C19. Only *G. inflata* inhibits CYP3A4 and 2D6.

5. *Curcuma longa* (turmeric) – Antiviral and anti-inflammatory

5.1 Turmeric in COVID-19

Turmeric has antiviral and anti-inflammatory effects that might benefit COVID-19 patients [10, 13]. It has also been hypothesized that the antioxidant effects of turmeric benefit diabetic patients during COVID-19 infection [67]. However, some has expressed concerns that curcumin, the main active component of turmeric, might increase the expression of ACE2 and worsen COVID-19 infection as well as increase pro-inflammatory cytokines and worsen COVID-19 in patients with cytokine storm [26]. In the contrary, curcumin binds to viral S protein and the viral attachment sites of the ACE2 receptor protein to inhibit the entry of SARS-CoV2 [18, 68]. In addition, curcumin has shown to reduce inflammatory cytokines in COVID-19 patients. In a clinical study with 40 COVID-19 patients, curcumin given as nano-curcumin at 160 mg/day for 14 days reduced the inflammatory cytokines IL-6 and IL-1β as well as clinical manifestations (fever, cough, dyspnea, headache, chest radiography, lymphocyte, white blood cells, and platelets count) in comparison to placebo-treated group. Both groups were taking atorvastatin, bromhexine, and betaferon concomitantly with 5–15% of them having diabetes, cardiovascular disease or renal disease. These results suggest the effectiveness and safety of curcumin in COVID-19 patients with underlying medical conditions [69].

5.2 Turmeric in diabetes

In clinical trials with type 2 diabetic patients, curcuminoids from 250 mg/day for 9 months to 1 g/day for 3 months improved glycemic control, β-cell function, insulin resistance, and reduced inflammatory cytokines with no major adverse effects. Minor side effects included diarrhea, constipation, vertigo, and itching. Some clinical and preclinical studies also showed that curcumin improve biomarkers of liver and kidney damage [70]. In a clinical trial on 46 patients with diabetic nephropathy, 1.5 g curcumin for 16 weeks improved 24-h urine analysis for albuminuria with no change in blood urea nitrogen, creatinine, fasting blood sugar, 2-h postprandial blood sugar, lipid profile, serum albumin, and hemoglobin A1C in comparison to placebo and baseline [71].

5.3 Turmeric in heart disease

A recent meta-analysis found that turmeric or curcumin have no effect on diastolic blood pressure and minor effect on systolic blood pressure when taken for longer than 12 weeks [72]. A meta-analysis that included 7 randomized, placebo-controlled clinical trials in patients with cardiovascular risk factors (i.e., non-alcoholic fatty liver disease, metabolic syndrome, type 2 diabetes, prehypertension, and dyslipidemia) found turmeric powder at 2–2.4 g/day for 1–2 months, turmeric extract with 0.6–1.9 g curcuminoids/day for 2–6 months, or curcumin at 70–80 mg/day for 2–3 months were effective in reducing serum LDL-cholesterol and triglycerides levels. Adverse events reported were abdominal pain, nausea, dyspepsia, constipation, and hot flushes. Hot flushes were also reported in the placebo group. In 3 of the trials patients were kept on their medications during the study; however, only one trial disclosed the name of the concomitant drug treatment (metformin) [73].

5.4 Turmeric toxicity

Turmeric has GRAS status in the United States [74]. Through a toxicological assessment, the European Food Safety Authority (EFSA) has recommended curcumin daily intake be ≤3 mg/kg body weight per day (180 mg/day in 60 kg individuals) [75]. In 2–year oral feed studies, turmeric oil at 79–85% curcumin showed no biological significantly differences in hematology, clinical chemistry (liver and kidney function markers), and urinalysis parameters, but showed to potentially cause carcinogenicity in mice and rats especially in females at doses ≥100 mg/kg body weight in rats and 300 mg/kg body weight in mice [76]. However, the EFSA concluded that curcumin is not carcinogenic and studies have demonstrated the benefits of curcumin as an adjunct treatment of cancer [77]. High daily dose of curcumin might cause hepatotoxicity. In rats, 25 and 100 mg/kg body weight for 90 days of curcumin induced liver injury through the generation of reactive oxygen species and pro-inflammatory cytokines as well as reduced antioxidant and detoxifying enzymes SOD and GST [78]. Similarly, 5% turmeric via diet for 90 days in female Wistar rats and 0.2% turmeric via diet in female Swiss mice was hepatotoxic. Human equivalent dose for these rodent studies ranged from 250 mg curcumin/day to 1 g–50 g turmeric/day [79, 80].

5.5 Turmeric pharmacokinetics

Turmeric constituents have shown to inhibit p-glycoprotein in vitro and in vivo models [81]. Inhibition of p-glycoprotein can lead to increased bioavailability of drugs [82]. Curcumin is primarily eliminated in the feces with little renal excretion in a rat study [76]. In a pharmacokinetics study with healthy adults, turmeric reduced the bioavailability of the beta-blocker talinolol [83]. Curcumin was safe and effective when combined with glyburide in patients with type 2 diabetes. Better cholesterol and glycemia control without hypoglycemic side effects were observed. Curcumin increased AUC but did not change C_{max} of glyburide [84]. In rats, curcumin increased the C_{max}, $AUC_{0\text{-}t}$ and half-life of amlodipine – an antihypertensive drug [85]. Amlodipine is metabolized by CYP3A4 in humans [86]. Curcumin inhibits several hepatic CYP enzymes including 3A4, 1A2, 2B6 (competitive type of inhibition), 2D6 and 2C9 (non-competitive inhibition) in human recombinant cytochrome P450s [87]. However, it is been suggested that these effects are not clinically significant due to poor bioavailability of curcumin. In fact, in a pharmacokinetics study in healthy volunteers, 4 g curcuminoids +24 mg piperine to enhance bioavailability did not affect C_{max}, AUC, clearance, or half-life of drugs metabolized by CYP3A, CYP2C9, and UGT, SULT conjugation enzymes [88].

5.6 Turmeric safety summary

Turmeric is safe and effective at doses ≤250 mg curcumin/day. Higher doses are associated with hepatotoxicity and potentially carcinogenicity. Doses as low as 70–250 mg curcuminoids/day has shown to be effective in metabolic disorders and COVID-19. Although turmeric inhibits cytochrome P450 enzymes, these effects seem to be clinically negligible. Caution when taken with drugs that are substrates of p-glycoprotein in order to avoid drug overdose. Although turmeric has hypoglycemic effects and might cause side effects such as fainting when combined with antidiabetic medications, this combination has shown to be safe in clinical trials.

6. *Nigella sativa* (black seed) – Anti-inflammatory and respiratory support

6.1 *Nigella sativa* in COVID-19

N. sativa is a plant native to South East Asia with several pharmacological effects including bronchodilation, antitussive, and anti-inflammatory and used as treatments of respiratory conditions, diabetes, cardiovascular diseases, among others [89, 90]. For example, in a clinical trial with 90 obese women, 3 g/day of *N. sativa* oil for 8 weeks reduced serum levels of TNFα and hsCRP in comparison to placebo with no adverse events reported [91]. In patients with asthma, 1 g *N. sativa* oil per day for 4 weeks reduced several inflammatory markers and improved pulmonary function [92]. Preclinical studies have shown that constituents in the methanolic extract of *N. sativa* seeds are responsible for the bronchodilator effect [93]. Recently, *N. sativa* has been regarded as a potential therapy for COVID-19 [13, 18, 94, 95]. For example, in a molecular docking-based study *N. sativa* inhibited SARS-CoV2 [94].

6.2 *Nigella sativa* in diabetes

Several clinical trials have been conducted to evaluate *N. sativa* in patients with type 2 diabetes [96]. For example, three controlled studies investigated the adjuvant use of 1–3 g/day *N. sativa* seeds powder or 2.5 ml/day *N. sativa* oil for 12 weeks in patients with type-2 diabetes. Significant and similar effects were observed with doses of 2 and 3 g/d on the reductions in fasting blood glucose, 2-hour postprandial glucose, HbA1C levels, and insulin resistance. Treatments were not associated with any adverse renal or hepatic functions throughout the study period. Patients were concomitantly taking oral hypoglycemic drugs (glibenclamide, metformin, rosiglitazone) but not insulin. Patients with coronary artery disease, valvular heart disease, heart failure, uncontrolled hypertension, renal failure and hepatic failure were excluded [97–99].

6.3 *Nigella sativa* in heart disease

In a meta-analysis including 11 randomized clinical trials with 860 hypertensive or normotensive individuals, *N. sativa* seeds versus placebo and one versus standard treatment significantly reduced systolic blood pressure by −3.60 mmHg and diastolic blood pressure by −2.80 mmHg. [100]. A similar meta-analysis including 17 randomized clinical trials with 1185 individuals with hyperlipidemia, obesity, hypertension, type 2 diabetes, or others, found that *N. sativa* seed powder or oil 1–3 g/day for up to 3 months reduces total cholesterol, LDL-cholesterol, and triglycerides [101]. No adverse events were reported by the subjects [102, 103]. One study in elderly with hypertension reported mild adverse events including dyspepsia in 6 subjects (15.7%), nausea in 3 subjects (7.8%), and constipation in 2 subjects (5.2%). No electrolyte abnormalities, liver and renal toxicities, or orthostatic hypotension were observed [104].

6.4 *Nigella sativa* toxicity

N. sativa has GRAS status in the United States [74]. A randomized, placebo-controlled study with 40 healthy elderly investigated the safety profile after daily intake of 1 g *N. sativa* seed powder for 9 weeks. Results found no statistical changes in any of the biochemical markers of cardiac, liver, and kidney function [105]. In 70 patients with chronic renal disease, 2.5 ml/day *N. sativa* oil for 12 weeks was safe and effective in improving clinical and biochemical parameters of kidney function without adverse events [106]. In a clinical trial with obese women, *N. sativa* oil (which is present in whole seeds and polar seed extracts), reduced body weight, VLDL cholesterol and triglycerides [107]. Traditional toxicity studies in rodents have been performed. In mice, hepatotoxicity was observed after 14-days of oral dosing at 6–21 g/kg body weight of *N. sativa* seeds water extract. No signs of hepatotoxicity were observed with methanolic and chloroform extracts. Body weight reductions were seen in methanolic extracts [108]. Similar findings were observed with the water extract in rats with increases in serum gamma-glutamyl transferase and alanine aminotransferase, but no changes in alkaline phosphatase and degeneration of hepatocytes [109]. In another rat study, 1 g/day for 6 weeks of whole *N. sativa* seeds were protective against hyperlipidemia to a similar extent as simvastatin without adverse effects to liver markers [110]. Human equivalent doses are 30–100 g *N. sativa* extracts per day for the mice study, and 10 g whole seeds per day for the rat study [80].

6.5 *Nigella sativa* pharmacokinetics

One of the main active constituents in *N. sativa* seeds and oil is thymoquinone. Thymoquinone has shown to bind to human α(1)-acid glycoprotein in the plasma [111] and inhibit CYP2C9 > 2D6 > 1A2 > 3A4 liver enzymes [112]. In hypertensive rats, *N. sativa* + alompidine showed greater reduction in blood pressure and heart rate than *N. sativa* alone, but no effect on alompidine pharmacokinetics (C_{max}, $AUC_{0\text{-}t}$, K_{el}, and terminal half-life) [113]. In another study in hypertensive rats, *N. sativa* + losartan showed greater reduction in blood pressure than *N. sativa* alone. *N. sativa* slightly reduced losartan C_{max} and $AUC_{0\text{-}t}$ [114]. Alompidine is metabolized by CYP3A4 and losartan by CYP2C9 and 3A4 in humans. These data suggest that *N. sativa* has minimal effect on CYP3A4 but inhibits CYP2C9. In other words, *N. sativa* has antihypertensive effects on its own but potentiates the effect of drugs metabolized by CYP2C9 which can cause further drop in blood pressure and lead to side effects such as fainting.

6.6 *Nigella sativa* safety summary

N. sativa whole seeds, oil or polar extracts (i.e., non-aqueous) at human doses up to 3 g/day for 12 weeks beneficially affect inflammatory and metabolic markers without adverse effects on heart, liver, or kidneys in healthy adults as well as in patients with heart disease and diabetes. *N. sativa* reduces blood glucose and blood pressure. Thus, caution when combining with hypoglycemic and antihypertensive drugs to avoid side effects. *N. sativa* can increase the bioavailability of drugs metabolized by CYP2C9 leading to higher risks of their side effects. Some diabetes and heart medications metabolized by CYP2C9 are losartan, fluvastatin, glipizide [25].

7. Interactions summary

The combination of several dietary ingredients might be desirable when their main mechanisms of action and clinical effects differ. For example, combination of an anti-inflammatory, antiviral, immunostimulant, and bronchodilator herbs might be recommended. Safety combination of black seed and turmeric has been demonstrated in a clinical study. *N. sativa* seed (1.5 g/d) and turmeric (2.4 g/d) in patients with metabolic syndrome for 4 weeks was safe and effective in reducing blood glucose, cholesterol, and blood pressure despite both ingredients having hypoglycemic and antihypertensive effects alone [115]. Any effect of echinacea on blood glucose and blood pressure is insufficient to evaluate. Licorice can increase blood pressure depending on the dose. Since turmeric and *N. sativa* have anti-hypertensive effects, the addition of licorice might be safe. All of the four dietary ingredients described here inhibit CYP2C9. All except *N. sativa* also inhibit CYP1A2. Turmeric and licorice also inhibit CYP2B6. Turmeric inhibits CYP3A4 and echinacea induces

Dietary ingredient	Level of evidence in COVID-19	Level of evidence in heart disease	Level of evidence in diabetes	Main interactions	References
Echinacea	None	None	Scarce: positive effect in 1 preclinical study	Induces CYP3A, inhibits CYP1A2, and CYP2C9	[27–29, 36, 38]
Licorice	Positive effects in vitro and 2 ongoing clinical trials	Negative effects in several clinical trials showing hypertension and hyperkalemia	Scarce: positive effects in 1 clinical, 1 preclinical, and 1 in vitro study	Safe dose <100 mg glycyrrhizin. Inhibits CYP1A2, 2B6, 2C8, 2C9, and 2C19. Only *G. inflata* inhibits CYP3A4 and 2D6	[45–61, 63–66]
Turmeric	Positive effects in vitro and 1 completed clinical trial	Positive effects in several clinical trials	Positive effects in several clinical trials	Safe dose ≤250 mg. Inhibits p-glycoprotein and not clinically significant inhibition of P450 enzymes	[68–77, 88]
Black seed	Positive effects in vitro	Positive effects in several clinical trials	Positive effects in several clinical trials	Inhibits CYP2C9	[94, 96–104, 111–114]

Table 2.
Summary of level of evidence for efficacy and safety of echinacea, licorice, turmeric, and black seed in COVID-19, heart disease, and diabetes.

it. Thus, caution should be taken when combining these dietary ingredients with drugs metabolized by CYP3A4, 2C9, 1A2 and 2B6. As presented in **Table 1**, many drugs used in COVID-19, diabetes, and heart disease are metabolized by CYP3A4 and 2C9. Caution should be taken with echinacea and turmeric because they induce or inhibit CYP3A4, respectively. Lastly, many drug examples presented in **Table 2** are excreted via urine. Turmeric and black seed are likely safe when combined with medications that are excreted by the kidneys. Caution when combining with licorice due to its potential to cause hyperkalemia. No sufficient evidence to evaluate echinacea's effect on kidney function.

8. Conclusions

All the four dietary ingredients discussed herein are safe for use short-term as in a setting of treating a disease. However, some might not be safe when taken long-term. For example, no safety data was found for echinacea in heart disease and diabetes. Long-term use of low dose or short-term use of high dose licorice can cause reversible hypertension. Hepatotoxicity might occur with long-term use of turmeric >250 mg/day. Lastly, all of these four dietary ingredients are metabolized by cytochrome P450 enzymes to some extent. Mostly they inhibit CYP2C9, 1A2 and 2B6. Caution with echinacea because it induces CYP3A4 and turmeric because it inhibits it.

Conflict of interest

The author declares no conflict of interest.

Author details

Paula Vieira-Brock
Independent Pharmacologist and Toxicologist, Salt Lake City, USA

*Address all correspondence to: paulavbrock@gmail.com

References

[1] Congress, *Dietary Supplement Health and Education Act of 1994*, in *S.784*, U.S.o. America, Editor. 1994: https://www.congress.gov/bill/103rd-congress/senate-bill/784/text.

[2] Dwyer, J.T., P.M. Coates, and M.J. Smith, *Dietary Supplements: Regulatory Challenges and Research Resources.* Nutrients, 2018. **10**(1).

[3] Bailey, R.L., et al., *Why US adults use dietary supplements.* JAMA Intern Med, 2013. **173**(5): p. 355-361.

[4] Fields, J.M., *Dangers of scientific bias against herbal drugs for coronavirus disease 2019.* J Integr Med, 2020. **18**(6): p. 459-461.

[5] Posadzki, P., L.K. Watson, and E. Ernst, *Adverse effects of herbal medicines: an overview of systematic reviews.* Clin Med (Lond), 2013. **13**(1): p. 7-12.

[6] Miller, L.G., *Herbal medicinals: selected clinical considerations focusing on known or potential drug-herb interactions.* Arch Intern Med, 1998. **158**(20): p. 2200-2211.

[7] Liu, M.Z., et al., *Pharmacogenomics and herb-drug interactions: merge of future and tradition.* Evid Based Complement Alternat Med, 2015. **2015**: p. 321091.

[8] WHO. *Naming the Coronavirus Disease (COVID-19) and the Virus that Causes it.* 2020 January 31st.]; Available from: https://www.who.int/emergencies/diseases/novel-coronavirus-2019/technical-guidance/naming-the-coronavirus-disease-(covid-2019)-and-the-virus-that-causes-it

[9] Fuzimoto, A.D. and C. Isidoro, *The antiviral and coronavirus-host protein pathways inhibiting properties of herbs and natural compounds - Additional weapons in the fight against the COVID-19 pandemic?* J Tradit Complement Med, 2020. **10**(4): p. 405-419.

[10] Jalali, A., et al., *A pharmacology-based comprehensive review on medicinal plants and phytoactive constituents possibly effective in the management of COVID-19.* Phytother Res, 2020.

[11] WHO. *Weekly epidemiological update - 27 January 2021.* 2021 January, 31st.]; Available from: https://www.who.int/publications/m/item/weekly-epidemiological-update---27-january-2021.

[12] Wyganowska-Swiatkowska, M., et al., *Influence of Herbal Medicines on HMGB1 Release, SARS-CoV-2 Viral Attachment, Acute Respiratory Failure, and Sepsis. A Literature Review.* Int J Mol Sci, 2020. **21**(13).

[13] Brendler, T., et al., *Botanical drugs and supplements affecting the immune response in the time of COVID-19: Implications for research and clinical practice.* Phytother Res, 2020.

[14] Boukhatem, M.N. and W.N. Setzer, *Aromatic Herbs, Medicinal Plant-Derived Essential Oils, and Phytochemical Extracts as Potential Therapies for Coronaviruses: Future Perspectives.* Plants (Basel), 2020. **9**(6).

[15] Fang, L., G. Karakiulakis, and M. Roth, *Are patients with hypertension and diabetes mellitus at increased risk for COVID-19 infection?* Lancet Respir Med, 2020. **8**(4): p. e21.

[16] Hushmandi, K., et al., *A review of medications used to control and improve the signs and symptoms of COVID-19 patients.* Eur J Pharmacol, 2020. **887**: p. 173568.

[17] Mehta, N., et al., *Pharmacotherapy in COVID-19; A narrative review for emergency providers.* Am J Emerg Med, 2020. **38**(7): p. 1488-1493.

[18] Ho, P., et al., *Perspective Adjunctive Therapies for COVID-19: Beyond*

Antiviral Therapy. Int J Med Sci, 2021. **18**(2): p. 314-324.

[19] AHA. *Types of Heart Medications*. 2020 January 31st]; Available from: https://www.heart.org/en/health-topics/heart-attack/treatment-of-a-heart-attack/cardiac-medications#ARB.

[20] ADA. *Oral medications*. 2021 January 31st]; Available from: https://www.diabetes.org/healthy-living/medication-treatments/oral-medication.

[21] ASHP. *AHFS Drug Information Monographs*. 2021 January 31st]; Available from: https://www.drugs.com/monograph/.

[22] FDA. *Professional Drug Information*. 2021 January 31st]; Available from: https://www.drugs.com/pro/.

[23] Hsu, A., G.R. Granneman, and R.J. Bertz, *Ritonavir. Clinical pharmacokinetics and interactions with other anti-HIV agents.* Clin Pharmacokinet, 1998. **35**(4): p. 275-291.

[24] FDA. *Full prescribing information for Actemra*. 2010 January 31st]; Available from: https://www.accessdata.fda.gov/drugsatfda_docs/label/2016/125276s107_125472s018lbl.pdf.

[25] Van Booven, D., et al., *Cytochrome P450 2C9-CYP2C9.* Pharmacogenet Genomics, 2010. **20**(4): p. 277-281.

[26] Nugraha, R.V., et al., *Traditional Herbal Medicine Candidates as Complementary Treatments for COVID-19: A Review of Their Mechanisms, Pros and Cons.* Evid Based Complement Alternat Med, 2020. **2020**: p. 2560645.

[27] FDA, *Echinacea*, U.S.D.o.H.a.H. Services, Editor. 2020: https://www.nccih.nih.gov/health/echinacea.

[28] Aucoin, M., et al., *The effect of Echinacea spp. on the prevention or treatment of COVID-19 and other respiratory tract infections in humans: A rapid review.* Adv Integr Med, 2020. **7**(4): p. 203-217.

[29] Aarland, R.C., et al., *Studies on phytochemical, antioxidant, anti-inflammatory, hypoglycaemic and antiproliferative activities of Echinacea purpurea and Echinacea angustifolia extracts.* Pharm Biol, 2017. **55**(1): p. 649-656.

[30] Chiou, S.Y., et al., *Antioxidant, Antidiabetic, and Antihypertensive Properties of Echinacea purpurea Flower Extract and Caffeic Acid Derivatives Using In Vitro Models.* J Med Food, 2017. **20**(2): p. 171-179.

[31] Albert, N.M., et al., *Predictors of over-the-counter drug and herbal therapies use in elderly patients with heart failure.* J Card Fail, 2009. **15**(7): p. 600-606.

[32] Silveira, D., et al., *COVID-19: Is There Evidence for the Use of Herbal Medicines as Adjuvant Symptomatic Therapy?* Front Pharmacol, 2020. **11**: p. 581840.

[33] Rezaie, A., et al., *Effects of Echinacea purpurea on Hepatic and Renal Toxicity Induced by Diethylnitrosamine in Rats.* Jundishapur J Nat Pharm Prod, 2013. **8**(2): p. 60-64.

[34] Mengs, U., C.B. Clare, and J.A. Poiley, *Toxicity of Echinacea purpurea. Acute, subacute and genotoxicity studies.* Arzneimittelforschung, 1991. **41**(10): p. 1076-1081.

[35] EMA, *Assessment report on Echinacea purpurea (L.) Moench., herba recens*, E.M. Agency, Editor. 2014: https://www.ema.europa.eu/en/documents/herbal-report/final-assessment-report-echinacea-purpurea-l-moench-herba-recens_en.pdf.

[36] Gorski, J.C., et al., *The effect of echinacea (Echinacea purpurea root) on*

cytochrome P450 activity in vivo. Clin Pharmacol Ther, 2004. **75**(1): p. 89-100.

[37] Penzak, S.R., et al., *Echinacea purpurea significantly induces cytochrome P450 3A activity but does not alter lopinavir-ritonavir exposure in healthy subjects.* Pharmacotherapy, 2010. **30**(8): p. 797-805.

[38] Yale, S.H. and I. Glurich, *Analysis of the inhibitory potential of Ginkgo biloba, Echinacea purpurea, and Serenoa repens on the metabolic activity of cytochrome P450 3A4, 2D6, and 2C9.* J Altern Complement Med, 2005. **11**(3): p. 433-439.

[39] Tsai, H.H., et al., *Evaluation of documented drug interactions and contraindications associated with herbs and dietary supplements: a systematic literature review.* Int J Clin Pract, 2012. **66**(11): p. 1056-1078.

[40] FDA, *Licorice Root*, U.S.D.o.H.a.H. Services, Editor. 2020: www.nccih.nih.gov/health/licorice-root.

[41] Ang, L., et al., *Herbal medicine for treatment of children diagnosed with COVID-19: A review of guidelines.* Complement Ther Clin Pract, 2020. **39**: p. 101174.

[42] Ang, L., et al., *Herbal medicine for the management of COVID-19 during the medical observation period: A review of guidelines.* Integr Med Res, 2020. **9**(3): p. 100465.

[43] Cinatl, J., et al., *Glycyrrhizin, an active component of liquorice roots, and replication of SARS-associated coronavirus.* Lancet, 2003. **361**(9374): p. 2045-2046.

[44] Hoever, G., et al., *Antiviral activity of glycyrrhizic acid derivatives against SARS-coronavirus.* J Med Chem, 2005. **48**(4): p. 1256-1259.

[45] Chen, H. and Q. Du, *Potential natural compounds for preventing SARS-CoV-2 (2019-nCoV) infection.* Preprints, 2020. **2020010358**.

[46] Vardhan, S. and S.K. Sahoo, *In silico ADMET and molecular docking study on searching potential inhibitors from limonoids and triterpenoids for COVID-19.* Comput Biol Med, 2020. **124**: p. 103936.

[47] Network, E.B.R., *Evaluation of The Potential Therapeutic Effects of Licorice and Boswellia Serrata Gum in Egyptian Patients With COVID-19 as a Complementary Medicine.* Identifier: NCT04487964, 2020, January - 2021, January. **https://clinicaltrials.gov/ct2/show/NCT04487964**.

[48] Safa, O., et al., *Effects of Licorice on clinical symptoms and laboratory signs in moderately ill patients with pneumonia from COVID-19: A structured summary of a study protocol for a randomized controlled trial.* Trials, 2020. **21**(1): p. 790.

[49] Alizadeh, M., et al., *Changes of Insulin Resistance and Adipokines Following Supplementation with Glycyrrhiza Glabra L. Extract in Combination with a Low-Calorie Diet in Overweight and Obese Subjects: a Randomized Double Blind Clinical Trial.* Adv Pharm Bull, 2018. **8**(1): p. 123-130.

[50] Hsu, Y.C., et al., *De-Glycyrrhizinated Licorice Extract Attenuates High Glucose-Stimulated Renal Tubular Epithelial-Mesenchymal Transition via Suppressing the Notch2 Signaling Pathway.* Cells, 2020. **9**(1).

[51] Nakagawa, K., et al., *Licorice flavonoids suppress abdominal fat accumulation and increase in blood glucose level in obese diabetic KK-A(y) mice.* Biol Pharm Bull, 2004. **27**(11): p. 1775-1778.

[52] Kwon, Y.E., D.J. Oh, and H.M. Choi, *Severe asymptomatic hypokalemia associated with prolonged licorice ingestion: A case report.* Medicine (Baltimore), 2020. **99**(30): p. e21094.

[53] Yang, L.Y., et al., *Liquorice-induced severe hypokalemic rhabdomyolysis with Gitelman syndrome and diabetes: A case report.* World J Clin Cases, 2019. **7**(10): p. 1200-1205.

[54] Smedegaard, S.B. and M.V. Svart, *Licorice induced pseudohyperaldosteronism, severe hypertension, and long QT.* Endocrinol Diabetes Metab Case Rep, 2019. **2019**.

[55] Benge, E., et al., *Trick or Treat? Licorice-Induced Hypokalemia: A Case Report.* Cureus, 2020. **12**(11): p. e11656.

[56] Kwon, Y.J., et al., *A Review of the Pharmacological Efficacy and Safety of Licorice Root from Corroborative Clinical Trial Findings.* J Med Food, 2020. **23**(1): p. 12-20.

[57] Penninkilampi, R., E.M. Eslick, and G.D. Eslick, *The association between consistent licorice ingestion, hypertension and hypokalaemia: a systematic review and meta-analysis.* J Hum Hypertens, 2017. **31**(11): p. 699-707.

[58] Luís, Â., F. Domingues, and L. Pereira, *Metabolic changes after licorice consumption: A systematic review with meta-analysis and trial sequential analysis of clinical trials.* Phytomedicine, 2018. **39**: p. 17-24.

[59] Bernardi, M., et al., *Effects of prolonged ingestion of graded doses of licorice by healthy volunteers.* Life Sci, 1994. **55**(11): p. 863-872.

[60] Sigurjonsdottir, H.A., et al., *Subjects with essential hypertension are more sensitive to the inhibition of 11 beta-HSD by liquorice.* J Hum Hypertens, 2003. **17**(2): p. 125-131.

[61] CFR, *Listing of Specific Substances Affirmed as GRAS: Licorice and licorice derivatives*, in *21 Code of Federal Regulations 184.1408*, U.S.o. America, Editor. 2020.

[62] Jung, W., et al., *Influence of herbal complexes containing licorice on potassium levels: a retrospective study.* Evid Based Complement Alternat Med, 2014. **2014**: p. 970385.

[63] van Gelderen, C.E., et al., *Glycyrrhizic acid: the assessment of a no effect level.* Hum Exp Toxicol, 2000. **19**(8): p. 434-439.

[64] Isbrucker, R.A. and G.A. Burdock, *Risk and safety assessment on the consumption of Licorice root (Glycyrrhiza sp.), its extract and powder as a food ingredient, with emphasis on the pharmacology and toxicology of glycyrrhizin.* Regul Toxicol Pharmacol, 2006. **46**(3): p. 167-192.

[65] Li, G., et al., *Cytochrome P450 inhibition by three licorice species and fourteen licorice constituents.* Eur J Pharm Sci, 2017. **109**: p. 182-190.

[66] Pandit, S., et al., *Exploring the possible metabolism mediated interaction of Glycyrrhiza glabra extract with CYP3A4 and CYP2D6.* Phytother Res, 2011. **25**(10): p. 1429-1434.

[67] Tabatabaei-Malazy, O., M. Abdollahi, and B. Larijani, *Beneficial Effects of Anti-Oxidative Herbal Medicines in Diabetic Patients Infected with COVID-19: A Hypothesis.* Diabetes Metab Syndr Obes, 2020. **13**: p. 3113-3116.

[68] Noor, H., et al., *Immunomodulatory and anti-cytokine therapeutic potential of curcumin and its derivatives for treating COVID-19 - a computational modeling.* J Biomol Struct Dyn, 2021: p. 1-16.

[69] Valizadeh, H., et al., *Nano-curcumin therapy, a promising method in modulating inflammatory cytokines in COVID-19 patients.* Int Immunopharmacol, 2020. **89**(Pt B): p. 107088.

[70] Pivari, F., et al., *Curcumin and Type 2 Diabetes Mellitus: Prevention and Treatment.* Nutrients, 2019. **11**(8).

[71] Vanaie, A., et al., *Curcumin as a major active component of turmeric attenuates proteinuria in patients with overt diabetic nephropathy.* J Res Med Sci, 2019. **24**: p. 77.

[72] Hadi, A., et al., *The effect of Curcumin/Turmeric on blood pressure modulation: A systematic review and meta-analysis.* Pharmacol Res, 2019. **150**: p. 104505.

[73] Qin, S., et al., *Efficacy and safety of turmeric and curcumin in lowering blood lipid levels in patients with cardiovascular risk factors: a meta-analysis of randomized controlled trials.* Nutr J, 2017. **16**(1): p. 68.

[74] CFR, *Substances generally recognized as safe: Spices and other natural seasonings and flavorings.*, in *21 Code of Federal Regulations 182.10*, U.S. government, Editor. 2021.

[75] EFSA, *Scientific Opinion on the re-evaluation of curcumin (E 100) as a food additive.* European Food Safety Authority Journal, 2010. **8**(9): p. 1679.

[76] NTP, *NTP Toxicology and Carcinogenesis Studies of Turmeric Oleoresin (CAS No. 8024-37-1) (Major Component 79%-85% Curcumin, CAS No. 458-37-7) in F344/N Rats and B6C3F1 Mice (Feed Studies).* Natl Toxicol Program Tech Rep Ser, 1993. **427**: p. 1-275.

[77] Adiwidjaja, J., A.J. McLachlan, and A.V. Boddy, *Curcumin as a clinically-promising anti-cancer agent: pharmacokinetics and drug interactions.* Expert Opin Drug Metab Toxicol, 2017. **13**(9): p. 953-972.

[78] Qiu, P., et al., *Overdose Intake of Curcumin Initiates the Unbalanced State of Bodies.* J Agric Food Chem, 2016. **64**(13): p. 2765-2771.

[79] EPA, *Recommendations for and Documentation of Biological Values for Use in Risk Assessment.*, U.S.E.P. Agency, Editor. 1988: https://cfpub.epa.gov/ncea/risk/recordisplay.cfm?deid=34855.

[80] Reagan-Shaw, S., M. Nihal, and N. Ahmad, *Dose translation from animal to human studies revisited.* Faseb j, 2008. **22**(3): p. 659-661.

[81] Lopes-Rodrigues, V., E. Sousa, and M.H. Vasconcelos, *Curcumin as a Modulator of P-Glycoprotein in Cancer: Challenges and Perspectives.* Pharmaceuticals (Basel), 2016. **9**(4).

[82] Lin, J.H. and M. Yamazaki, *Role of P-glycoprotein in pharmacokinetics: clinical implications.* Clin Pharmacokinet, 2003. **42**(1): p. 59-98.

[83] Juan, H., et al., *Unexpected effect of concomitantly administered curcumin on the pharmacokinetics of talinolol in healthy Chinese volunteers.* Eur J Clin Pharmacol, 2007. **63**(7): p. 663-668.

[84] Neerati, P., R. Devde, and A.K. Gangi, *Evaluation of the effect of curcumin capsules on glyburide therapy in patients with type-2 diabetes mellitus.* Phytother Res, 2014. **28**(12): p. 1796-1800.

[85] Jiang, N., et al., *Effects of curcumin on the pharmacokinetics of amlodipine in rats and its potential mechanism.* Pharm Biol, 2020. **58**(1): p. 465-468.

[86] Zhu, Y., et al., *Amlodipine metabolism in human liver microsomes and roles of CYP3A4/5 in the dihydropyridine dehydrogenation.* Drug Metab Dispos, 2014. **42**(2): p. 245-249.

[87] Appiah-Opong, R., et al., *Inhibition of human recombinant cytochrome P450s by curcumin and curcumin decomposition*

products. Toxicology, 2007. **235**(1-2): p. 83-91.

[88] Volak, L.P., et al., *Effect of a herbal extract containing curcumin and piperine on midazolam, flurbiprofen and paracetamol (acetaminophen) pharmacokinetics in healthy volunteers.* Br J Clin Pharmacol, 2013. **75**(2): p. 450-462.

[89] Clarke, R., F. Lundy, and L. McGarvey, *Herbal treatments in asthma and COPD - current evidence.* Clinical Phytoscience, 2015. **1**(4).

[90] Dajani, E.Z., T.G. Shahwan, and N.E. Dajani, *Overview of the preclinical pharmacological properties of Nigella sativa (black seeds): a complementary drug with historical and clinical significance.* J Physiol Pharmacol, 2016. **67**(6): p. 801-817.

[91] Mahdavi, R., et al., *Nigella sativa oil with a calorie-restricted diet can improve biomarkers of systemic inflammation in obese women: A randomized double-blind, placebo-controlled clinical trial.* J Clin Lipidol, 2016. **10**(5): p. 1203-1211.

[92] Koshak, A., et al., *Nigella sativa Supplementation Improves Asthma Control and Biomarkers: A Randomized, Double-Blind, Placebo-Controlled Trial.* Phytother Res, 2017. **31**(3): p. 403-409.

[93] Keyhanmanesh, R., et al., *The Relaxant Effects of Different Methanolic Fractions of Nigella sativa on Guinea Pig Tracheal Chains.* Iran J Basic Med Sci, 2013. **16**(2): p. 123-128.

[94] Ahmad, S., et al., *Molecular docking, simulation and MM-PBSA studies of Nigella sativa compounds: a computational quest to identify potential natural antiviral for COVID-19 treatment.* J Biomol Struct Dyn, 2020: p. 1-9.

[95] Islam, M.N., et al., *Revisiting pharmacological potentials of Nigella sativa seed: A promising option for COVID-19 prevention and cure.* Phytother Res, 2020.

[96] Hamdan, A., R. Haji Idrus, and M.H. Mokhtar, *Effects of Nigella sativa on Type-2 Diabetes Mellitus: A Systematic Review.* Int J Environ Res Public Health, 2019. **16**(24).

[97] Bamosa, A.O., et al., *Effect of Nigella sativa seeds on the glycemic control of patients with type 2 diabetes mellitus.* Indian J Physiol Pharmacol, 2010. **54**(4): p. 344-354.

[98] Hosseini, M., et al., *Effects of Nigella sativa L. Seed Oil in Type II Diabetic Patients: a Randomized, Double-Blind, Placebo - Controlled Clinical Trial.* Journal of Medicinal Plants, 2013. **12**(47): p. 93-99.

[99] Kaatabi, H., et al., *Nigella sativa improves glycemic control and ameliorates oxidative stress in patients with type 2 diabetes mellitus: placebo controlled participant blinded clinical trial.* PLoS One, 2015. **10**(2): p. e0113486.

[100] Sahebkar, A., et al., *A systematic review and meta-analysis of randomized controlled trials investigating the effects of supplementation with Nigella sativa (black seed) on blood pressure.* J Hypertens, 2016. **34**(11): p. 2127-2135.

[101] Sahebkar, A., et al., *Nigella sativa (black seed) effects on plasma lipid concentrations in humans: A systematic review and meta-analysis of randomized placebo-controlled trials.* Pharmacol Res, 2016. **106**: p. 37-50.

[102] Sabzghabaee, A.M., et al., *Clinical evaluation of Nigella sativa seeds for the treatment of hyperlipidemia: a randomized, placebo controlled clinical trial.* Med Arch, 2012. **66**(3): p. 198-200.

[103] Dehkordi, F.R. and A.F. Kamkhah, *Antihypertensive effect of Nigella*

sativa seed extract in patients with mild hypertension. Fundam Clin Pharmacol, 2008. **22**(4): p. 447-452.

[104] Rizka, A., et al., *Effect of Nigella sativa Seed Extract for Hypertension in Elderly: a Double-blind, Randomized Controlled Trial.* Acta Med Indones, 2017. **49**(4): p. 307-313.

[105] Bin Sayeed, M.S., et al., *The effect of Nigella sativa Linn. seed on memory, attention and cognition in healthy human volunteers.* J Ethnopharmacol, 2013. **148**(3): p. 780-786.

[106] Alam, M.A., et al., *Evaluation of safety and efficacy profile of Nigella sativa oil as an add-on therapy, in addition to alpha-keto analogue of essential amino acids in patients with chronic kidney disease.* Saudi J Kidney Dis Transpl, 2020. **31**(1): p. 21-31.

[107] Mahdavi, R., et al., *Effects of Nigella sativa oil with a low-calorie diet on cardiometabolic risk factors in obese women: a randomized controlled clinical trial.* Food Funct, 2015. **6**(6): p. 2041-2048.

[108] Vahdati-Mashhadian, N., H. Rakhshandeh, and A. Omidi, *An investigation on LD50 and subacute hepatic toxicity of Nigella sativa seed extracts in mice.* Pharmazie, 2005. **60**(7): p. 544-7.

[109] Tennekoon, K.H., et al., *Possible hepatotoxicity of Nigella sativa seeds and Dregea volubilis leaves.* J Ethnopharmacol, 1991. **31**(3): p. 283-289.

[110] Muneera, K.E., A. Majeed, and A.K. Naveed, *Comparative evaluation of Nigella sativa (Kalonji) and simvastatin for the treatment of hyperlipidemia and in the induction of hepatotoxicity.* Pak J Pharm Sci, 2015. **28**(2): p. 493-498.

[111] Lupidi, G., et al., *Characterization of thymoquinone binding to human* α_1*-acid glycoprotein.* J Pharm Sci, 2012. **101**(7): p. 2564-2573.

[112] Albassam, A.A., et al., *Inhibition of cytochrome P450 enzymes by thymoquinone in human liver microsomes.* Saudi Pharm J, 2018. **26**(5): p. 673-677.

[113] Alam, M.A., et al., *Effect of Nigella sativa and Fenugreek on the Pharmacokinetics and Pharmacodynamics of Amlodipine in Hypertensive Rats.* Curr Drug Metab, 2020. **21**(4): p. 318-325.

[114] Ahad, A., et al., *Potential pharmacodynamic and pharmacokinetic interactions of Nigella Sativa and Trigonella Foenum-graecum with losartan in L-NAME induced hypertensive rats.* Saudi J Biol Sci, 2020. **27**(10): p. 2544-2550.

[115] Amin, F., et al., *Clinical efficacy of the co-administration of Turmeric and Black seeds (Kalongi) in metabolic syndrome - a double blind randomized controlled trial - TAK-MetS trial.* Complement Ther Med, 2015. **23**(2): p. 165-174.

Chapter 7

Pharmacological Investigation of Genus *Pistacia*

Abdur Rauf, Yahya S. Al-Awthan, Naveed Muhammad, Muhammad Mukarram Shah, Saikat Mitra, Talha Bin Emran, Omar Bahattab and Mohammad S. Mubarak

Abstract

Several plants in the genus *Pistacia* are used in the treatment of various pathogenic and non-pathogenic disorders. Especially important are the major species belonging to this genus such as *Pistacia lentiscus, Pistacia atlantica, Pistacia vera, Pistacia terebinthus,* and *Pistacia khinjuk,* among others; these have been reported for their potential benefits both in medical and commercial purposes. In addition, members of this genus exhibit numerous ethnomedicinal uses, such as analgesic, anti-inflammatory, anticancer, antimicrobial, antihypertension, antihyperlipidemic, antiviral, and antiasthma. In light of these potential uses, the present chapter aimed to collect and summarize the literature about all of this medicinal information. Accordingly, this chapter focuses on the pharmacological uses and benefits of the genus Pistacia, especially those related to health issues.

Keywords: *Pistacia*; *Pistacia lentiscus*, *Pistacia atlantica*, *Pistacia vera*, *Pistacia terebinthus*, *Pistacia khinjuk*, pharmacological activities

1. Introduction

Pistacia, a genus that belongs to the family and order of Anacardiaceae and Sapindales, respectively, includes almost twenty species five of which have been classified and characterized as significant and economically important [1]. Flowers of this genus are in panicles or racemes, unisexual, small, apetalous, subtended by 1–3 small bracts and wind-pollinated, and 2–7 bracteoles. Deciduous, alternative or evergreen leaves are typically pinnate, sometimes simple or trifoliate, leathery, or membranous [2]. *Pistacia vera, P. khinjuk, P. atlantica, P. terebinthus,* and *P. lentiscus* are the foremost species of the genus *Pistacia*, where studies carried out by numerous researchers showed that the *Pistacia vera L.* as the utmost economically valuable species [3]. Cultivated pistachio, which is scientifically known as *Pistacia vera* has continued to rise to an annual estimated value of around $2 billion over the last two decades [4]. It has comestible seeds and a commercially important influence. Pistachios, often utilized in the shell, are fresh to consume; baked products, fruit, and ice cream are used for manufacturing purposes. Their applications as traditional, medical, and non-food products, such as toothache relief, are also

IntechOpen

available. In addition, Pistachio has been documented as a solution for sclerosis and scirrhus of the liver, abscesses, impaired circulation, and other health-related issues [2, 5]. Furthermore, the *Pistacia* genus has been tested for multiple ethnomedicinal ailments, including inflammation, cancer, microbial attack, hypertension, and asthma, among others. The frequent usage of representatives of this genus rendered it as core plants in natural medicines. For instance, several health problems and disorders caused by free radicals may be can be mitigated by means of antioxidants.

Antioxidants are the strongest protective agents against free radicals. In this respect, members of genus *Pistacia* have been documented to display variable degrees of free radical scavenging potential. Leaf extracts obtained from *Pistacia lenticus* and *P. atlantica* exert antioxidant effects with 14.16% and 19.3%, respectively [6]. In addition, research findings indicated that genus *Pistacia* gens has been established as natural antimicrobial agents. Fungal growth was substantially decreased by the crude leaf extract of *P. atalantica* and *P. lenticus*, but the growth of bacteria was not significantly suppressed [6]. Similarly, the mouthwash of *P. atalantica* has an impressive antimicrobial effect on the microorganism of gingivae and has been recommended as reliable and effective [7]. In addition, essential oil from *P. vera* with an effective effect against some pathogenic bacteria particularly *S. aureus* and *E. coli* [8]. On the other hand, the lipophilic extract of *P. vera* demonstrated potential antiviral effect [9]. In addition, *P. lenticus, P. atlantica, P. palaestina,* and *P. vera*, among others exhibited anticancer activity in numerous experimental studies. In this context, the crude extract of leaves and fruits of *P. lenticus* substantially suppressed the growth in the cell line of the growing melanoma [10], where inhibitory potential against BHK21 cell line has been identified in the seed oil of *P. lenticus* [11]. Furthermore, the ethanol extract of *P. atlantica* showed significant activity against gastric and cervical carcinoma [12]. Besides, the essential oil obtained from *P. palaestina* is efficient in inhibiting malignant colorectal cancer [13]. Mansouri et al. [79] evaluated in vivo the neuroprotective effect of *P. vera L.* gum extract on oxidative damage during cerebral ischemia–reperfusion in rats and concluded that the neuroprotective potency may be due to cumulative antioxidant defense as well as suppression of free radical production [14]. Besides, anthelmintic role has been observed for different extract and essential oil of *P. khinjuk* particularly against *Echinococcus granulosus*, which develops hydatid cyst [15]. Except for all of these biological activities, members of genera *Pistacia* exert high therapeutic activity against numerous health issues, including peptic ulcer, colitis, Hypoglycemia, obesity, hypertension, Nephritic disorders, hepatic disorders, and other toxicological problems. Based on the previous discussion, the aim of the present work is to collect and summarize the medicinal information along with recent references pertaining to members of the genus *Pistacia*, which would be helpful to and further researchers in the field. Below are details about documented biological activities related to the members of the genus *Pistacia*.

2. Biological activity

2.1 Antioxidant effect

Free radicals are responsible for ample of disorder in human medicines. Blockage, neutralization or complexation of these noxious radicals can prevent or mitigate numerous health issues. In this respect, synthetic antioxidants might be responsible for several side effects; therefore, natural antioxidants are preferred. Antioxidants are the best preventive agents against free radicals responsible for various diseases. Within this context, different members of the genus *Pistacia* demonstrated variable degree of free radicals scavenging potential. The leaf extracts of *P. lenticus* and *P. atlantica* showed a week antioxidant effect (14.16 and 19.3% respectively) [1]. On the other hand, the methanol extract of *P. linticus* at the flowering season was tested for antioxidant effect using DPPH and PRAP assays; results showed a high significant antioxidant (131 mmol/L) effect [2]. Similarly, the crude leaf extract of *P. linticus* demonstrated significant antioxidant potential [3]. The methanol leaf extracts of *P. atlantica* of 34 collected samples were tested for antioxidant effect, and results revealed significant antioxidant [4]. In a similar fashion, the seeds and skin of pistachio (*P. vera* L) were subjected to antioxidant effect (DPPH, TEAC, and SOD mimetic assays), and the phenolic contents quantification (HPLC) were determined. The best antioxidant effect of skin as compared to seeds was attributed to the highest phenolic contents [5]. Additionally, the hydrophilic extract of pistachio nut showed antioxidant effects due to the presence of polyphenolic compounds [6]. The acetone and methanol extracts of *P. terebinthus* demonstrated good antioxidant effects, attributed to the presence of various phenolic contents and flavones [7]. *P. weinmannifolia* is a shrub and widely distributed in the Yunnan area of China. The leaves of this plant are used traditionally by the herbalist. The leaves are rich in phenolic constitutes, among which Gallotannins, Pistafolin A and B were confirmed. The protection of lipid, proteins and DNA damage from the reactive oxygen species (ROS) by Pistafolin A and B through antioxidant effect was reported. The free radical scavenging effect of Pistofoli A was more potent than Pistofolin B due to structural changes [1]. Taken all together, *Pistacia* plants could be excellent free radical scavengers, which could help to cure or mitigate several diseases. The ROS or RNS etc., as free radicals interact with the cell membrane, such as the free radicals interact with hemoglobin and making them denatured, the denatured hemoglobin accumulating at the surface of RBC and making the cell membrane non-flexible, which leads to the rupturing of RBC known as hemolytic anemia. The use of antioxidanta, especially the plants-based antioxidants, can prevent a lot of health problems.

2.2 Anti-microbial effect

The list of antibiotics is supplementing day by day due to antimicrobial resistance issues. These antibiotics are helpful and have extended spectrum but are responsible for various adverse effects. These adverse effects minimize the patient compliance, and, therefore, the search for new, effective and affordable antibiotic is a big challenge to phytochemical researchers. In this respect, natural antibiotics could have multiple uses in addition to the antibiotic effect; therefore, the use of natural antibiotic can minimize the polypharmacy. Within this context, research findings indicated that the crude leaf extract of *P. lenticus* and *P. atalantica* significantly reduce the fungal growth, whereas weak bacterial inhibitory effect was reported [1]. Roozegar investigated the effect of *P. atalantica* leaf extract against mouth and saliva bacterial load, and reported a significant effect against *S. mutans* and *S. mitis* in disk diffusion method with zone of inhibition of 19 and 25 mm, respectively; no significant effect was observed against *S. salivarius* [8]. The mouthwash of *P. atalantica* exhibited excellent antimicrobial effect against gingival microorganism. Therefore, this mouthwash was recommended as effective and safe [9].

Similarly, the hydro extract of *P. atalantica* was tested against different bacteria *in vitro* and was found effective against *E. coli*, *P. aeruginosa* and *S. aureus*, except for *H. pylori* [10]. Additionally, the hydro distilled essential oil from the stem of *P. vera* was tested against some pathogenic bacteria, and exhibited significant effect against *E. coli* and *S. aureus* [11]. Furthermore, the antibacterial potential of *P. lenticus* extract was tested against gram positive and gram-negative bacteria. Results demonstrated that the extract exerts significant effect against gram positive as compared to gram negative bacteria [12]. The leaf extract of *P. khinjuk* when screened for the antibacterial and antifungal potential exhibited significant activity [13]. On the other hand, the essential oil of *P. khinjuk* was found to contain, through GC–MS analyses γ-terpinene (81.14%) (w/w), β-pinene (3.93%) (w/w), and α-terpinolene (2.38%) (w/w). This essential oil was tested for activity against *P. aeruginosa* and *S. subtilis.* Chemical constituents of the essential oil might be responsible for the antibacterial effect against the tested pathogenic bacteria [14]. Similarly, the essential oil from the leaves of *P. lenticus* was also tested against different gram positive and gram negative pathogenic bacteria. The major chemical constitutes in essential oil were α-pinene and β-pinene, and a variable degree of antibacterial effects were observed [15]. Volatile compounds from the essential of leaves and fruits of *P. lenticus* exhibited best antibacterial effect [16]. Likewise, the antimicrobial effect of *P. integerrima* has been reported against various pathogenic microbes. The oil was found rich in 1-terpinen-4-ol (28.82%), p-menth-1-en-8-ol, (43.38%), *n*-octyl acetate (19.91%), and β-farnesene (7.88%). The concentration of α-terpinolene, limonene and α-thujene were less than 1%. The tested oils exhibited promising antibacterial activities. The zone on inhibition against *E. coli, S. aureus, K. pneumonia, Straptodirimu, B. stearothermophilus* and *S. typhimurium* was 16, 18, 26, 22, 18 and 20 mm, respectively [17]. The essential oil of *P. terebinthus* (collected from Tunisia and Italy) was reported along with chemical composition (GC and GC–MS). The oil was isolated through hydrodistillation. The oil consisted of monoterpene hydrocarbons (86.3% and 90.9%, respectively), α-pinene (62.4 vs. 35.0)%, camphene (3.0 vs. 2.4)%, β-pinene (12.1 vs. 4.5)%, terpinolene (1.7 vs. 35.2)% and β-phellandrene (3.8 vs. 4.5)% as the main components. The oil demonstrated significant effect against *T. rubrum*, *M. canis* and *E. floccosum*, with MIC and MLC values in the range (0.16–0.32) μL/mL [18]. In view of the above discussion, these plants might help against different pathogenic infections. Plants accumulate numerous phytochemicals that interact with the micro-organisms. The inhibition or the killing of these micro-organisms might be due to cell wall inhibition or protein synthesis inhibition or might be due to the antimetabolite action of constitutes. These plants' use for the above infections needs to explore the exact mechanism on related microbes and clinical trials.

2.3 Antiviral effect

The antiviral effect of natural products cannot be ignored. The non-polar extract of *P. vera* is antiviral. *Herpes simplex* (DNA) and *Parainfluenza virus* (RNA) was used for confirmation of antiviral effect [19]. The extracts demonstrated antiviral effect at a concentration of 128–256 μg/mL. Different antiviral compounds have been identified in *P. lenticus.* The HSV-2, Coxasakievirus-3 and adenovirus-5 were used. The methanolic extract of *P. lenticus* demonstrated antiviral action against HSV-2 [20]. The polyphenolic rich extract at concentration range of 0.4, 0.6, 0.8 mg/mL of *P. vera* has been used against the HSV-1 with significant results [21]. Further study is needed to confirm the antiviral action of *Pistacia* against a wide range of viruses, including coronavirus.

2.4 Antiemetic effect

Emesis is one of the common side effects of numerous drugs. Emesis is also a common problem of other associated diseases. Natural products used for this purpose are well developed traditionally. The copper sulfate and ipecac-induced emesis has been blocked by the *P. vera* leaves and nut extract [22]. Copper sulfate induces emesis through GIT irritation, and ipecac induces emesis through GIT irritation and chemically stimulated the CTZ. The chemical constitutes of ipecac get readily absorbed and interact with 5HT3 and dopamine receptors. A mechanistic study is needed to confirm the extract mode of action.

2.5 Anticancer effect

The crude extract of *P. lenticus* leaves and fruits significantly inhibited the growth of melanoma cell line (B16F10 cells). The leaves and fruits significantly inhibited the B16F10 cells (IC_{50} = 56.40 and 58.04 μg/mL, respectively) [3], whereas the seed oil and phenolic compounds fraction of *P. lenticus* showed inhibitory potential against BHK21 cell line. The IC_{50} was 0.029 g/mL and the percent effect was 42.4 at the concentration of 0.09 g/mL [23]. The essential oils extracted from the fruits and leaves of *P. lenticus* were tested for anticancer effect. The oil of leaves exhibited interesting anticancer activity as compared to fruits [16] on RD and L20B cell lines with IC_{50} values of 26.43 ± 2.18 and 33.02 ± 2.84 μg/mL, respectively. A protective effect of the oil of *P. lenticus* has been reported in bleomycin-induced lung fibrosis and oxidative stress in rats [24]. A significant anti- proliferation potential of *P. atlantica* against COLO205 has been noticed along with good antioxidant effect [25]. Colorectal cancer is one of the major malignant forms of cancer, which has been blocked by the essential oil from the *P. palaestina* [26]. The ethanol extract of *P. atlantica* demonstrated a significant effect against the gastric and cervical carcinoma along with antioxidant effect, which is attributed to the presence of phenolic compounds in the extract [27]. The ethyl acetate extract of *P. vera* L. also attenuated the growth of MCF-7 human breast cancer [28]. The anticancer effect of these plants against various cells line is well established. Interestingly, the antiemetic effect of these plants is very good regarding the anticancer effect because most anticancer drugs have emesis as a significant side effect. If a natural remedy has anticancer and antiemetic potential, it might be one of the best therapeutic mixtures.

2.6 Cytotoxic effect

The crude extract and fractions of *P. integerrima* of gall, root, bark, and leaves have been reported with cytotoxic effect against brine shrimp [29]. The extract and fractions were tested at various concentrations (10, 100, and 1000 ppm). This preliminary study is a pathway toward anticancer potentials.

2.7 Antiparasitic effect

The essential oil (EO) from the leaves and fruits of *P. lenticus* were tested against leishmanial species (*L. infantum*, *L. major*, and *L. tropica*) using MTT assay. Both of the tested samples demonstrated a variable degree of cytotoxic effect. The major constituents of leaves were myrcene, α pinene, while limonene and α-pinene constituted fruits essential oil. The EO of leaves demonstrated significant effect against *L. major* (IC50 = 17.52 ± 1.26 μg/mL) as compared to the EO of fruits (IC50 = 21.42 ± 2.92 μg/mL), while the EO of fruits exhibited more effect than leaves against *L. infantum* (IC50 = 08 ± 0.83 μg/mL) than *P. lentiscus* leaves essential oils (IC50 = 11.28 ± 1.63 μg/mL) [16]. The extract of *P. khinjuk* demonstrated a significant *in vivo* and *in vitro* effects against *L. major* and *L. tropica* [30]. The fruits and leaves extract of *P. atlantica* and *P. vera* demonstrated a significant inhibition against hydatid cyst protoscolices [31, 32]. The essential oil of *P. vera* inhibited *in vitro* and *in vivo* leishmanial effect [33]. The essential oil of *P. lenticus* also demonstrated *anti-trichomonas vaginalis* trophozoites. The EO were tested at concentration range of 15, 10 and 5 mg/mL with different time duration of incubation. The morphological changes were monitored through TEM [34, 35]. A new antiparasitic agent (pistagremic acid) has been reported from the *P. integerrima* with IC_{50}: 6.71 ± 0.09 μM against *Leishmania major* [36].

2.8 Antidiarrheal effect

The crude methanolic extract of *P. integerrima* significantly reduces the castor oil-induced diarrhea in mice, where maximum relaxation of smooth muscle was noticed. The induced contraction was exerted through calcium and muscarinic receptor agonist. Contraction was inhibited by the plant extract, and this inhibition reflects the plant's antimuscarinic action as well as the calcium channel blocking properties [37]. The crude methanolic effect of *P. integerrima* bark has been tested for its GIT motility effect and showed a significant reduction in induced loos motion [38].

2.9 Antispasmodic effect

The crude methanolic extract of *P. integerrima* demonstrated a significant inhibition in spontaneous contraction of rabbit jejunum [37]. This calcium-induced contraction was reversed by the plant extract. This relaxing effect on GIT smooth muscle reflects the constipating effect of plant extract.

2.10 Bronchodilator effect

The induced contraction of tracheal section was completely relaxed with the application of methanolic extract of *P. integerrima* [37]. The essential oil of *P. integerrima* has been reported with antiasthematic effect [39].

2.11 Analgesic effect

P. integerrima bark's methanolic extract significantly reduced the induced writhing in mice representing the painkiller potential in this plant. This attenuation of acetic acid-induced writhing at the dose of 100 mg/kg reflects the peripheral analgesic effect of plant extract [38]. *P. integerrima* gall's extract demonstrated significant analgesic effect against acid-induced writhing, formalin-induced pain, and thermal-induced central algesia. The extract also attenuated the thermal-induced pain [40]. The gall analgesic effect was due to the presence of analgesic flavonoids [41]. The *P. vera* leaves extract proved central and peripheral analgesic in animal models [42]. The oil of *P. atalantica* fruits attenuated acetic acid-induced writhing in rats [43]. The *P. atlantica* was also reported to have a good painker in another study [44, 45]. Pestagremic acid is one of the potential analgesic constitutes of the *P. integerrima* bark [46]. The oleoresin demonstrated the anti-inflammatory effect while rest of the samples were devoid of analgesic potential [47]. The gold nanoparticles *P. integerrima* gall have been tested for analgesic effect at the tested doses of 10 and 20 in acetic acid-induced pain model. Results demonstrated significant analgesic potential [48]. These research data reflect that this genus has central and peripheral analgesic potential. The opioids receptors mediate the central pain, while peripheral pain receptors or COX inhibition are responsible for the peripheral analgesic effect. The available synthetic drugs having a good analgesic effect but are associated with side effects like a peptic ulcer. To find the analgesic remedy free of side effects is a big challenge to the researcher in the current modern era. The above-tested extract or constitutes needs to inter in the clinical trial to find more useful analgesic drugs.

2.12 Anti-osteoarthritis effect

Osteoarthritis (OA) is one of the chronic health problems around the globe. The patient of OA commonly uses NSAID as self-medication, especially in developing countries. To develop or discover new effective and safe medication for OA, plants' screening is essential. The oleoresin from *P. atlantica* demonstrated a comparable effect with diclofenac in knee osteoarthritis [49]. The *P. atlantica* cream might inhibit various enzymes involved in inflammation. The formulated cream significantly inhibited the OA induced condition. The topical anti-OA is far better than systemic use for the elimination of severe side effects.

3. Anti-inflammatory activity

The faction of the leaves of *P. lenticus* significantly attenuated the induced edema as compared to acetylsalicylic acid [3]. The crude extract of the gall of *P. integerrima* also demonstrated anti-inflammatory effect in various doses [40]. In another study, the anti-inflammatory effect of gall was attributed to the presence of flavonoids. The isolated flavonoids were tested for carrageenan-induced edema and provided significant anti-inflammatory [41]. The EO from the fruits of *P. lenticus* attenuated the carrageenan-induced edema (inflammation) at various tested doses [50]. The crude leaves extract of *P. vera* demonstrated anti-inflammatory effect both in acute and chronic inflammatory models [42]. The *P. atlantica* has been proven significant anti-inflammatory in animal model [44, 51]. The bark of *P. integerrima* accumulated anti-inflammatory constitutes like pestagremic acid [46, 52]. The nano particles of *P. integerrima* gall also showed significant anti-inflammatory effect [48]. The ethanolic and aqueous extracts of different parts of *P. vera* as its oleoresin have been tested for anti-inflammatory effect. The oleoresin demonstrated the anti-inflammatory effect while the rest of the samples were devoid of anti-inflammatory potential [47]. In another study, the significant anti-inflammatory effect (*in-vivo* and *in-vitro*) of *P. vera* has been reported [53]. The extract and triterpene from the *P. terebinthus* gall demonstrated significant acute and chronic anti-inflammatory effects [54]. The aqueous extract of *P. khinjuk* demonstrated anti-inflammatory effect [54, 55]. The above data mean that the genus has the best anti-inflammatory plants. Inflammation is caused by prostaglandin (PG) production. The PGs are the product of arachidonic acid through COX. Inhibition of COX is responsible for the anti-inflammatory effect. These COX are widely distributed in the body. The extract or constitutes blocking COX are considered as anti-inflammatory drugs.

3.1 Anti-gout effect

The leaves of *P. integerrima* demonstrated uric acid (UA) lowering effect in fructose induced hyperuricemia animal model [56]. The chemical constituents such as quercetin-3-*O*-β-d-glucopyranoside, kaempferol-3-*O*-β-d-glucopyranoside, quercetin-3-*O*-(6″-*O*-syringyl)-β-d-glucopyranoside, kaempferol-3-*O*-(4″-*O*-galloyl)-α-l-arabinopyranoside, rutin together with aglycons, quercetin, kaempferol and apigenin inhibited the XO up to a variable degree. The inhibition of XO is a strong indicator of *P. integerrima* as a significant anti-gout. Hyperuricemia is also a chronic pain condition and needs to prolong treatment.

3.2 Anti- epileptic effect

Epilepsy is one of the most common, serious neurological conditions, affecting more than 50 million people worldwide. The hydroalcoholic extract of *P. vera* demonstrated a significant anti-epileptic effect in pentylenetetrazole (PTZ) chronic induced kidding in male rats [57]. The epileptic condition was induced by PTZ (40 mg/kg, IP), and the induced condition was significantly inhibited by the extract of *P. vera* at the tested doses of 50 and 100 mg/kg. The inhibition of chronic induced seizure indicate that *P. vera* is a significant antiepileptic. The petroleum ether extract of *P. integerrima* attenuated the PTZ-induced jerks in zebrafish and mice models at the dose of 50 and 100 mg/kg. The antiepileptic effect was further confirmed through maximum electroshock (MES) in a rat model. The tested extract significantly attenuated various aspects of induced jurks [58].

4. Sedative and hypnotic activity

Insomnia is a worldwide health issue with different etiology. This condition is treated as self-medication through benzodiazepines, which have a potential side effect of addiction. Once the patient gets addicted, then these medicines are used for life. The natural plant's based tranquilizers might be free of such addiction due to the accumulation of agonist and antagonistic chemical constituents. The hydroalcoholic extract of *P. vera* gum showed a sedative effect in the locomotor test. The extract at the dose of 0.25, 0.5, 1 g/kg showed the increased duration of sleep and shortened sleep latency hypnotic effect in phenobarbital-induced sleep [59].

4.1 Muscle relaxation

The hydroalcoholic extract of *P. vera* gum acted as muscle relaxant in traction and rotarod test. When tested at 0.25, 0.5, 1 g/kg and only the higher dose (1 g/kg) demonstrated this muscle relaxation effect [59]. No further syudies are available.

4.2 Effect on memory

The essential oil of *P. lenticus* attenuated memory dysfunction in rats. *P. lentiscus* oil (PLo) at a dose of 3.3 mL/kg for 15 days reversed LPS-induced memory deficits in rats. Besides, the increased acetylcholinesterase activity in brain structures of LPS-treated rats was reduced by PLo. Additionally, PLo significantly attenuated the increased oxidative stress in the brain of LPS-treated rats [60]. The chemical induced memory impairment was regulated with *P. vera* fruit [61].

4.3 Anti-fatigue effect

The hydro-alcoholic extract of *P. vera* seed is significant anti-fatigue. The extract was tested at the dose of 10, 100 and 1000 mg/kg in male rats. Animals were allowed to run at the speed of 20 m/min on treadmills. The extract tested animals demonstrated less fatigue as compared to a negative control [62].

4.4 Anxiolytic effect

A significant population of the world is affected by anxiety and depression. The chronic use of anxiolytics is responsible for the physical dependence and withdrawal syndrome. To minimize or avoid such harmful effects, the natural plants based treatment might be helpful. The fruits extract of *P. atlantica* demonstrated significant anxiolytic effect in intact and gonadectomized rates [63]. In elevated puls maze animal model the extract of *P. vera* gum showed anxiolytic effect at higher dose (1 g/kg) [59].

5. Wound healing

The treatment of wounds is directly related to the use of antibiotics. The systemic use of antibiotics is associated with various side effects in addition to resistance. The topical use is far better than systematic use to avoid side effects. The natural products based topical application have more positive aspects as compared to the available synthetic chemical molecules. The beauty of plant-based topical wound healing dosage form is that these remedies accumulated various synergetic chemicals in addition to phytosterols. The development of the natural product-based topical dosage might provide analgesic, anti-inflammatory and antibiotic effects. The wound-healing effect of these valuable plants is also outstanding. The fruits oil of *P. lenticus* has been reported with significant healing of laser burn [64]. In another study the fruits oil of *P. lenticus* has accelerated the cutaneous wound healing [65]. The *P. lenticus* resin also shorten the duration of skin burn in rats in a dose-dependent manner [66]. The *P. atlantica* and *P. khinjuk* extracts increased the curing rate of skin wound in experimental animals [67]. The methanolic extract of *P. khinjuk* is also a worthy topical anti- wound agent [68]. The mastic extract, seed oil and resin oil of *P. lenticus* have been tested as significant wound healing agent in different experimental models [69–71]. Bioassay -guided isolation and identification of various chemical constitutes of *P. vera* has been reported [72]. The topical wound healing gel has been formulated with significant effect of *P. atlantica* [73].

5.1 Diabetic wound healing effect

Healing of the diabetic wound is a big problem around the globe. There is no specific treatment for a diabetic foot or wound. Therefore, medicinal plants are the best option to screen for the said action. The *P. atlantica* resin oil is the best wound healing agent in STZ-induced diabetic experimental rat [74]. No more studies are available for this activity.

5.2 Anti-second degree burn

The curing of second-degree burns is also not so much easier by standard antibiotic treatment. Therefore, the search for a new, effective and safe anti-burn therapeutic agent is essential. The topical application of *P. vera* oil on the second-degree burn accelerated the wound healing effect [75].

5.3 Anti-colitis effect

The oil of *P. lenticus* has been reported as a significant curative and preventive agent in colitis induced animals [76]. In addition to this plant, no further work is perform in this regard.

5.4 Anti-peptic ulcer effect

H. pylori is the main cause of peptic ulcer. The ulcer duration is shortened by the triple therapy of metronidazole, clarithromycin and omeprazole for 15 days. But the complete eradication of peptic ulcers takes years. The anti-peptic ulcer effect of oil of *P. atlantica* is also worth mentioning [77]. A limited study is available of this genus on anti-peptic ulcer action.

5.5 Neuroprotective effect

The *Pistacia* genus is one of the best neuroprotective natural products [78]. The neuroprotective effect of *P. vera* gum in induced ischemia animal is worth mentioning [79]. The significant inhibition of acetylcholinesterase and related enzymes is responsible for the neuroprotective effect of *P. terebinthus* [80]. The leaf extract and its major phenolic compounds of *P. lenticus* reversed the aluminum-induced neurotoxic effect in mice [81]. The toxic effect of mercury on brain was regulated by the *P. atlantica* indicating the neuroprotective role [82].

5.6 Hypoglycemic effect

The antidiabetic effect of *P. atlantica* has been reported [44]. In a study, the *n*-hexane extract of *P. atlantica* significantly improved the streptozotocin (STZ)- induced hyperglycemic condition. The same extract also improved the beta cell of pancreas [83]. The leaf extract of this plant also inhibited the α-amylase and α-glucosidase enzymes responsible for the diabetic disorders [84]. The leaf and fruit extract of *P. lenticus* significantly attenuated the induced diabetic condition [85]. The alloxan-induced diabetic condition has been normalized by *P. lenticus* crude extract [86]. The STZ induced hyperglycemic condition in experimental animal was normalized by the crude extract of *P. terebinthus* [87]. The crude methanolic extract of *P. vera* fruit stem metabolites are week antidiabetic [88]. The potential inhibition of 11β-hydroxysteroid dehydrogenase 1 by the oleoresin of *P. lenticus* demonstrate a good antidiabetic property [89]. Pistagremic acid, one of the potential constitutes of *P. integerrima*, is also α-glucosidase inhibitor [90]. Interestingly, the plants in this genus can cure diabetic patients mostly suffering from diabetic neuropathy as well as from wounds. So the treatment of all these conditions at a time resulting from the polypharmacy situation. This polypharmacy situation leads to poor patient compliance. These plants at a time are antidiabetic, antidiabetic wound healers and neuroprotective. So further work is highly recommended to test these plants on such patients who suffer from all these conditions.

5.7 Effect on GLUT

This effect is also directly linked with the anti-diabetic effect. The body has different types of glucose transporters (GLUT). These GLUT are responsible for the influx of glucose molecules and keeping glucose concentration in the blood flow. Among these transporters, the GLUT-II is bi-directional, and the rest are unidirectional. The extract of *P. thlantica* improved the GLUT-IV transporter expression indicating the improved function of insulin [91]. Other plants of this genus are highly recommended to be tested on these GLUTs.

5.8 Lipid lowering effect

The genus looks quite interesting with a particular aspect of diabetic treatment. Because the lipid-lowering activity is highly adjuvant to diabetic patients, only two genus-species have been tested on the lipid-lowering effect, and it is highly recommended to test the rest of the spp. For this effect, *P. atlantica* subsp. *kurdica* has been reported as the best lipid lowering medicinal plant in STZ-induced diabetic animals. The lipid-lowering effect is helpful in diabetic condition [92]. This effect has been shown by the *P. lenticus* fatty oil in egg yolk fed rabbit [93].

5.9 Anti-obesity effect

The bioactive compounds mainly protocatechuic acid (452 μg/g dw) and quinic acid (960 μg/g dry weight dw) derived from *P. atlantica* root have been established to possess a notable lipase inhibition effect on porcine pancreatic lipase [94]. No further studies are available.

5.10 Antihypertensive effect

The genus is very limitedly explored for the antihypertensive (HTN) effect. The leaf extract of *P. atlantica* strongly inhibited the angiotensin-converting enzyme–I (ACE-I), indicating the antihypertensive effect [84]. The HTN is mostly associated with DM. So if a clinical trial is conducted on patients suffering from HTN and DM, it will be very fruitful.

5.11 Acetyl cholinesterase

P. atlantica exhibited a significant acetylcholinesterase inhibition effect [44]. The crude extract and different fractions and fruit stem metabolites of *P. vera* caused the significant acetylcholinesterase inhibition [88]. The inhibition of acetylcholinesterase by the *P. khinjuk* has also been reported [95].

5.12 Nephroprotective effect

Nephrotoxicity is related to chronic consumption of NSAID, DM, and even with HTN. The plants are analgesic, anti-inflammatory and nephroprotective. This is the beauty of natural products that they have multiple indications at a time. Ehsani et al. [96] established the protective effects of *Pistacia vera*-derived hydroalcoholic extract against rat nephrotoxicity induced by gentamicin. Nephrotoxicity in rats was caused by intraperitoneal gentamicin injection at a dose of 100 mg/kg/day. Pistachio hydroalcoholic extracts (10, 50, and 100 mg/kg) were administered for seven days. The findings from this study reported that treatment with pistachio could ameliorate renal failure and structural damage by mitigating inflammation and oxidative stress in the kidney [96].

5.13 Hepatoprotective effect

A significant hepatoprotective effect was observed in carbon tetrachloride-induced hepatitis by the hydroalcoholic extract of *P. vera*. Hepatoprotective effects were observed against CCl_4-triggered liver damage in 40 male rats when treated with *P. vera* hydroalcoholic extract. The antioxidant properties of hydroalcoholic extract potentially supported hepatic cells to suppress inflammation and necrosis caused by CCl_4. Findings from this study along with earlier studies confirm that Pistachio extract can act as a potential candidate for liver damage treatment [97]. Another study has been undertaken to ascertain the hepatoprotective effect of the fruit and leaf extracts of *P. lentiscus* on acute hepatitis induced by paracetamol, as evidenced by lowering tissue necrosis, reducing transaminase as well as MDA serum levels. Hepatoprotective capacity against paracetamol (165 mg/kg body weight) toxicity was found in mice pretreated with the same dosage of PL (*Pistacia* leaves) or PF (*Pistacia* fruits) extract (125 mg/kg) or a mixture of both. These findings were verified via histological analysis of the liver, which revealed substantial defense against hepatic necrosis triggered by paracetamol [85].

5.14 Anti-melanogenic effect

The methanol extract derived from seeds of *P. vera* has been documented to have anti-melanogenic effects against human Melanoma SKMEL-3 cells. The consequence of MPH on the content of melanin, the activity of cellular tyrosinase as well as cytotoxicity (MTT assay) of the SKMEL-3 human melanoma cell was assessed, followed by 72 hours incubation. Findings demonstrated that MPH has powerful radical DPPH (2,2-diphenyl-1-picryl-hydrazyl-hydrate) scavenging activity and low anti-tyrosinase function in comparison with the prominent antioxidant (BHT) and tyrosinase (kojic acid) inhibitors, respectively. MPH demonstrated substantial cytotoxic activity (~63 percent) with a large dose (0.5 mg/mL) and powerful anti-melanogenic influence (~57 percent) in SKMEL-3 cells. The consequence of MPH on melanin reduction may be attributed to its cytotoxicity. Thus it can be concluded from the findings that MPH may be used as a potential agent to treat hyperpigmentation conditions such as melanoma [98].

5.15 Anti-nipple fissure effect

Painful nipple fissure is a severe concern for breastfeeding mothers. In breastfeeding mothers, nipple fissures are typically induced by improper positioning when breastfeeding or complications with latching or suction. They may also be triggered by breast engorgement. In athletes, nipple fissures are started by nipple chaffing to assess the effectiveness of saqez (*Pistacia atlantica*) on breastfeeding women's improvement of nipple fissure. A randomized clinical trial was performed on 100 suitable women who accessed the health centers in their post-partum period at Shahid Beheshti University of Medical Sciences in Tehran, Iran. A total of 100 participants were divided randomly into two equal groups of 50 women, divided into breast-milk and saqez ointments group. The findings revealed that the demographic and obstetric characteristics of the two classes were matched. Additionally, it can be concluded that saqez ointment is comparatively effective than breast milk in curing and managing nipple fissures during one-month follow-up, without culminating in any adverse effects [99].

5.16 Anti-oral mucositis effect

Oral mucositis refers to the ulcerative and lesions of the oral mucosa found in people having cancer when treated with chemotherapy and radiation therapy of areas, including the oral cavity. Oral mucositis lesions also are extremely painful and impair diet and oral health and raise the severity of the local and systemic infection. An experimental analysis conducted by Tanideh et al. [100] verified that the essential oil of *P. atlantica* (bene) accelerated the healing status of oral mucositis induced by 5-fluorouracil in hamsters. The healing influence of bene oil could predominantly be local and due to possessing antioxidants and fatty acids in saponified and non-saponified fractions, respectively [100].

5.17 Anthelmintic effect

Different extract and essential oil of *P. khinjuk* are significant anthelmintics, especially against *Echinococcus granulosus*, which causes hydatid cyst [101]. The *P. lenticus* is also the best anthelmintic [102]. The exsheathment of gastro-intestinal nematode larvae is impaired by polyphenols of *Pistacia lentiscus* [103]. Additionally, The *P. lenticus* along with other plants in mixture form killed the nematodes in naturally infected sheep [104].

5.18 Toxicological effect

In an acute toxicity study, the methanolic extract of *P. integerrima* bark proved to be safe [38]. Besides, the *P. atalantica* fruits also proved safe in acute toxicological studies where the acute toxicity was evaluated for two days. Antinociceptive action was conducted with tail-flick, hot plate, and rotarod test. The *P. atlantica* fruit extract levels for LD50 were 1.66 g/kg with a cumulative non-lethal dosage of 0.93 g/kg. The fruit extract derived from *P. atlantica* at the doses range of 50–350 mg/kg conferred analgesic effects dose-dependently 30 minutes after administration during the hot plate and tail-flick tests so that a substantial difference between the groups obtaining saline and the extract was observed ($p < 0.05$). Results also revealed no significant differences in a sensory-motor assessment with *P. atlantica* fruit extract's administration at doses ranging from 50 to 350 mg/kg. Additionally, findings revealed robust antinociceptive behavior of the *P. atlantica* fruit extract in mice [45].

6. Conclusions and future perspectives

Medicinal plants are potential source of various chemical constitutes which are responsible for the cure of different diseases. Scientific work of these plants is based on the ethnopharmacological use, largely based on trial and error, which may cause harm to humans. In addition, there is a false public perception that natural remedies are free of side or toxic effects. Although this claim is correct to some extent due to the presence of agonist and antagonist molecules in the same plant or extract, however, use of such chemical constitutes without scientific knowledge could lead to serious health problems. For this reason, researchers have tested these alternative medicines for various disorders. Within this context, the genus *Pistacia* has been screened for different diseases based on ethnomedicinal uses. In the present work, we tried to collect all pharmacological data related to *Pistacia*. The wide spread use of members of this genus made it a key source of natural medicines. Furthermore, the purpose of this data collection was to encourage researchers for development and commercialization of these valuable members into various dosage forms.

The genus *Pistacia* accumulated many potential plants with significantly correlated activities such as analgesic, anti-inflammatory, nephroprotective, hepatoprotective, and anti-peptic ulcers. These activities are positively correlated because most of the NSAIDs cause hepatic, renal and stomach problems. So plants in this genus are tested on such an experimental model. The same animal is subjected to pain, inflammation, peptic ulcers, hepatitis and nephrotic damages and then treated with these plants individually or in a mixture with the hope to cure with time. If the researcher succeeded in such a study, it would be a breakthrough in pharmaceutical sciences to minimize polypharmacy. It is worth mentioning that the plants of this genus are anti-diabetic, neuroprotective, anti-diabetic wound, GLUT enhancer, lipid-lowering and anti-HTN. All these conditions have a significant correlation. A substantial number of patients worldwide suffered at times with these conditions. Therefore, we strongly recommend these plants be tested up to the clinical trial level for curing such diseases. The significant curing of such correlated disorders can abolish the problem of polypharmacy. Polypharmacy is one of the major factors leading to poor patient compliance. Moreover, chronic toxicological profiling of these plants is needed on all vital organs.

Author details

Abdur Rauf[1*], Yahya S. Al-Awthan[2,3], Naveed Muhammad[4], Muhammad Mukarram Shah[5], Saikat Mitra[6], Talha Bin Emran[7], Omar Bahattab[2] and Mohammad S. Mubarak[8]

1 Department of Chemistry, University of Swabi, Anbar, Khyber Pakhtunkhwa, Pakistan

2 Department of Biology, Faculty of Sciences, University of Tabuk, Tabuk, Saudi Arabia

3 Department of Biology, Faculty of Science, Ibb University, Ibb, Yemen

4 Department of Pharmacy, Abdul Wali Khan University, Mardan, Khyber Pakhtunkhwa, Pakistan

5 Department of Pharmacy, University of Swabi, Anbar, Khyber Pakhtunkhwa, Pakistan

6 Department of Pharmacy, Faculty of Pharmacy, University of Dhaka, Dhaka, Bangladesh

7 Department of Pharmacy, BGC Trust University Bangladesh, Chittagong, Bangladesh

8 Department of Chemistry, The University of Jordan, Amman, Jordan

*Address all correspondence to: mashaljcs@yahho.com

References

[1] Benhammou, N., F.A. Bekkara, and T.K. Panovska, *Antioxidant and antimicrobial activities of the Pistacia lentiscus and Pistacia atlantica extracts.* African Journal of Pharmacy and Pharmacology, 2008. **2**(2): p. 022-028.

[2] Gardeli, C., et al., *Essential oil composition of Pistacia lentiscus L. and Myrtus communis L.: Evaluation of antioxidant capacity of methanolic extracts.* Food chemistry, 2008. **107**(3): p. 1120-1130.

[3] Remila, S., et al., *Antioxidant, cytoprotective, anti-inflammatory and anticancer activities of Pistacia lentiscus (Anacardiaceae) leaf and fruit extracts.* European Journal of Integrative Medicine, 2015. **7**(3): p. 274-286.

[4] Gourine, N., et al., *Antioxidant activities and chemical composition of essential oil of Pistacia atlantica from Algeria.* Industrial Crops and Products, 2010. **31**(2): p. 203-208.

[5] Tomaino, A., et al., *Antioxidant activity and phenolic profile of pistachio (Pistacia vera L., variety Bronte) seeds and skins.* Biochimie, 2010. **92**(9): p. 1115-1122.

[6] Gentile, C., et al., *Antioxidant activity of Sicilian pistachio (Pistacia vera L. var. Bronte) nut extract and its bioactive components.* Journal of Agricultural and Food Chemistry, 2007. **55**(3): p. 643-648.

[7] Topçu, G., et al., *A new flavone from antioxidant extracts of Pistacia terebinthus.* Food chemistry, 2007. **103**(3): p. 816-822.

[8] Roozegar, M.A., et al., *Antimicrobial effect of Pistacia atlantica leaf extract.* Bioinformation, 2016. **12**(1): p. 19.

[9] Arami, S., et al., *The effect of Pistacia atlantica var. mutica mouthwash on dental plaque bacteria and subgingival microorganisms: a randomized and controlled triple-blind study.* Drug research, 2015. **65**(09): p. 463-467.

[10] Ahmed, Z.B., et al., *Four Pistacia atlantica subspecies (atlantica, cabulica, kurdica and mutica): A review of their botany, ethnobotany, phytochemistry and pharmacology.* Journal of Ethnopharmacology, 2020: p. 113329.

[11] Ghalem, B. and B. Mohamed, *Antimicrobial activity evaluation of the oleoresin oil of Pistacia vera L.* African Journal of Pharmacy and Pharmacology, 2009. **3**(3): p. 092-096.

[12] Tassou, C.C. and G. Nychas, *Antimicrobial activity of the essential oil of mastic gum (Pistacia lentiscus var. chia) on Gram positive and Gram negative bacteria in broth and in Model Food System.* International biodeterioration & biodegradation, 1995. **36**(3-4): p. 411-420.

[13] Taran, M., et al., *Antimicrobial activity of the leaves of Pistacia khinjuk.* Journal of Medicinal Plants, 2010. **9**(6):81-85.

[14] Tahvilian, R., et al., *Chemical composition and screening of antibacterial activity of essential oil of Pistacia khinjuk against two selected pathogenic bacteria.* Annals of Tropical Medicine and Public Health, 2017. **10**(5): p. 1159.

[15] Derwich, E., et al., *GC/MS analysis and in vitro antibacterial activity of the essential oil isolated from leaf of Pistacia lentiscus growing in Morocoo.* World Applied Sciences Journal, 2010. **8**(10): p. 1267-1276.

[16] Bouyahya, A., et al., *Could volatile compounds from leaves and fruits of Pistacia lentiscus constitute a novel source of anticancer, antioxidant, antiparasitic and antibacterial drugs?* Industrial Crops and Products, 2019. **128**: p. 62-69.

[17] Rauf, A., et al., *Chemical composition and biological screening of essential oils from Pistacia integerrima*. African Journal of Pharmacy and Pharmacology, 2013. **7**(20): p. 1220-1224.

[18] Piras, A., et al., *Chemical characterisation and biological activity of leaf essential oils obtained from Pistacia terebinthus growing wild in Tunisia and Sardinia Island*. Natural product research, 2017. **31**(22): p. 2684-2689.

[19] Özçelik, B., et al., *Antibacterial, antifungal, and antiviral activities of the lipophylic extracts of Pistacia vera*. Microbiological Research, 2005. **160**(2): p. 159-164.

[20] Bouslama, L., et al., *Identification of an antiviral compound isolated from Pistacia lentiscus*. Archives of Microbiology, 2020. **202**(9): p. 2569-2578.

[21] Musarra-Pizzo, M., et al., *In vitro anti-HSV-1 activity of polyphenol-rich extracts and pure polyphenol compounds derived from pistachios kernels (Pistacia vera L.)*. Plants, 2020. **9**(2): p. 267.

[22] Hosseinzadeh, H., M. Mirshojaeian, and B.M. Razavi, *Antiemetic effect of Pistacia vera L.(Pistachio) leaves and nuts aqueous extracts in young chicken*. Pharmacol online, 2008. **2**: p. 568-571.

[23] Mezni, F., et al., *Evaluation of Pistacia lentiscus seed oil and phenolic compounds for in vitro antiproliferative effects against BHK21 cells*. Pharmaceutical biology, 2016. **54**(5): p. 747-751.

[24] Abidi, A., et al., *Protective effect of Pistacia lentiscus oil against bleomycin-induced lung fibrosis and oxidative stress in rat*. Nutrition and cancer, 2017. **69**(3): p. 490-497.

[25] Rahman, H.S., *Phytochemical analysis and antioxidant and anticancer activities of mastic gum resin from Pistacia atlantica subspecies kurdica*. OncoTargets and therapy, 2018. **11**: p. 4559.

[26] Awwad, O., et al., *Effect of Pistacia palaestina Boiss. Essential Oil on Colorectal Cancer Cells: Inhibition of Proliferation and Migration*. Journal of Essential Oil Bearing Plants, 2020. **23**(1): p. 26-37.

[27] Hashemi, L., et al., *Anticancer activity and phenolic compounds of Pistacia atlantica extract*. International Journal of Pharmaceutical and Phytopharmacological Research, 2017. **7**(2): p. 26-31.

[28] Seifaddinipour, M., et al., *Cytotoxic effects and anti-angiogenesis potential of pistachio (Pistacia vera L.) hulls against MCF-7 human breast cancer cells*. Molecules, 2018. **23**(1): p. 110.

[29] Uddin, G., et al., *Cytotoxic activity of extracts/fractions of various parts of Pistacia integerrima stewart*. Transl Med, 2013. **3**(118): p. 2161-1025.100011.

[30] Ezatpour, B., et al., *In vitro and in vivo antileishmanial effects of Pistacia khinjuk against Leishmania tropica and Leishmania major*. Evidence-Based Complementary and Alternative Medicine, 2015. **2015**.

[31] Zibaei, M., R. Rostamipour, and H. Nayebzadeh, *Effect of Pistacia atlantica fruit and leaf extracts on hydatid cyst protoscolices*. Recent patents on anti-infective drug discovery, 2016. **11**(1): p. 53-58.

[32] Mahmoudvand, H., et al., *Chemical composition, efficacy and safety of Pistacia vera (var. Fandoghi) to inactivate protoscoleces during hydatid cyst surgery*. Biomedicine & Pharmacotherapy, 2016. **82**: p. 393-398.

[33] Mahmoudvand, H., et al., *In vitro and in vivo antileishmanial activities of Pistacia vera essential oil*. Planta medica, 2016. **82**(4).

[34] Eldin, H.M.E. and A.F. Badawy, *In vitro anti-Trichomonas vaginalis activity of Pistacia lentiscus mastic and Ocimum basilicum essential oil.* Journal of Parasitic Diseases, 2015. **39**(3): p. 465-473.

[35] Hasheminya, S.-M. and J. Dehghannya, *Composition, phenolic content, antioxidant and antimicrobial activity of Pistacia atlantica subsp. kurdica hulls' essential oil.* Food Bioscience, 2020. **34**: p. 100510.

[36] Uddin, G., et al., *Pistagremic acid a new leishmanicidal triterpene isolated from Pistacia integerrima Stewart.* Journal of enzyme inhibition and medicinal chemistry, 2012. **27**(5): p. 646-648.

[37] Janbaz, K.H., et al., *Antidiarrheal, antispasmodic and bronchodilator activities of Pistacia integerrima are mediated through dual inhibition of muscarinic receptors and Ca++ influx.* Science, Technology and Development, 2015. **34**(1): p. 52.

[38] Ismail, M., et al., *Analgesic, anti GIT motility and toxicological activities of Pistacia integerrima Stewart ex Brandis bark in mice.* Journal of Medicinal Plants Research, 2012. **6**(14): p. 2827-2831.

[39] Shirole, R., et al., *Investigation into the mechanism of action of essential oil of Pistacia integerrima for its antiasthmatic activity.* Journal of Ethnopharmacology, 2014. **153**(3): p. 541-551.

[40] Ahmad, N.S., et al., *Analgesic and anti-inflammatory effects of Pistacia integerrima extracts in mice.* Journal of Ethnopharmacology, 2010. **129**(2): p. 250-253.

[41] Rauf, A., et al., *Antinociceptive and anti-inflammatory activities of flavonoids isolated from Pistacia integerrima galls.* Complementary Therapies in Medicine, 2016. **25**: p. 132-138.

[42] Hosseinzadeh, H., E. Behravan, and M.M. Soleimani, *Antinociceptive and Anti-inflammatory Effects of Pistacia vera LeafExtract in Mice.* Iranian journal of pharmaceutical research: IJPR, 2011. **10**(4): p. 821.

[43] Tanideh, N., et al., *Healing effect of pistacia atlantica fruit oil extract in acetic Acid-induced colitis in rats.* Iranian journal of medical sciences, 2014. **39**(6): p. 522.

[44] Bahmani, M., et al., *The effects of nutritional and medicinal mastic herb (Pistacia atlantica).* Journal of Chemical and Pharmaceutical Research, 2015(1): p. 646-653.

[45] Nadri, S., et al., *Chemical composition, antinociceptive and acute toxicity of Pistacia atlantica fruit extract.* Entomol Appl Sci Letters, 2018. **5**(3): p. 8-12.

[46] Rauf, A., et al., *In-vivo antinociceptive, anti-inflammatory and antipyretic activity of pistagremic acid isolated from Pistacia integerrima.* Phytomedicine, 2014. **21**(12): p. 1509-1515.

[47] Orhan, I., et al., *Bioassay-guided evaluation of anti-inflammatory and antinociceptive activities of pistachio, Pistacia vera L.* Journal of Ethnopharmacology, 2006. **105**(1-2): p. 235-240.

[48] Islam, N.U., et al., *Pistacia integerrima gall extract mediated green synthesis of gold nanoparticles and their biological activities.* Arabian Journal of Chemistry, 2019. **12**(8): p. 2310-2319.

[49] Peivastegan, M., et al., *Comparing the Effects of Oleoresin of Pistacia atlantica Tree and Diclofenac Gel on the Knee Osteoarthritis Improvement.* Shiraz E-Medical Journal, 2020. **21**(10).

[50] Ben Khedir, S., et al., *In vivo evaluation of the anti-inflammatory effect*

of Pistacia lentiscus fruit oil and its effects on oxidative stress. Evidence-Based Complementary and Alternative Medicine, 2016. **2016**.

[51] Karimi, F., M. Minaiyan, and A. Ghannadi, *Anti-inflammatory effect of Pistacia atlantica subsp. kurdica* volatile oil and gum on acetic acid-induced acute colitis in rats. 2015.

[52] Rauf, A., et al., *Phytochemical, ethnomedicinal uses and pharmacological profile of genus Pistacia.* Biomedicine & Pharmacotherapy, 2017. **86**: p. 393-404.

[53] Paterniti, I., et al., *The anti-inflammatory and antioxidant potential of pistachios (Pistacia vera L.) in vitro and in vivo.* Nutrients, 2017. **9**(8): p. 915.

[54] Giner-Larza, E.M., et al., *Anti-inflammatory triterpenes from Pistacia terebinthus galls.* Planta medica, 2002. **68**(04): p. 311-315.

[55] Esmat, A., et al., *Anti-inflammatory activity of Pistacia khinjuk in different experimental models: isolation and characterization of its flavonoids and galloylated sugars.* Journal of medicinal food, 2012. **15**(3): p. 278-287.

[56] Ahmad, N.S., et al., *Pharmacological basis for use of Pistacia integerrima leaves in hyperuricemia and gout.* Journal of Ethnopharmacology, 2008. **117**(3): p. 478-482.

[57] Fatehi, F., et al., *The effect of hydroalcoholic extract of Pistacia vera on pentylenetetrazole-induced kindling in rat.* Research Journal of Pharmacognosy, 2017. **4**(2): p. 45-51.

[58] Jain, P.D., et al., *Screening of Pistacia integerrima extracts for their anticonvulsant activity in acute zebrafish and rodent models of epilepsy.* International Journal of Nutrition, Pharmacology, Neurological Diseases, 2015. **5**(2): p. 56.

[59] Ziaee, T. and H. Hosseinzadeh, *Muscle relaxant, hypnotic and anti-anxiety effects of Pistacia vera gum hydroalcoholic extract in mice.* Journal of Medicinal Plants, 2010. **9**(36): p. 96-207.

[60] Ammari, M., et al., *Pistacia lentiscus oil attenuates memory dysfunction and decreases levels of biomarkers of oxidative stress induced by lipopolysaccharide in rats.* Brain research bulletin, 2018. **140**: p. 140-147.

[61] Singh, S. and M. Kulshreshtha, Pharmacological approach of Pistacia Vera fruit to assess learning and memory potential in chemically-induced memory impairment in mice. Central Nervous System Agents in Medicinal Chemistry (Formerly Current Medicinal Chemistry-Central Nervous System Agents), 2019. 19(2): p. 125-132.

[62] Khatami, F., et al., The anti-fatigue effects of the hydro-alcoholic extract of Pistacia vera seeds (pistachios) on male Wistar rats. Pistachio and Health Journal, 2018. 1(2): p. 17-21.

[63] Rashidi, S., N. Askari, and M. Abbasnejad, Anxiolytic-like effect of Pistacia atlantica fruit in intact and gonadectomized rats subjected to chronic stress. Journal of Occupational Health and Epidemiology, 2014. 3(3): p. 152-159.

[64] Khedir, S.B., et al., The healing effect of Pistacia lentiscus fruit oil on laser burn. Pharmaceutical biology, 2017. 55(1): p. 1407-1414.

[65] Boulebda, N., et al., Dermal Wound Healing Effect of Pistacia Lentiscus Fruit's Fatty Oil. Pharmacognosy Research, 2009. 1(2): p. 66.

[66] Haghdoost, F., et al., Pistacia atlantica resin has a dose-dependent effect on angiogenesis and skin burn wound healing in rat. Evidence-Based Complementary and Alternative Medicine, 2013. 2013.

[67] Tohidi, M., et al., Evaluation of antibacterial activity and wound healing of Pistacia atlantica and Pistacia khinjuk. Journal of Medicinal Plants Research, 2011. 5(17): p. 4310-4314.

[68] Azadpour, M., et al., *Antioxidant, antibacterial, and wound-healing properties of methanolic extract of Pistacia khinjuk.* Comparative Clinical Pathology, 2015. **24**(2): p. 379-385.

[69] Mezni, F., et al., *Wound healing effect of Pistacia lentiscus L. seed oil: confirmation of its uses in Mediterranean traditional medicine.* Boletín Latinoamericano y del Caribe de Plantas Medicinales y Aromáticas, 2020. **19**(3).

[70] Shahouzehi, B., et al., *Effect of Pistacia atlantica resin oil on anti-oxidant, hydroxyprolin and VEGF changes in experimentally-induced skin burn in rat.* World Journal of Plastic Surgery, 2018. **7**(3): p. 357.

[71] Fakour, S., et al., *Effect of Pistacia atlantica mastic extract on experimental wound healing and various biochemical parameters of blood serum in rabbit models.* J. Med. Plants 2017, **16**(63): 78-91.

[72] Sarkhail, P., et al., *Bioassay-guided fractionation and identification of wound healing active compound from Pistacia vera L. hull extract.* Journal of Ethnopharmacology, 2020. **248**: p. 112335.

[73] Hamidi, S.A., et al., *Cutaneous wound healing after topical application of pistacia atlantica gel formulation in rats.* Turkish Journal of Pharmaceutical Sciences, 2017. **14**(1): p. 65.

[74] Shahouzehi, B., et al., *Effects of Pistacia atlantica resin oil on the level of VEGF, hydroxyproline, antioxidant and wound healing activity in STZ-induced diabetic rats.* The Ukrainian Biochemical Journal, 2018. **90**(1): p. 34-41.

[75] Taghipour, Z., et al., *The effects of the topical administration of Pistacia vera oil on the second-degree burn model in rats.* Pistachio and Health Journal, 2018. **1**(2): p. 7-11.

[76] Naouar, M.S., et al., *Preventive and curative effect of Pistacia lentiscus oil in experimental colitis.* Biomedicine & Pharmacotherapy, 2016. **83**: p. 577-583.

[77] Memariani, Z., et al., *Protective effect of essential oil of Pistacia atlantica Desf. On peptic ulcer: role of α-pinene.* Journal of Traditional Chinese Medicine, 2017. **37**(1): p. 57-63.

[78] Moeini, R., et al., *Pistacia genus as a potential source of neuroprotective natural products.* Planta medica, 2019. **85**(17): p. 1326-1350.

[79] Mansouri, S.M.T., B. Naghizadeh, and H. Hosseinzadeh, *The effect of Pistacia vera L. gum extract on oxidative damage during experimental cerebral ischemia-reperfusion in rats.* 2005.

[80] Orhan, I.E., et al., *Neuroprotective potential of some terebinth coffee brands and the unprocessed fruits of Pistacia terebinthus L. and their fatty and essential oil analyses.* Food Chemistry, 2012. **130**(4): p. 882-888.

[81] Azib, L., et al., *Pistacia lentiscus L. leaves extract and its major phenolic compounds reverse aluminium-induced neurotoxicity in mice.* Industrial Crops and Products, 2019. **137**: p. 576-584.

[82] Fatiha, B., et al., *Toxicity of mercury on the brain: ability of extract of Pistacia atlantica regulated effect.* Journal of Drug Delivery and Therapeutics, 2020. **10**(4-s): p. 17-24.

[83] Hashemnia, M., Z. Nikousefat, and M. Yazdani-Rostam, *Antidiabetic effect of Pistacia atlantica and Amygdalus scoparia in streptozotocin-induced diabetic mice.* Comparative Clinical Pathology, 2015. **24**(6): p. 1301-1306.

[84] Ahmed, Z.B., et al., *Potentially antidiabetic and antihypertensive compounds identified from Pistacia atlantica leaf extracts by LC fingerprinting.* Journal of pharmaceutical and biomedical analysis, 2018. **149**: p. 547-556.

[85] Mehenni, C., et al., *Hepatoprotective and antidiabetic effects of Pistacia lentiscus leaf and fruit extracts.* journal of food and drug analysis, 2016. **24**(3): p. 653-669.

[86] Rehman, M.S.U., et al., *Anti-diabetic activity of crude Pistacia lentiscus in alloxan-induced diabetes in rats.* Bangladesh Journal of Pharmacology, 2015. **10**(3): p. 543-547.

[87] Uyar, A. and N. Abdulrahman, *A histopathological, immunohistochemical and biochemical investigation of the antidiabetic effects of the Pistacia terebinthus in diabetic rats.* Biotechnic & Histochemistry, 2020. **95**(2): p. 92-104.

[88] Lawali, Y.D., et al., *Antidiabetic and Anticholinesterase Properties of Extracts and Pure Metabolites of Fruit Stems of Pistachio (Pistacia vera L.).* Current Organic Chemistry, 2020. **24**(7): p. 785-797.

[89] Vuorinen, A., et al., *Pistacia lentiscus oleoresin: Virtual screening and identification of masticadienonic and isomasticadienonic acids as inhibitors of 11β-hydroxysteroid dehydrogenase 1.* Planta medica, 2015. **81**(06): p. 525-532.

[90] Uddin, G., et al., *Pistagremic acid, a glucosidase inhibitor from Pistacia integerrima.* Fitoterapia, 2012. **83**(8): p. 1648-1652.

[91] Zarekar, M., et al., *Combined effect of aerobic training and pistacia athlantica extract on GLUT-4 protein expression and muscle glycogen in diabetic rats.* Iranian Journal of Endocrinology and Metabolism, 2014. **16**(4): p. 245-253.

[92] Hosseini, S., et al., *Antihyperlipidemic and antioxidative properties of Pistacia atlantica subsp. kurdica in streptozotocin-induced diabetic mice.* Diabetes, Metabolic Syndrome and Obesity: Targets and Therapy, 2020. **13**: p. 1231.

[93] Djerrou, Z., *Anti-hypercholesterolemic effect of Pistacia lentiscus fatty oil in egg yolk-fed rabbits: A comparative study with simvastatin.* Chinese Journal of Natural Medicines, 2014. **12**(8): p. 561-566.

[94] Ben Hmed, M., et al., *Antiobesity and Inhibitory Pancreatic Lipase Effects of Bioactive Compounds of Pistacia atlantica Roots Extract.* Austin Pancreat Disord, 2019. **3**(1): p. 1013.

[95] Ghajarbeygi, P., et al., *An In Vitro and In Vivo Cholinesterase Inhibitory Activity of Pistacia khinjuk and Allium sativum Essential Oils.* Journal of Pharmacopuncture, 2019. **22**(4): p. 231.

[96] Ehsani, V., et al., *Protective effect of hydroalcoholic extract of Pistacia vera against gentamicin-induced nephrotoxicity in rats.* Renal failure, 2017. **39**(1): p. 519-525.

[97] Iranmanesh, F., et al., *Effects of Pistacia vera hydro-alcoholic extract on carbon tetrachloride-induced hepatotoxicity in male rats.* Iranian Journal of Pharmacology and Therapeutics, 2016. **14**(2): p. 35-0.

[98] Sarkhail, P., et al., *Anti-melanogenic activity and cytotoxicity of Pistacia vera hull on human melanoma SKMEL-3 cells.* Acta Medica Iranica, 2017: p. 422-428.

[99] As'adi, N., et al., *The effect of Saqez (Pistacia atlantica) ointment on the treatment of nipple fissure and nipple pain in breastfeeding women.* Electronic physician, 2017. **9**(8): p. 4952.

[100] Tanideh, N., et al., *Healing acceleration of oral mucositis induced by*

5-fluorouracil with Pistacia atlantica (bene) essential oil in hamsters. Journal of Oral Pathology & Medicine, 2017. **46**(9): p. 725-730.

[101] Taran, M., et al., *The anthelmintic effect of Pistacia khinjuk against protoscoleces of Echinococcus granulosus.* World Journal of Zoology, 2009. **4**(4): p. 291-295.

[102] Landau, S., et al., *Anthelmintic activity of Pistacia lentiscus foliage in two Middle Eastern breeds of goats differing in their propensity to consume tannin-rich browse.* Veterinary parasitology, 2010. **173**(3-4): p. 280-286.

[103] Azaizeh, H., et al., *Polyphenols from Pistacia lentiscus and Phillyrea latifolia impair the exsheathment of gastro-intestinal nematode larvae.* Veterinary parasitology, 2013. **191**(1-2): p. 44-50.

[104] Saric, T., et al., *Anthelmintic effect of three tannin-rich Mediterranean shrubs in naturally infected sheep.* Small Ruminant Research, 2015. **123**(1): p. 179-182.

Chapter 8

Medicinal Plants and Traditional Practices of Baiga Tribe in Amarkantak Region of Eastern Madhya Pradesh

Ramesh Kumar Ahirwar

Abstract

The present ethnobotanical study was carried out in Amarkantak region eastern part of Madhya Pradesh during January 2018 to January 2019 to document the medicinal plants used by the Baiga tribes. Traditional medicinal plants used by the Baiga tribes of 37 plant species belonging to 35 genera and 28 families used to menstrual disorder, piles, sore throat, respiratory disorder, haematuria, miscarriage, jaundice, fever, insanity, leucorrhoea, bleeding during pregnancy, spermatorrhea, infertility in women, motiabind, scorpion bite, wounds of animals, stomach disorder, intestinal worms, diabetes, leukoderma, rheumatism, scabies, wart and easy delivery etc. and other various unreported medicinal plants are reported here.

Keywords: Ethnomedicine, Baiga tribe, Madhya Pradesh

1. Introduction

The district Anuppur in Madhya Pradesh located between 23°15′ to 24°N Latitude and 81°0′ to 81°45′E Longitude, covering an area of 3701 sq. km. The district is surrounded by Korea district (Chhattisgarh) in the East, Dindori district in the West, Shahdol district in North and Northwest district in Umaria (**Figure 1**). This region is popularly known as the Plateau of Beghel-Khand for its rich and diverse flora. The Pushprajgarh block of Anuppur district mostly inhabited by Gond, Baiga, Panika, Kol, Agaria tribes in sporadic remote hill tracts. The total population of the study site is 194,574. The maximum temperature goes up to 45°C in the month of May and minimum recorded is 20°C in the month of January. The area has been categorised as Central India sub-tropical forest endowed with various forest as natural resources. The holy river '*Narmada*' origin in Amarkantak in '*Mai ki Bagiya*' passes through the district Anuppur, Madhya Pradesh. The *Baigas* are one of the oldest aboriginal tribes and classified as one of the primitive tribes of Madhya Pradesh based on pre-agricultural technology, low literacy and stagnant and diminishing population [1]. The area has been categorised as Central India sub-tropical forest endowed with various forest as natural resources. The *Baiga* tribes still practice on herbal medicines. Hence, the use of herbs to treatment of various health disorders is being done at a very low cost. A number of valuable research papers on ethno-medicinal plants of the Amarkantak region have been published by various

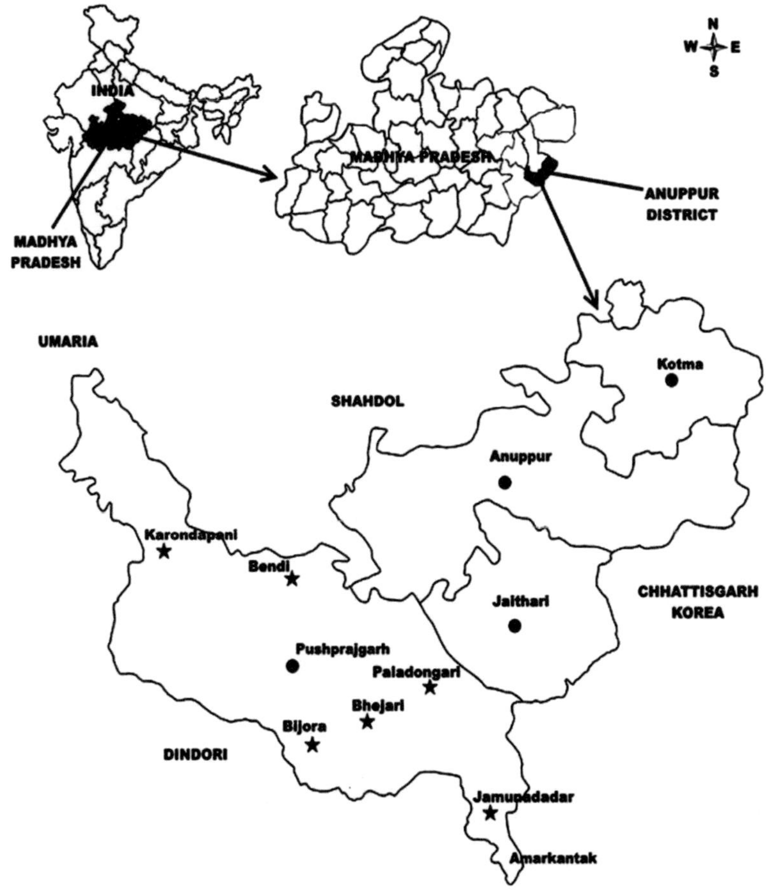

Figure 1.
Location map of study area in Amarkantak region (District Anuppur), Eastern Madhya Pradesh.

researchers [2–7]. However, the vast storage of ethno-medicinal information of these study areas has not been fully documented. In the present paper, an attempt has been made to present indigenous knowledge and uses of wild plants, which are used by *Baiga* tribes for treatment of various ailment diseases.

2. Methodology

Intensive ethnobotanical explorations were conducted in seven villages, namely Pushprajgarh, Jamunadadar, Bijora, Bhejari, Paladongari, Bendi and Karondapani of district Anuppur from January 2018 to January 2019. The method adopted for collection of data was interview with *Baiga* tribes, local medicine men (*Vaidyas*) and one to one discussion about therapeutic uses of local plants in the treatment of various diseases. A questionnaire was prepared to gather data for this purpose. The herbarium specimens were prepared by following the standard method [8]. Plants used by the tribal were identified with the help of Flora of Madhya Pradesh [9–11] and identification was confirmed by consulting the herbaria of Botanical Survey of India, Central Regional Circle, Allahabad (BSA). These voucher specimens are prepared and deposited in the herbarium of Department of Botany, Pt. S.N.S. Govt. Post Graduate College, Shahdol, Madhya Pradesh. The plants are arranged alphabetically according to their botanical name followed by family, local name and mode of administration for different diseases as given in the (**Table 1**).

S.No.	Botanical name	Family	Local name	Uses
1	*Abrus precatorius* L.	Fabaceae	Ghumchi	Two spoonful of seed paste (red variety) is given orally once daily before breakfast for 3 days to cure menstrual disorder.
2	*Abutilon indicum* (L.) Sweet.	Malvaceae	Kanghi	Leaves are boiled in coconut oil and the oil is externally applied on head once daily for one week to cure cold and scabies on the head.
3	*Acacia nilotica* (L.) Willd. ex Delile	Mimosaceae	Bamoor	Fresh leaf paste (20 gm) is externally applied on anus daily in the morning after bath for 15 days to cure piles.
4	*Achyranthes aspera* L.	Amaranthaceae	Lathjira	Root paste is externally applied on affected area immediately after scorpion sting.
5	*Alpinia calcarata* Roscoe	Zingiberaceae	Kulanjan	Root in small pieces, chewed once in a day for 4 times to cure sore throat.
6	*Andrographis paniculata* (Burm. f.) Wall. ex Nees	Acanthaceae	Chirayta	Decoction of the whole plant (15 ml) is given orally thrice in a day for 7 days to cure respiratory disorder.
7	*Asparagus racemosus* Willd.	Liliaceae	Satavar	Fresh tuber juice (15 ml) mixed with a cup of cow's milk is given orally twice in a day for 10 days to cure haematuria.
8	*Azadirachta indica* A. Juss.	Meliaceae	Neem	Leaf juice (10 ml) mixed with a cup of water is given orally once in a day before breakfast for 7 days to prevent miscarriage.
9	*Bauhinia variegata* L.	Caesalpiniaceae	Kachnar	A spoonful of sun-dried flower bud powder mixed with a cup of water is given orally twice in a day for 21 days to cure piles.
10	*Boerhavia diffusa* L.	Nyctaginaceae	Punarnava	Root decoction (10 ml) mixed with a cup of water is given thrice in a day for 7 days to cure jaundice.
11	*Bryonia laciniosa* L.	Cucurbitaceae	Shivlingi	A teaspoonful crushed seed with a glass of water is given orally once in the morning before breakfast for 3 months to cure sterility in women.
12	*Buchanania lanzan* Spreng.	Anacardiaceae	Chironji	Leaf juice (15 ml) mixed a cup of water taken twice in a day for 2 days to prevent dysentery.
13	*Butea monosperma* (Lam.) Taub.	Fabaceae	Chheula	Flowers are boiled in water and the water is used to take bath for 3 days to cure mild fever among children.

S.No.	Botanical name	Family	Local name	Uses
14	*Cordia macleodii* Hook. f. & Thomson	Boraginaceae	Dahiman	Seed paste (20 gm) is given orally with added sugar lump (*Misri*) 10 gm once in a day in the morning before breakfast for 40 days to get relief from insanity.
15	*Curculigo orchioides* Gaertn.	Hypoxidaceae	Kali musali	Decoction of fresh rhizome (10 ml) mixed with a cup of water is given orally twice in a day for 7 days to cure leucorrhoea.
16	*Cuscuta reflexa* Roxb.	Cuscutaceae	Amerbel	Stem paste (20 gm) with 2-3 seeds of black pepper powder (*Piper nigrum* L.) is mixed and the paste is given orally once in the morning on empty stomach for 3 days to cure jaundice.
17	*Cynodon dactylon* (L.) Pers.	Poaceae	Doobi	Whole plant juice (30 ml) mixed with a glass of cow's milk is given orally once at the bedtime for 3 days to cure bleeding during pregnancy.
18	*Cyperus rotundus* L.	Cyperaceae	Nagarmotha	Rhizome extract (5 ml) with one teaspoonful of honey is to take orally thrice daily for 3 days to cure diarrhoea.
19	*Ferula assa-foetida* L.	Apiaceae	Heeng	Oleo-gum-resin (5 gm) (It's obtained from the rhizome and root of the plant) is mixed with a cup of lukewarm water and applied on the stomach of the child, twice in a day for 3 days to cure flatulence.
20	*Ficus benghalensis* L.	Moraceae	Bargad	A spongy sugar-cake (*Batasa*) filled with latex (4 drops) is eaten once in a day in the morning after breakfast for 20 days to cure spermatorrhea.
21	*Ficus racemosa* L.	Moraceae	Dumer	Flower buds (7 buds at a time) which is ground well and mixed with a glass of cow's milk is consumed after dinner for 4 months to cure infertility in women.
22	*Gloriosa superba* L.	Liliaceae	Kalihari	Fresh root paste (20 gm) and 3 fruits of black pepper (*Piper nigrum* L.) are mixed in a glass of lukewarm goat's milk and it is given orally once at the bedtime for 3 days of pregnancy, which can be up to 3 months, for abortifacient.
23	*Hedychium coronarium* J. Koeing	Zingiberaceae	Gulbakavali	Two drops of flower *arrack* is dripped into human eyes thrice in a day for 15 days to prevent cataract (*Motiabind*).

S.No.	Botanical name	Family	Local name	Uses
24	*Helicteres isora* L.	Sterculiaceae	Marorphali	Seed powder (5 gm) mixed in a glass of water is given orally twice in a day for 3 days in case of gastroenteritis.
25	*Ipomea pes-tigridis* L.	Convolvulaceae	Panchpatiya	Leaf paste is externally applied on the affected area just after scorpion bite. Meanwhile, two spoonsful are given orally with a cup of water.
26	*Jatropha curcas* L.	Euphorbiaceae	Bhakrenda	Latex of stem is externally applied on wounds twice in a day for 5 days to cure wounds of animals.
27	*Kigelia pinnata* (Jacq.) DC.	Bignoniaceae	Balamkhira	Fruit juice (5 ml) is mixed in a cup of water and it is given orally twice in a day for 3 days to cure stomach disorder.
28	*Madhuca longifolia* (J. Koeing) Macbr. var. *latifolia* (Roxb.) Chevalier	Sapotaceae	Mahua	Decoction of bark (10 ml), mixed with a glass of water, is given orally twice in a day for 3 days to cure stomachache.
29	*Mucuna pruriens* (L.) DC.	Fabaceae	Kevanch	A teaspoonful of sun-dried powder, mixed with a glass of water, is given orally once daily for 7 days to remove intestinal worms.
30	*Nyctanthes arbor-tristis* L.	Oleaceae	Harsingar	Leaf juice (5 ml) mixed with a cup of water is given orally twice in a day for 20 days as a cure for excessive thirst and loss of weight caused by diabetes.
31	*Plumbago zeylanica* L.	Plumbaginaceae	Chitrak	A teaspoonful of sun-dried root powder is diluted in a cup of boiled water to make a paste and the paste is externally applied twice in a day on affected area for 3 months to cure leukoderma.
32	*Pongamia pinnata* (L.) Pierre	Fabaceae	Karanj	Seed oil is applied twice in a day for one month on the affected area to cure rheumatism.
33	*Pterocarpus marsupium* Roxburgh	Fabaceae	Beeja	A cup of filtrate, which is filtered from the water used for soaking its bark for a night, is consumed on empty stomach once in a day for 3 months to cure diabetes.
34	*Ricinus communis* L.	Euphorbiaceae	Rendi	Leaves are boiled in water and the water is used for bathing once in a day for one month to cure scabies.
35	*Terminalia arjuna* (Roxb. ex DC.) Wight & Arn.	Combretaceae	Arjun	Leaf juice (5 ml) mixed with a cup of water is given orally twice in a day for 15 days to cure wart.

S.No.	Botanical name	Family	Local name	Uses
36	*Terminalia bellirica* (Gaertn.) Roxb.	Combretaceae	Bahera	Fruit decoction (100 ml) is given orally once daily in the morning before breakfast for a week to cure piles.
37	*Ziziphus nummularia* (Burm. f.) Wight & Arn.	Rhamnaceae	Jharberi	A teaspoonful of root paste (10 ml), mixed with a cup of water, is given orally for easy delivery.

Table 1.
Ethnomedicinal uses of plants by Baiga *tribe in Amarkantak region.*

3. Results and discussion

The ethnobotanical research reports 37 plant species belonging to 35 genera and 28 families used for curing various diseases by the Baiga tribes in the Amarkantak region. The representing plants are mostly used to cure various diseases viz. menstrual disorder, piles, sore throat, respiratory disorder, haematuria, miscarriage, jaundice, fever, insanity, leucorrhoea, bleeding during pregnancy, spermatorrhea, infertility in women, abortifacient, Motiabind, scorpion bite, wounds of animals, stomach disorder, intestinal worms, diabetes, leukoderma, rheumatism, scabies, wart and easy delivery. The presence of such a large number of medicinal plants indicates that the area has a very rich diversity of medicinal plant species and is a site for different indigenous knowledge. The present ethno-medicinal information provided in this paper, is compared with well-known Indian medicinal literature [12–14]. The result of the present study continues to play a vital role in the healthcare system of the tribal people and paves the way for the development and discovery of new drugs.

Acknowledgements

The author is grateful to the Baiga tribal people, who have shared unreservedly valuable information in the field during to study area.

Conflict of interest

The author declares no conflict of interest.

Author details

Ramesh Kumar Ahirwar
Department of Botany, Guru Ghasidas Vishwavidyalaya, Bilaspur, Chhattisgarh, India

*Address all correspondence to: dr.rkahirwar@gmail.com

References

[1] Tewari DN: Primitive Tribes of Madhya Pradesh, Ministry of Home Affairs, Govt. of India, New Delhi, 1984.

[2] Parna IC, Ahirwar RK: Socio-Economic importance of some plant's species used by the tribes of Chanda forest district Dindori Madhya Pradesh, India, International Journal Science and Research, 4 (3), 2015, 1733-1735.

[3] Ahirwar RK, Kapale R: A survey of traditional health care practices of the tribals of Dindori district, Madhya Pradesh, Indian Journal Applied & Pure Biology, 29 (1), 2014, 77-80.

[4] Parna IC, Ahirwar RK, Singh GK: Traditional medicinal knowledge about some herbaceous plants used by Baiga tribes of Bajag forest, Dindori district Madhya Pradesh, India, International Journal Science and Research, 3 (12), 2014, 2232-2236.

[5] Soni V, Prakash A, Nema M: Study on ethno-medico-botany of some plants of Dindori district of Madhya Pradesh, India, International Journal Pharmacy & Life Science, 3 (8), 2012, 1926-1929.

[6] Ahirwar RK, Singh GK: some antidiabetic plants from Dindori district of Madhya Pradesh (India), Indian Journal Applied & Pure Biology, 26 (2), 2011, 269-271.

[7] Mudaiya RK, Lale SK, Shankar R, Dhiman KS: Medicinal wealth of Dindori forest division of Madhya Pradesh, India needs conservation and systematic collection, World Journal Pharmaceutical Research, 5 (2), 2016, 347-372.

[8] Jain SK, Rao RR: A Handbook of Field Herbarium Methods, Today and Tomorrow's Printers and Publishers, New Delhi, 1976.

[9] Verma DM, Balakrishnan NP, Dixit RD: Flora of Madhya Pradesh, Vol. I, Botanical Survey of India, Calcutta, 1993.

[10] Singh NP, Khanna KK, Mudgal V, Dixit RD: Flora of Madhya Pradesh, Vol. III, Botanical Survey of India, Calcutta, 2001.

[11] Mudgal V, Khanna KK, Hajra PK: Flora of Madhya Pradesh, Vol, II, Botanical Survey of India, Calcutta, 1997.

[12] Jain SK: Dictionary of Indian Folk Medicine and Ethnobotany, Deep Publications, New Delhi, 1991.

[13] Kirtikar KR, Basu BD: Indian Medicinal Plants, Part I to IV, Reprint (Bishen Singh Mahendra Pal Singh, Dehradun), 1998.

[14] Chopra RN, Nayer SL, Chopra IC: Glossary of Indian Medicinal Plants of India, CSIR, New Delhi, 1956.

Chapter 9

Extraction of Bioactive Compounds from Medicinal Plants and Herbs

Fongang Fotsing Yannick Stéphane, Bankeu Kezetas Jean Jules, Gaber El-Saber Batiha, Iftikhar Ali and Lenta Ndjakou Bruno

Abstract

Human beings have relied on herbs and medicinal plants as sources of food and remedy from time immemorial. Bioactive compounds from plants are currently the subject of much research interest, but their extraction as part of phytochemical and/or biological investigations present specific challenges. Herbalists or scientists have developed many protocols of extraction of bioactive ingredients to ensure the effectiveness and the efficacy of crude drugs that were used to get relief from sickness. With the advent of new leads from plants such as morphine, quinine, taxol, artemisinin, and alkaloids from *Voacanga* species, a lot of attention is paid to the mode of extraction of active phytochemicals to limit the cost linked to the synthesis and isolation. Thus, the extraction of active compounds from plants needs appropriate extraction methods and techniques that provide bioactive ingredients-rich extracts and fractions. The extraction procedures, therefore, play a critical role in the yield, the nature of phytochemical content, etc. This chapter aims to present, describe, and compare extraction procedures of bioactive compounds from herbs and medicinal plants.

Keywords: Herbs, Medicinal plants, Plants extracts, Extraction, Bioactive ingredients, Phytoconstituents, Secondary metabolites, Phytochemicals

1. Introduction

With the increasing demand for herbal medicinal products, nutraceuticals, and natural products for primary healthcare worldwide, medicinal plant extract manufacturers and essential oil producers have started using the most appropriate extraction techniques. Different methods are used to produce extracts and essential oil of defined quality with the least variations.

Herbs and medicinal plants have been used for centuries as source of a wide variety of biologically active compounds. The plant crude material or its pure compounds are extensively used to treat diverse ailments by generations of indigenous practitioners [1, 2]. They are currently the subject of much research interest, but their extraction as part of phytochemical and biological investigations presents specific challenges that must be addressed throughout the solvent extraction [3]. Natural products provide unlimited opportunities for new drug discovery because of the unmatched availability of chemical diversity [4]. Thanks to two drugs

Galanthus woronowii Losinsk

Alzheimer's disease

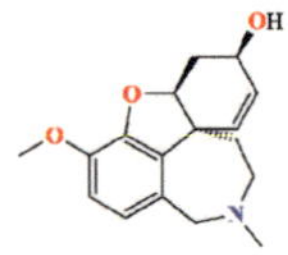

Galantamine

Galantamine is an alkaloid in the Amaryllidaceae family widely used to treat Alzheimer's Disease (AD). Since its approval for clinical use in 2001, its effectiveness has been attested in numerous clinical trials. Galantamine is an effective well-tolerated symptomatic treatment which improves cognition, function and activities of daily living in the short term (up to 6 months) in patients with mild to moderate AD [9,10].

Strychnopsis thouarsii bark

Malaria

Tazopsine

Tazopsine Malaria treatment (in development) isolated from *Strychnopsis thouarsii* bark from Madagascar [A]. Tazopsine, a derivative of a related morphinan, was reported to be active against the liver stages of *Plasmodium falciparum* K1 parasite. This may have the potential as anti-malarial leads [11].

Vicia faba L.

Parkinson's disease

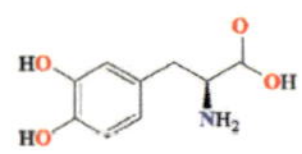

Levodopa

Parkinson's disease (PD) is a common neurodegenerative disease typified by a movement disorder consisting of bradykinesia, rest tremor, rigidity, and postural instability [12].
Levodopa, the precursor of dopamine, was first developed to treat PD in the 1960s and continues to be the most-effective therapeutic agent for PD in 2020 [D]. Treatment of PD involves pharmacologic approaches (typically with levodopa preparations prescribed with or without other medications) and non-pharmacologic approaches (such as exercise and physical, occupational, and speech therapies) [12].

Papaver somniferum

Pain

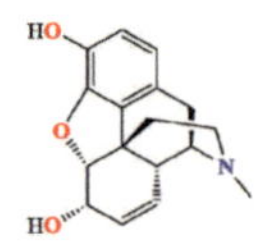

Morphine

The isolation of morphine from *Papaver somniferum* (Poppy straw) in its pure form resulted from years of research and testing, occurring most prominently between the years 1803 and 1817 [B]. Morphine is one of the foremost opioid agents due to its easier access in the hospital system in treating patients with extremity trauma and moderate to severe pain [C]. It is an effective analgesic and recommended to treat cancer-related pain [13].

Artemisia annua

Antimalarial

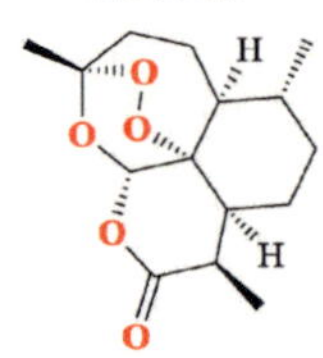

Artemisinin

The 2015 Nobel Prize in Physiology or Medicine was awarded to Professor Youyou Tu for her critical contributions to the discovery of artemisinin. Artemisinin has saved millions of lives and represents one of the significant contributions of China to global health [E]. Artemisinin was isolated from the leaves of a Chinese medicinal plant, *Artemisia annua*. The use of artemisinin-based combination therapy (ACT) with a competent partner drug and having multiple ACT as first-line treatment choice sustained high levels of effectiveness [14].

Taxus brevifolia
(pacific yew or Western yew)

Apoptotic effect on cancer cells

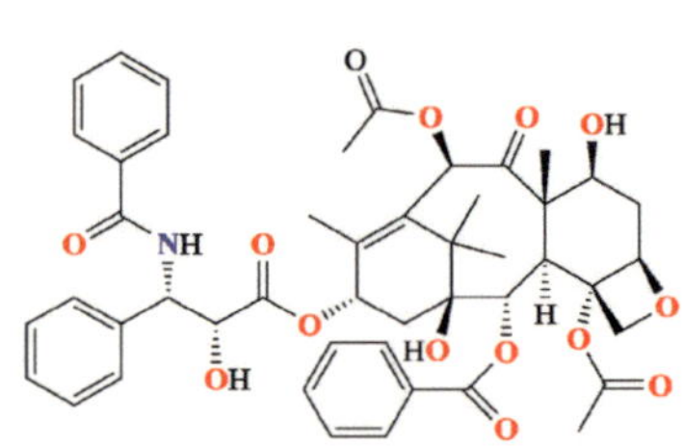

Paclitaxel (Taxol)

Paclitaxel, sold under the brand name Taxol (isolated from the bark of *Taxus brevifolia*) among others, is a chemotherapy medication used to treat a number of types of cancer. This includes ovarian cancer, esophageal cancer, breast cancer, lung cancer, Kaposi sarcoma, cervical cancer, and pancreatic cancer. It is given by injection into a vein. Taxol is classified as a "plant alkaloid," a "taxane" and an "antimicrotubule agent.".

Figure 1.
Chemical structures of a few important bioactive compounds isolated from plants.

derived from alkaloids of Madagascar's rosy periwinkle (*Catharanthus roseus*), the likelihood of remission for a child who has leukemia increased by 85 percent between 1960 and 1997 [5, 6]. New compounds, such as one recently discovered in a plant in Madagascar, are likely to provide novel antibiotics and help to curb the epidemic of antibiotic-resistant diseases [7].

Natural products are currently of considerable significance due to their unique attributes as a significant source of therapeutic phytochemicals and their efficacy, safety, and minimal side effects [2, 8]. Bioactive compounds in plants include alkaloids, terpenoids, coumarins, flavonoids, nitrogen-containing compounds, organosulfur compounds, phenolics, etc. A wide spectrum of bioactivities is exhibited by these compounds such as anti-inflammatory, immunostimulatory, anticancer, antioxidant, antimicrobial, etc.

Research on medicinal plants is particularly important as that on conventional drugs due to the beneficial phytochemicals from plants and the shift towards natural products in pharmaceutical and cosmeceutical industries. Chemical structures of a few essential bioactive compounds isolated from plants are presented in **Figure 1** [9–14].

Extraction of the bioactive constituents from plants has always been challenging for researchers [15]. As the target compounds may be non-polar to polar and thermally labile, the suitability of the extraction methods must be considered. The study on medicinal plants starts with extraction procedures that play a critical role in the extraction outcomes and the consequent assays.

Hence, this chapter aims to provide an overview of the process of plant extraction, describe, and compare extraction methods based on their principle, the effect of solvent on extraction procedures, strength, limitations, and economic feasibility, with their advantages and disadvantages. This chapter shall also emphasize the common problems encountered and methods for reducing or eliminating these problems. Since millions of natural products derived from plants are known, only selected groups and compounds are presented.

2. Medicinal plants and herbs

The term "medicinal" as applied to a plant indicates that it contains a substance or substances which modulate beneficially the physiology of sick mammals, and man has used it for healthful purpose [16]. Medicinal plants were described by Farnsworth and Soejarto as: "all higher plants with medicinal effects that relate to health, or which are proven as drugs by Western standards, or which contain constituents that are defined as hits." [17].

Medicinal plant (MP) refers to any plant which, in one or more of its organs, contains substances that can be used for therapeutic purposes or which are precursors of the synthesis of valuable drugs. A whole plant or plant parts may be medicinally active [18–22]. Medicinal plants (MPs) are becoming very important due to their uses mainly as a source of therapeutic compounds that may lead to novel drugs. MPs are plants that are used for healthcare purposes in both allopathic and traditional medicine systems. MPs cover various species used including condiments, food aromatic and cosmetics [23–26].

Herbs may be defined as the dried leaves of aromatic plants used to impart flavor and odor to foods with, sometimes, the addition of color. The leaves are commonly traded separately from the plant stems and leaf stalks [27].

Herbal medicine is referred to as medicinal preparations comprising active ingredients obtained from the herbal plant. The product can be made from the whole plant or any part. Preparations from by-product herbal plants such as oil, gum, and other secretions are also considered herbal medicines [18, 19, 22].

3. Primary and secondary metabolites

Metabolites are intermediate processes in nature and are small molecules. Primary metabolites are known vital or essential compounds and are directly involved in the average growth, development, and reproduction of plants [28]. Primary metabolites include cell constituents (e.g. carbohydrates, polysaccharides, amino acids, sugars, proteins, and lipids) and fermentation products (ethanol, acetic acid, citric acid, and lactic acid), and are mainly used during their growth and development stages [19, 22, 29, 30].

Secondary metabolites are not directly involved in those processes and usually have a function but are not that important for the organism (e.g. phenolic, steroids, lignans, etc.). They are found only in specific organisms or groups of organisms, and express of the individuality of species [19, 30, 31]. They are not necessarily produced under all conditions, and most often, the function of these compounds and their benefit to the organism is not yet known. Some are undoubtedly made for readily appreciated reasons, e.g., as toxic material providing defense against predators, as volatile attractants towards the same or other species, but it is logical to assume that all do not play some vital role for the well-being of the producer [27, 30]. Secondary metabolites are produced after the growing stage and are used to increase the ability of plants to survive and overcome their local challenges. Bioactive compounds are classified as terpenoids, alkaloids, nitrogen-containing compounds, organosulfur compounds, and phenolic compounds [29].

Bioactive compounds are reported to possess diverse bioactivities such as antioxidant, anticancer, antimalarial, antiulcer, antimicrobial, anti-inflammatory activity [32–36].

4. Bioactive compounds

The definition of bioactive compounds remained ambiguous and unclear for a long time. Very few references describe the term "bioactive". It is composed of two words *bio-* and *-active*. In etymology *bio-* is from the Greek (βίο-) "bios" that means life while *-active* is derived from the Latin word "activus" that refers to dynamic, full of energy, with energy, or involved in activity [37–39]. The term "bioactive" is an alternative term for "biologically active" [40]. Hence, a bioactive compound is simply a substance with biological activity [41, 42].

A plant extract is a substance or an active substance with desirable properties removed from the tissues of a plant, frequently by treating it with a solvent, to be used for a particular purpose. The term "bioactive compounds" is generally referred to as biologically significant chemicals but not established as essential nutrients [43]. Bioactive compounds are essential (e.g., vitamins) and non-essential (e.g., polyphenols, alkaloids, etc.) compounds that occur in nature, are part of the food chain, and can affect human health [44]. They are derived from various natural sources such as plants, animals, microorganisms (e.g., fungi) and marine organisms (e.g., lichens) [2]. The amount of bioactive natural products in natural sources is always fairly low [45, 46]. Plant active compounds are usually contained inside plant matrixes. Active compounds are synthesized in small quantities and different concentrations in all plant organs or parts such as leaves, roots, barks, tubers, woods, gums or oleoresin exudations, fruits, figs, flowers, rhizomes, berries, twigs, as well as the whole plant. Further processes may be required after extraction to purify or isolate the desired compounds.

5. Fresh or dried plant materials

Fresh and dried samples are used and are reported in the literature in the preparation of medicinal remedies. Ideally, fresh plant tissues should be used for phytochemical analysis, and the material should be plunged into boiling alcohol within minutes of its collection. Alternatively, plants may be dried before extraction [47]. In most reported cases, dried materials are preferred considering their long conservation time compared to fresh samples. Furthermore, fresh specimens are fragile and tend to deteriorate faster than dried ones. Phytoconstituents such as Essential Oils (EOs) are found in fewer dried samples than in fresh samples. In case of fresh plant material extraction using organic solvents such as methanol or ethanol, is required to deactivate enzymes present in the plant sample. The extractive might contain a substantial portion of water; hence it can be partitioned using specific immiscible organic solvents [3].

6. Drying procedures

Drying is the most common method to preserve the plant material from enzymatic degradation, such as hydrolysis of glucoside, etc. It should be dried as quickly as possible in the open room under primitive conditions at ambient room temperature with air circulation around the plant material to avoid heat and moisture [47]. However, they placed in shallow trays with good atmospheric air-up dryness either in the sunshine or in shade depending on nature of the indicated or identified constituents. However, direct sunlight is usually avoided to reduce the possibility of chemical reactions, responsible for forming of the artifact that may result from chemical transformations after exposure to ultraviolet radiation. Alternatively, plant materials should be dried under optimum temperature conditions between 40 and 50°C, or they can be dried in the oven if needed. Generally, plant material is dried at temperatures below 30°C to avoid the decomposition of thermolabile compounds [3]. Plants containing volatile or thermolabile components may be lyophilized (freeze-dried). In freeze-drying the frozen material is placed in an evacuated apparatus with a cold surface maintained at −60 to −80°C. Water vapors from the frozen material then pass rapidly to the cold surface to yield the dry material [8, 48].

7. Grinding or powdering plant materials

Lowering particle sizes increase surface contact between samples and extraction solvents and therefore, increase the yield rate and yield. Grinding resulted in coarse smaller samples, meanwhile, powdered samples gave a more homogenized and smaller particle, leading to better surface contact with solvents used for extraction. Before the extraction, pretreatments such as drying and grinding of plant materials are usually conducted to increase the extraction efficiency [48]. It is essential that the particles are of as uniform size as possible because larger particles take a longer time to complete the extraction process [49]. Usually, solvent molecules most contact the larger analytes, and particle size smaller than 05 mm is ideal for efficient extraction [8]. Conventional methods are usually used to reduce the particle size of dried plant samples viz. mortar and pestle or electric blenders and mills, etc.

8. Extraction techniques of actives compounds from plants and herbs

Extraction is separating the medicinally active mixture of many naturally active compounds usually contained inside plant materials (tissues) using selective solvents through the standard procedure [50]. It can also be defined as the treatment of the plant material with solvent, whereby the medicinally active constituents are dissolved and most of the inert matter remains undissolved. Thus, the purpose of all extraction is to separate the soluble plant metabolites, leaving behind the insoluble cellular marc known as residue [8]. The obtained product is a relatively complex mixture of metabolites, in liquid or semisolid state or (after removing water) in dried powder form, and are intended for oral and/or external uses. Extraction is based on the difference in solubility between the solute, other compounds in the matrix, and the solvent used to stabilize [29].

In general, there are three common type of extractions: liquid/solid, liquid/liquid and acid/base [51]. The extraction of these active compounds needs appropriate extraction methods that consider the plant parts used as starting material, the solvent used, extraction time, particle size and the stirring during extraction [52, 53]. Extraction methods include solvent extraction, distillation method, pressing, and sublimation according to the extraction principle. Solvent extraction is the most widely used method [47].

The solvent used, the plant part used as starting material and the extraction procedure are three basic parameters reported that influence the quality of an extract [15]. Proper extraction procure is the first step towards isolating and identifying the specific compounds in crude herbal material. It plays a significant and crucial role in the outcome. Successful extraction begins with careful selection and preparation of plant sample and thorough review of the appropriate literature for indications of which protocols are suitable for a particular class of compounds or plant species [3]. For instance, if the components are volatile or prone to degradation, they can first be frozen and homogenized with liquid nitrogen [29]. The extraction, in most cases, involves soaking the plant material in solvent for some specific time. Reported properties on an excellent extraction solvent include low toxicity, preservative action, ease of evaporation at low heat, promotion of rapid physiologic absorption of the extract, and inability to cause the extract to be complex or dissociate.

The principle of solid–liquid extraction is that when a solid material comes in contact with the solvent, the soluble components in the solid material are dissolved in, and move to the solvent. In solvent extraction, the mass transfer of soluble ingredients to the solvent takes place in a concentration gradient. The mass transfer rate depends on the concentration of ingredients, until equilibrium is reached. After that, there will no longer be a mass transfer from plant material to the solvent. In addition, heating the solvent can also enhance the mass transfer because of better solubility.

Moreover, the concentration gradient changes if fresh solvent replace the solvent equilibrium with the plant material [50]. Properties required for an excellent extracting solvent (or a mixture of solvents) include removal, inert, non-toxic, free from plasticizers, not easily inflammable, and no or less chemical interaction [53]. The selection of solvent is therefore crucial for solvent extraction. Solubility, selectivity, cost, and safety should be taken into account in selecting solvent [47]. The factors affecting the choice of solvent are quality of phytochemicals to be extracted, rate of extraction, diversity of metabolites extracted, the toxicity of the solvent in the bioassay process, and the potential health hazard of the extractants and ease of subsequent handling of the extract. Obtaining maximum yield and the highest quality of the targeted compounds is the central goal of the extraction process [29]. Extraction methods are usually chosen per the properties of targeted active compounds, the water content of the plant material, and the objectives of extraction. Initially, natural bioactive compounds are extracted using various extraction techniques, and their bioactivities are identified using *in vitro* and *in vivo* testing [45, 47]. A successful determination of biologically active compounds from plant material is largely dependent on the type of solvent used in the extraction. Since the extract will contain traces of residual solvent, the solvent should not interfere with the bioassay [15].

Various conventional (classical) and non-conventional (innovative) methods can extract plant materials. Variation in extraction procedures usually depends on key factors as extraction time, the temperature used, the particle size of tissues, the solvent-to-sample ratio, the pH of the solvent.

8.1 Classical and/or conventional techniques

The commonly employed extraction methods (long been used) are primarily based on liquid–solid extraction. They are ordinarily easy to operate and are based on heat and/or solvents with different polarities.

8.1.1 Maceration

This process is conducted by soaking the plant materials (coarse or powered) in a closed stoppered container in a solvent allowed to stand at room temperature for 2–3 days with frequent stirring to obtain plant extracts. A sealed extractor is used to avoid solvent evaporation at atmospheric pressure. The process is intended to soften and break the plant's cell walls to release the soluble phytoconstituents. The mixture is then pressed or strained by filtration or decantation after a specific time [8, 54]. Maceration is the simplest and still widely used procedure. The extraction procedure in this stationary process works on principle of molecular diffusion, which is a time-consuming process. Maceration ensures dispersal of the concentrated solution accumulation around the particles' surface and brings fresh solvent to the surface of particles for further extraction [46].

8.1.2 Digestion

This is a kind of maceration in which gentle heat is applied during the maceration extraction process. The temperature does not alter the active ingredients of plant material, so there is greater efficiency in the use of menstruum (solvent or mixture of solvent used for extraction). It is used when the moderately elevated temperature is not objectionable and the solvent efficiency of the menstruum is increased thereby [15]. The most used temperatures are between 35 and 40°C, although it can rise to no higher than 50°C. The plant part to be extracted is placed in a container with the pre-heated liquid to the indicated temperatures, is maintained for a period that may vary between half an hour to 24 hours, shaking the container regularly. This process is used for the herbal material or plant parts that contain poorly soluble substances or polyphenolic compounds [49].

8.1.3 Infusion

Infusion is a simple chemical process used to extract plant material that is volatile and dissolves readily or release its active ingredients easily in organic solvents [49]. Infusion and decoction use the same principle as maceration; both involve soaking the plant material in boiled or cold water which is then allowed to steep in the liquid. The maceration time for infusion is, however shorter. The liquid may then be separated and concentrated under a vacuum using a rotary evaporator.

Infusion finds its application in tea preparation and consumption prescribed in psychophysical asthenia, diarrhea, bronchitis, asthma, etc. In Tropical Africa, the infusion of the bark of *Prunus africana* (pygeum) is taken orally to increase the ease of urination and reduce inflammation and cholesterol deposits [30].

8.1.4 Lixiviation (elution)

The word "lixiviation" (comes from the Latin lixivium, "lessive".) The extraction is carried out with cold or boiled, fresh and new solvent, always. Extraction of components is done using water as solvent.

8.1.5 Decoction

The current process involves boiling the plant material in water to obtain plant extracts. Heat is transferred through convection and conduction, and the choice of solvents will determine the type of compound extracted from the plant material [8]. The sample is boiled in a specified volume of water for a defined time (15 to 60 minutes.) It is then cooled, strained, filtered, and added enough water through the drug to obtain the desired volume. This method is suitable for extracting thermostable (that does not modify with temperature) and water soluble compounds, hard plant materials and commonly resulted in more oil-soluble compounds than maceration.

8.1.6 Tincture

It is the extraction of plant material in alcohol. Usually, the plant material (fresh) and ethyl alcohol are taken at the ratio of 1:5. Because of the alcohol content, the tinctures can be stored at room temperatures without decomposing [55].

8.1.7 Percolation

It is conducted by passing the boiled solvent through the plant material at a controlled and moderate rate (e.g. 5–7 drops per min) until the extraction is complete before evaporation. The concentrated plant extracts are commonly collected at the bottom of the vessel. To obtain a significant amount of extract, successive percolations can be performed by refilling the percolator with fresh solvent and pooling all extracts together. This procedure is mostly used to extract active compounds in the preparation of tinctures and fluid extracts. Its major disadvantage is that large volumes of solvents are required, and the procedure can be time-consuming and may require skilled persons [49].

8.1.8 Steam distillation and hydrodistillation

Steam and hydrodistillation methods are usually used to extract volatile compounds, including essential oil, insoluble in water, from various aromatic and medicinal plants. This is conducted by boiling the plant materials in water to obtain EOs after vapor condensation. Steam distillation occurs at a temperature lower than the boiling point of the ingredients. The method is useful for thermos-sensitive bioactive compounds e.g., natural aromatic compounds. The heat leads to breakage in the sample's pores and then enables the release of the target compound from a matrix. As Raoult's law states that while mixing two immiscible liquids, the boiling point will be reduced. Therefore, in the mixture of volatile compounds having a boiling point between 150 and 300°C and water having a boiling point at about 100°C (at atmospheric pressure), the mixture evaporation will be getting closer to that of the water [29, 56].

There are similarities between the hydrodistillation and the steam distillation principles. In brief, plant material is immersed in water or a proper solvent followed by heating to boiling under atmospheric pressure in the alembic. In a condenser, EOs vapors and water undergo a liquefaction process, and EOS are then separates from water/solvent after collection of the condensate in the decanter. The principle of extraction is based on isotropic distillation. Hydrodistillation with water immersion, direct vapor injection, and water immersion and vapor injection are the three main types of hydrodistillation. The distillation time depends on the plant material being processed [56].

8.1.9 Hot continuous extraction or Sohxlet extraction, soxhletation

In this method, finely ground sample is placed in a porous bag or "thimble" made from a strong filter paper or cellulose, set in the thimble chamber of the Soxhlet apparatus. The first Soxhlet apparatus was developed in 1879 by Franz von Soxhlet (**Figure 2**) [58]. Extraction solvents are heated in a round bottom flask, vaporized into the sample thimble, condensed in the condenser, and dripped back. When the liquid content reaches the siphon arm, the liquid content is emptied into the bottom flask again, and the process is continued [8]. The disadvantages include no possibility of stirring, and a large amount of solvent is required. This method is unsuitable for thermolabile compounds as prolonged exposure (long extraction time) to heat may lead to their degradation. It constitutes an official classical method used to determine different foods' fat content [15, 29, 57].

Exposure to hazardous and flammable liquid organic solvents are the most noticed disadvantages in this method, and the high purity of extraction solvents needed may add to the cost. Also, shaking or stirring cannot be provided in the Soxhlet device to accelerate the process [57].

However, it requires a smaller quantity of solvent as compared to maceration. Besides, instead of many portions of warm solvent passing through the sample, just one batch of solvent is recycled. Other advantages of this technique include its simple operational mode, its applicability to a higher temperature that increases the kinetics process, its low capital cost, the absence of filtration, and the continuous contact of the solvent and the sample. It maintains a relatively high extraction temperature with heat from the distillation flask [29, 57, 59].

8.1.10 Serial exhaustive extraction

It is a standard extraction procedure that involves successive extraction with various solvents of increasing polarity from non-polar to polar ones. The aim is to ensure that a broad polarity range of compounds could be extracted [15].

8.1.11 Fermentation (aqueous-alcoholic extraction)

Some medicinal preparations adopt the technique of fermentation for extracting the active principles. The extraction procedure involves soaking the crude drug, either a powder or a decoction, for a specified period. Alcohol is generating *in situ* after fermentation occur; this eases the extraction of the active components contained within the plant material. The alcohol hence generated additionally serves as preservative. Water should be boiled first, if the fermentation is to be performed in an earthen vessel. Wooden vats, porcelain jars, or metal vessels are used in place of earthen vessels in large-scale manufacturing. This method is not yet standardized [50].

Hydrodistillation and steam distillation, hydrolytic maceration followed by distillation, expression and effleurage (cold fat extraction) may be employed for

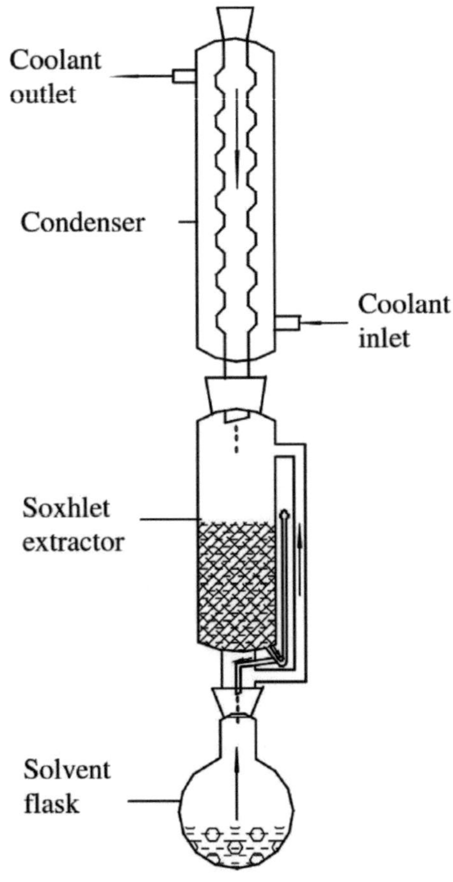

Figure 2.
Experimental Soxhlet extraction apparatus [57].

aromatic plants. Some of the latest extraction methods for aromatic plants include headspace trapping, solid phase micro extraction, protoplast extraction, micro distillation [15].

These techniques are the easiest and simplest methods. Despite the establishment of advanced extraction methods, the potential of conventional solid–liquid extractions is still being used to obtain active compounds from plants. These methods are criticized due to large solvent consumption and long extraction times that can destroy some metabolites. Solvents used in these techniques for soaking play a critical role. Many other advanced extraction methods that incorporate various technologies have been developed [8, 48].

8.2 Innovative (non-conventional) techniques

There is steady progress in the development of extraction technology in recent years. They are also known as advanced techniques with the most recently developed.

8.2.1 Microwave-assisted extraction (MAE)

Microwaves are part of the electromagnetic spectrum of light with a range of 300 MHz to 300 GHz, and wavelengths of these waves range from 1 cm^{-1} to 1 m^{-1} [60]. These waves are made up of two perpendicular oscillating fields which are used as energy and information carriers.

In this extraction process, the use of microwave energy results in faster heating. Due to the exposure of each molecule to the microwave field, its direct effects include, thermal gradients reduction, volume generation due to heat, equipment size reduction, because of the higher process rates, and thus increase in productivity, through better usage of the same equipment process volume [61]. MAE is a feasible green solvent extraction procedure as it uses water or alcohol at elevated temperature and controlled pressure conditions (**Figure 3**).

This procedure has demonstrated various benefits like ease to handle and understand steadiness. Many studies reported that MAE has higher yields and is significantly faster than conventional methods for extracting active substances from plant materials [48, 54, 62]. MAE can be presented as a potential alternative to the traditional soli-liquid extraction techniques. A few of the potential advantages are as follow:

i. a lesser amount of solvent is required (few milliliters of solvent can be used);

ii. shorter extraction time, from few seconds to few minutes (15–20 min);

iii. improved extraction yield;

iv. favorable for thermolabile constituents;

v. heavy metals and pesticides residue which is present in the trace can be extracted from a few milligrams of plant sample;

vi. during extraction, it provides a stirring, by which the mass transfer phenomenon is improved [54, 60, 62, 63].

MAE intensification needs special equipment to be functional, and electricity produces waves, leading to higher investments and higher operating costs than conventional methods [64]. Banar and collaborators extracted the bioactive compounds from *Urtica dioica* grown in Lebanon using conventional methods (maceration, reflux, Soxhlet, hydrodistillation, Ultrasound-Assisted Extraction (UAE) and

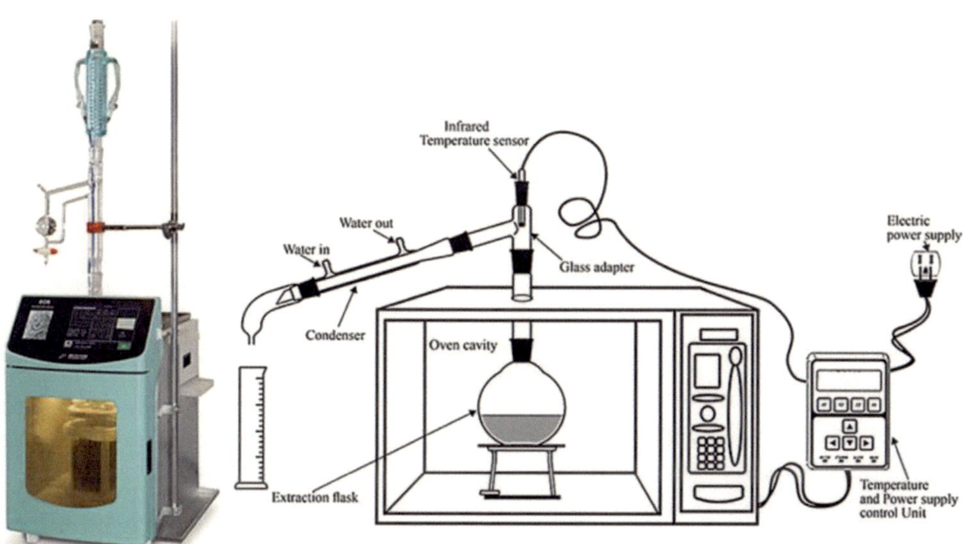

Figure 3.
Schematic representation of microwave-assisted extraction equipment [62].

Microwave-Assisted Extraction (MAE)) with different solvents. Their results revealed that MAE was the most effective technique. The extraction time was reduced, the lesser solvent was used and the amount of extracted compounds was increased [65].

8.2.2 Ultrasound-assisted extraction (UAE) or sonication extraction

This extraction method involves using ultrasound with frequencies ranging from 20 to 2000 KHz; this increases the permeability of cell walls and produce cavitation. Although the process is helpful in some cases, its large-scale application is limited due to its high cost. The most noticeable disadvantage of the procedure is the occasional but known deleterious effect of ultrasound energy on the active components of the medicinal plants through the formation of free radicals and consequently undesirable changes on the drug molecules [50]. The schematic representation of the equipment is given below (**Figure 4**).

Factors that affect the efficiency of UAE are extraction time, power, solvent, Liquid/Solid (L/S) ratio, plant material, frequency, amplitude, and intensity. UAE more advantageous than other advanced extraction methods and provided the best mass and heat transfer efficiency, lowest energy consumption and carbon emission. It was reported to yield high total phenolic content, antioxidant activity, or specific active compounds [62, 66].

8.2.3 Pressurized liquid extraction (PLE) or accelerated solvent extraction (ASE)

Pressurized liquid extraction (PLE) also known as pressurized fluid extraction (PFE), accelerated solvent extraction (ASE), and pressurized solvent extraction (PSE), or as enhanced solvent extraction system (ESE) [67].

Dionex Corporation introduced PLE in 1995 as an alternative to maceration, percolation, sonication, Soxhlet extraction, etc. It is an automated technique for extracting solid samples with liquid solvents (either aqueous or organic, single or mixtures) above their boiling point, combine high pressures (4–12 MPa) and moderate to high temperatures (50–300°C) [68]. When water is the extraction solvent,

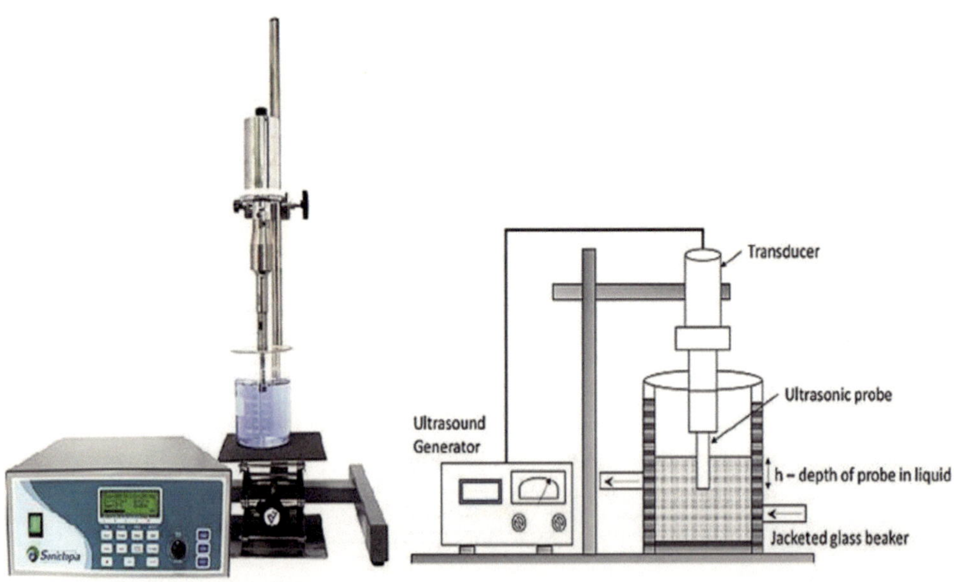

Figure 4.
Schematic representation of an ultrasound-assisted extraction equipment.

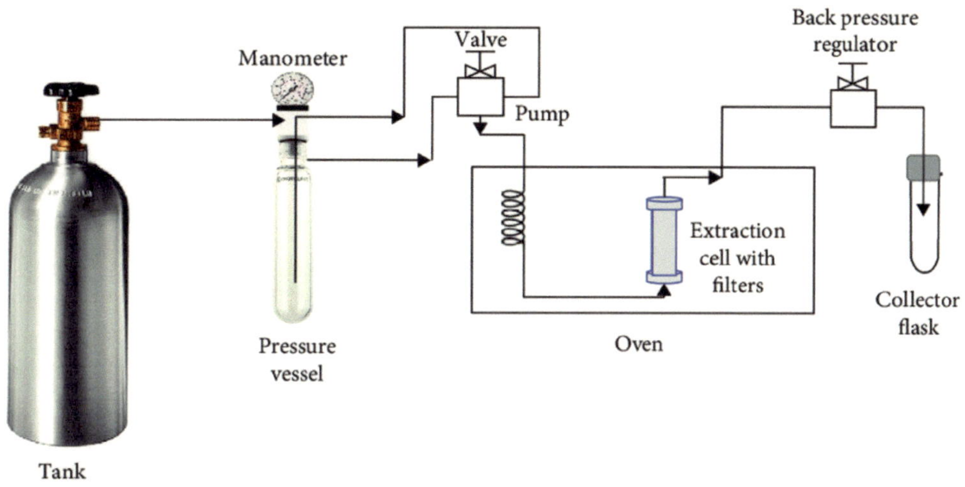

Figure 5.
Scheme of pressurized liquid extraction equipment [68].

different terms are used to define the method, that includes hot water extraction (HWE), subcritical water extraction (SWE), high-temperature water extraction (HTWE), hot water extract pressurized (PHWE), liquid water extraction or super-heated water extraction [67]. Sample size, solvent, pressure, temperature, pH, flow rate, extraction time are the standard parameters influencing the PLE process, with temperature and solvent type being the most significant ones [69–71].

In this process, for a short period of time (5–10 min), a cartridge in which the ample has been placed is filled with an extracting solvent and used to statically extract the sample under elevated temperature and pressure. To purge the sample extract from the extraction cell into a collector flask pressurized gas is used (**Figure 5**) [68].

To increase the efficiency of this extraction process, environmentally friendly liquid solvents are used at moderate to elevated temperature and pressure [72]. The increased temperature causes dramatic changes in the physical–chemical properties of water, enhances the analytes' solubility, breaks matrix-analyte interactions achieving a higher diffusion rate, and accelerates the extraction process by increasing the diffusivity of the solvent. The increased pressure in contrast, keeps the solvent in a liquid state without boiling and forces the solvent to penetrate the matrix pores [55, 73–75].

The main advantages of this technique are: (i) faster extraction from 15 to 50 min, (ii) low quantity of solvents (15–40 mL), and no filtration is required. However, costly equipment and the need for a throughout optimization of variables to avoid a matrix-dependent efficiency are the main demerits [72–74].

8.2.4 Supercritical fluid extraction (SFE)

SFE is used for separating components from the matrix with the application of supercritical fluids as the extracting solvent (**Figure 6**) [30].

Using CO2 as the extracting fluid has many advantages. Besides, its lower boiling point (31°C) and its critical pressure (74 bar). Moreover, carbon dioxide is abundant in nature, safe and inexpensive. But while carbon dioxide is the preferred fluid for SFE, it possesses several polarity limitations. When extracting polar solutes and when strong analyte-matrix interactions are present solvent polarity is crucial. Carbon dioxide fluid is usually mixed with organic solvents to alleviate the polarity limitations (**Figure 7**) [2].

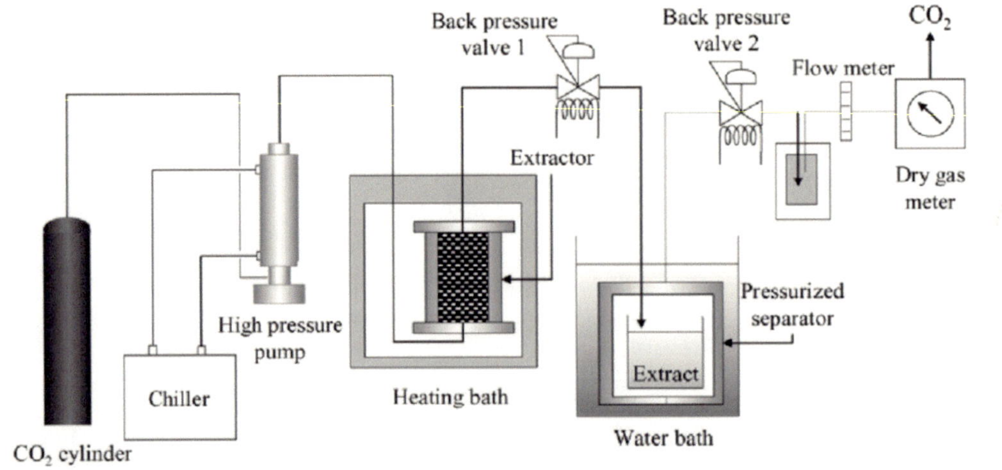

Figure 6.
Schematic diagram of supercritical fluid extraction (SFE) set-up [76].

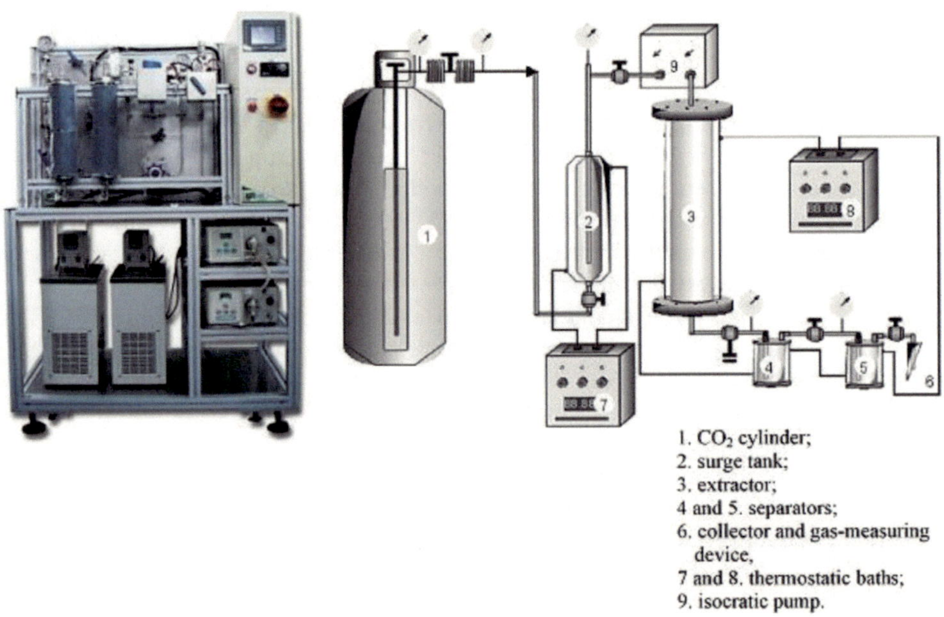

Figure 7.
Schematic representation of a supercritical fluid extraction (SFE) system [62].

i. The SFE extraction procedure possesses distinct advantages:

ii. the extraction of constituents is carried out at a low temperature, strictly avoiding damage from heat and some organic solvents. SFE offers gentle treatment for heat-sensitive material;

iii. fragrances and aroma remain unchanged;

iv. CO2 is an inexpensive solvent;

v. No solvent residues are left behind;

vi. possibility of direct coupling with analytical chromatographic techniques such as gas chromatography (GC) or supercritical fluid chromatography (SFC);

vii. environmentally friendly extraction procedure. CO2 as the solvent does not cause environmental problems and is physiologically harmless, germicidal, and non-flammable.

Some specific disadvantages of this method are:

i. high investment cost;

ii. the use of high pressures leads to capital costs for the plant, and operating costs may also be high, so the number of commercial processes utilizing supercritical fluid extraction is relatively small, due mainly to the existence of more economical methods;

iii. high polar substances (sugars, amino acids, inorganic salts, proteins, etc.) are soluble;

iv. phase equilibrium of the solvent/solute system is complex and making design of extraction conditions is difficult.

SFE finds extensive application in extracting pesticides, environmental samples, foods and fragrances, essential oils, polymers, and natural products [50, 77]. Conde-Hernández and collaborators extracted the essential oil of rosemary (*Rosmarinus officinalis*) by S-CO2 extraction, hydro distillation and steam distillation. They found that both yields of essential oil and antioxidant activity of SFC extract were higher than those from the other two methods [78, 79].

8.2.5 Pulsed electric field (PEF) extraction

Pulsed electric field extraction is a technique based on the exposure of vegetable matrix to an electrical potential. A transformer generates an electric pulse, increasing voltages from 140 or 220 V to 1000 V, or even greater than that (25000 V). A capacitor transforms this high voltage in a closed chamber with metallic electrodes. The general scheme of PEF equipment is presented in **Figure 8** [80].

This "cold" extraction assisted by PEF prevent the degradation of the cell and the extraction of components from the intracellular vacuoles [81]. It considerably increases the yield and decreases the time because it can increase mass transfer by destroying membrane structures during the extraction process.

Specific energy input, treatment temperature and field strength are considered among parameters that can influence the treatment efficacy of the PEF extraction. It is known as a non-thermal method which reduces the decomposition of the thermolabile components [47].

8.2.6 Enzyme-assisted extraction (EAE)

The EAE is an enzymatic pre-treatment that is carried out by the addition of specific hydrolyzing enzymes during the extraction step. In the cell membrane and cell wall structure, micelles are formed by macromolecules such as polysaccharides and protein. The coagulation and denaturation of proteins at high temperatures during extraction are the main barriers to extracting natural products. EAE enhance

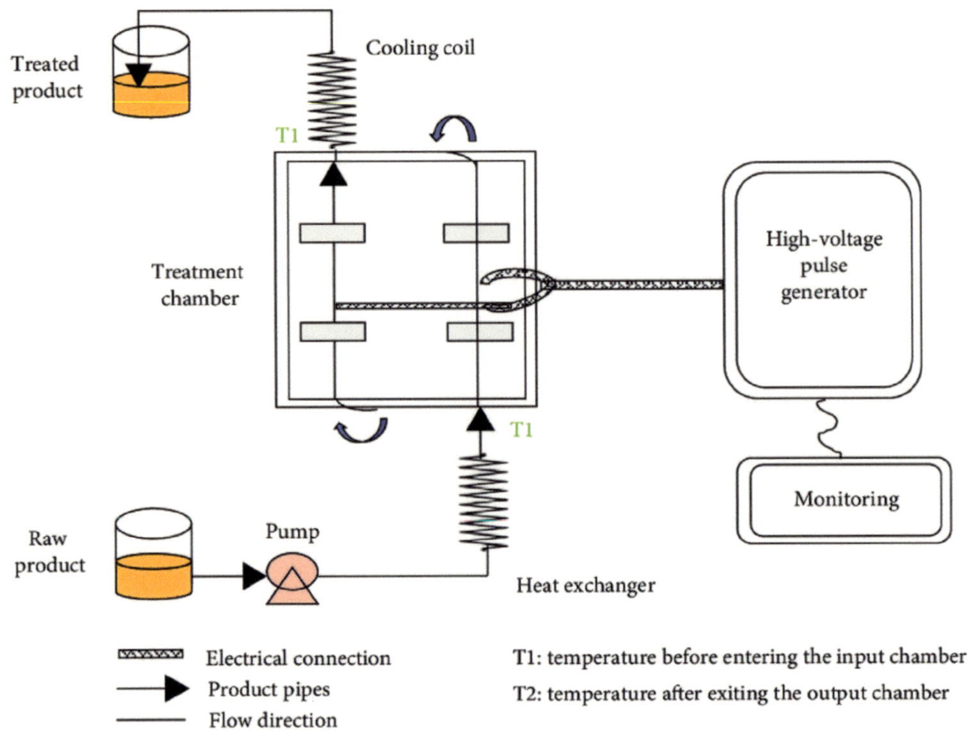

Figure 8.
General scheme of a PEF equipment process.

the extraction efficiency due to the hydrolytic action of the enzymes on the components of the cell wall and membrane and the macromolecules inside the cell, which facilitate the release of the natural products. Cellulose, α-amylase, and pectinase are hydrolyzing enzymes usually employed in EAE [47, 82]. This procedure is suitable for extracting various bioactive substances from plant matrices, but after filtration the obtained fraction is rich in small water-soluble molecules that include polyphenols and flavonoids [82].

8.2.7 Turbo-distillation extraction or turbo-extraction (turbolysis)

Turbo-distillation was patented in 1983 by Martel, and has been used in several companies as an industrial purpose for extracting EOs from hard matrixes (such as wood, bark, and seeds) [83]. The extraction process is similar to hydrodistillation with slight modifications [84]. The turbo-extraction or turbolysis is based on extraction with stirring and simultaneous reduction of particle size. Due to of high shearing force, cells disruption leads to rapid dissolution of the active constituents. It results in an extraction time of the order of minutes and the plant content is almost completely depleted [85]. Compare to hydrodistillation, turbo-distillation minimize extraction time and energy consumption and prevents the degradation of volatile constituents (**Figure 9**) [84].

In 2017, Martins and collaborators studied the turbo-extraction of stevioside and rebaudosideo A from *Stevia rebaudiana* dried and powdered leaves. The extraction is carried out by applying a fractional factorial design that allowed the evaluation of the main effects of drug powder size, solvent to drug ratio by weight, temperature, stirring and time on the yield of these glycosides. Their work demonstrated that turbo-extraction was promising for *Stevia rebaudiana* glycosides extraction.

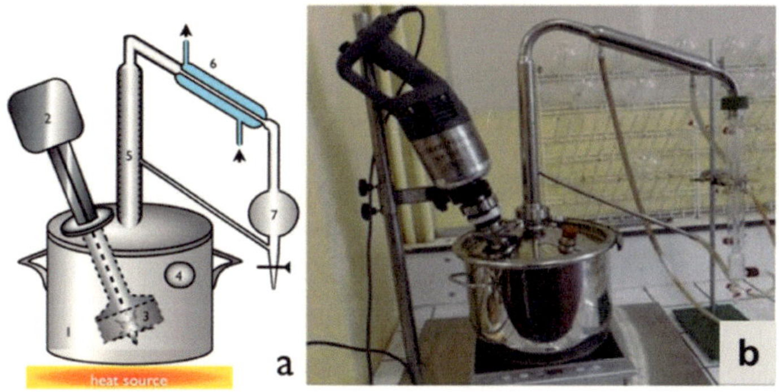

The vessel (1); the rotor (2); the turbo shredder (3); the thermometer (4); the distillation column (5); the condenser (6); the receiver-cum separator (7)

Figure 9.
Laboratory turbo-Clevenger: (a) schematic, (b) bench apparatus. The vessel (1); the rotor (2); the turbo shredder (3); the thermometer (4); the distillation column (5); the condenser (6); the receiver-cum separator (7) [46, 84].

It stimulated new research on the purification of these extracts, which became an exciting source of income for developing countries such as India and Brazil [86]. Perino and collaborators showed that the essential oil extracted by turbodistillation in 30 minutes were quantitatively (yield and kinetics profile) and qualitatively (aromatic profile) similar to those obtained using conventional hydrodistillation in 3 hours. They concluded that this process, which gave a reduced extraction time, was perfectly adapted to the extraction of hard matrixes [84]. It can be advantageous over dynamic maceration.

8.2.8 Counter-current extraction (CCE)

In this procedure, the wet raw material is pulverized to produce a fine slurry. The target material is moved in one direction (usually as a fine slurry) within a cylindrical extractor where it comes in contact with extracting solvent. Further, the starting material moves making more concentrated extract. Thus, complete extraction is possible when the amounts of material and the flow rate of solvent are optimized the complete extraction is possible. The process is extremely efficient, takes little time and poses no danger when high temperature is applied. Lastly, the extracts come out sufficiently concentrated at one end of the extractor, while the residue falls on the other end [50]. This extraction procedure has great advantages:

i. compared to other methods such as maceration, decoction, percolation a unit amount of the plant material cab be extracted with a much smaller volume of solvent;

ii. CCE is usually performed at room temperature, which avoids the thermolabile constituents from being exposed to heat which is used in most other techniques;

iii. Since the drug is pulverized under wet conditions, the heat generated during comminution is neutralized by water. This once more avoids the thermal degradation of components from heat exposure;

iv. Compare to continuous hot extraction, CCE is rated to be more efficient and effective.

8.2.9 Solid-phase extraction (SPE)

Solid-phase extraction (SPE) is a sample preparation technology using chromatographic packing material, solid particle, commonly found in a cartridge-type device, to chemically separate the different components. Samples are almost constantly in the liquid state (although special applications can be run with some samples in the gas phase). In this method, the dissolved or suspended compounds in a liquid mixture are separated from other compounds depending on their physical and chemical properties. The technically correct name for this technology is "Liquid–Solid Phase Extraction", since the chromatographic particles are solid and the sample is in the liquid state [87].

SPE has many benefits, but four significant benefits deserve special attention:

i. simplification of complex sample matrix along with compound purification;

ii. reduce ion suppression or enhancement in MS applications;

iii. capability to fractionate sample matrix to analyze compounds by class;

iv. trace concentration (enrichment) of very low-level compounds.

This rapid, economical and sensitive technique uses different types of cartridges and disks, with various sorbents, where the solute molecules are preferentially attached over the stationary phase.

8.2.10 High-voltage-assisted extraction

The principle of this equipment is similar to PEF, with the difference that electrical discharge is made through a small point. For this, a needle electrode is used from which the release is made in a plate ground electrode.

These methods are known as greener methods, are often better than conventional ones in terms of high yields, high selectivity, lower solvent consumption and shorter extraction time. They are also found to be environmentally ecofriendly since energy, and organic solvent consumption are reduced. The combination of extraction methods to obtain high purity extracts or high overall yields are described in the literature [40, 88–90]. Its main advantage is the operability in continuous mode, which is very important from an industrial and economic point of view [80].

8.2.11 Phytonics process

A new solvent-based on hydrofluorocarbon-134a and a new technology to optimize its remarkable properties in the extraction of plant material offer significant environmental advantages and health and safety benefits over traditional processes to produce advanced quality natural fragrant oil, flavors and biological extracts.

The technology known as "phytonics process" was developed and patented by Advanced Phytonics Limited (Manchester, UK). Fragrant components of EOs and biological or phytopharmacological extracts that can be used straightly without additional chemical or physical treatment are the products frequently extracted by this process. The properties of the new generation of fluorocarbon solvents have been applied to the extraction of plant material. The core of the solvent is 1,1,2,2-tetrafluoroethane, better known as hydrofluorocarbon-134a (HFC-134a) with a boiling point of – 25°C; a vapor pressure of 5.6 bar at ambient temperature. It is flammable and non-toxic. This product was developed as a replacement for chlorofluorocarbons and more importantly, it does not deplete the ozone layer. By most standards this is a poor solvent that is unable to break up (dissolve) plant waste.

The process is advantageous because the solvents can be customized: by using modified solvents with HFC-134a, the process can be made highly selective in extracting a specific class of phytoconstituents. Likewise, to withdraw a broader spectrum of constituents other modified solvents can be employed. The biological products obtained by this process contain extremely low residual solvent. Residuals are constantly below the levels of detection and are fewer than 20 parts per billion. Therefore, selected solvents have minimal potential reaction effects on the botanical material, and are neither acidic nor alkaline. At the end of each production cycle, the processing plant is sealed so that solvents are constantly recycled and totally recovered. Electricity is the unique utility required to perform these systems and, even then, they consume little energy. There is no scope for the escape of the solvents, and even if some solvents come to escape, they pose no threat to the ozone layer because they do not contain chlorine. The waste product (biomass) from these plants is dry and "ecofriendly" to handle.

As the benefits of this procedure, we have the following:

i. the phytonic process is soft and its products are never damaged by exposure to temperatures over ambient because relatively low temperatures are employed;

ii. vacuum stripping is necessary which, in other processes, leads to the loss of precious volatiles;

iii. the process is performed completely at neutral pH, and in without oxygen, the products never suffer acid hydrolysis damage or oxidation;

iv. the procedure is extremely selective, and offer a choice of operating conditions end products;

v. it requires a minimum amount of electrical energy;

vi. it is less threatening to the environment;

vii. no harmful emission in the atmosphere and the subsequent waste products (spent biomass) are inoffensive and pose no effluent disposal problems;

viii. the solvents employed are neither toxic, nor flammable, or ozone-depleting;

ix. the solvents are entirely recycled within the system.

In biotechnology, the utilization of the phytonics process is frequently employed to extract (e.g., for the production of antibiotics), herbal drug, food, EOs and flavor industries, and pharmacologically active products. It is particularly used to produce top-quality pharmaceutical-grade extracts, pharmacologically active intermediates, antibiotic extracts, and phytopharmaceuticals. However, the fact that it is used in all these areas prevents its use in other areas. The technique is being used to extract high-quality essential oils, oleoresins, natural food colors, flavors and aromatic oils from all types of plant material. The technique is also used in refining crude products obtained from other extraction processes. It provides extraction without wax or other contaminants. It helps in the removal of many biocides from contaminated biomass [50].

8.3 Liquid–Liquid extraction (partitioning)

Upon extraction of the solids and release of desired organics into the extraction solvent, the most common next step is a liquid–liquid extraction, taking advantage of mixing two (or sometimes three or even more that can establish two phases) non miscible solvents, for example, water and ether. The standard rule of thumb is that polar compounds go into polar solvents (e.g., amino acids, sugars, and proteins remain in water). To the contrary, the nonpolar components usually remain in the organic phase (e.g., steroids, terpenoids, waxes, and carotenoids are typically extracted into a solvent such as ethyl acetate).

It is important to minimize interference from compounds that may coextract with the target compounds during the extraction of plant material by conventional or by advanced methods. It is also needed to avoid contamination of the extract and to prevent decomposition of important metabolites or artifact formation as a result of extraction conditions or solvent impurities [3]. Regardless of the extracting procedure employed, the resulting solution should be filtered to withdraw whatever particulate matter. Due to the accompanying increased risk of formation of artifact and decomposition or isomerization of extract components plant extract should not be stored in the solvent for a long time at room temperature or in sunlight because [3].

9. Extraction of specific metabolites

The chemical investigation profile of a plant extract, fractionation of a crude extract is suitable to isolate the major classes of compounds from each other before further chromatographic analysis. One procedure based on varying polarity that might be used on an alkaloids-containing plant is indicated in **Figure 10**. The type and quantity of components to be separate into different fractions will, vary from plant to plant. Such procedure can be modified when labile substances are investigated [47].

9.1 Extraction of essential oils (EOs)

Essential oils (EOs) are concentrated aromatic hydrophobic oily volatile liquids characterized by a strong odor and produced by all plant organs [91]. They are obtained from raw material by several extraction techniques such as water or steam distillation, hydrodiffusion, solvent extraction, Soxhlet extraction, expression under pressure or cold pressing method, also known as scarification method, microwave-assisted extraction, microwave hydrodiffusion and gravity, supercritical fluid or subcritical water extractions. The best extraction method to use depends on the ease of evaporating (volatility) and the hydrophilicity or hydrophobicity (polarity) of the desired components [92–96]. However, the three most commonly applied techniques to extract EOs are Soxhlet, hydrodistillation, and SFE [97]. The extraction method chosen significantly affects the chemical composition of EOs [91]. Benmoussa and collaborators have recently found that the microwave hydrodiffusion and gravity (MHG) appeared like a rapid process, a green technology, and a desirable alternative protocol to enhance both the quality and the quantity of the EOs extracted from medicinal and aromatic plants [92].

9.2 Extraction of fats and oils

Lipids contain a broad category of non-polar molecules that are barely soluble or completely insoluble in water, but soluble in an organic solvent such

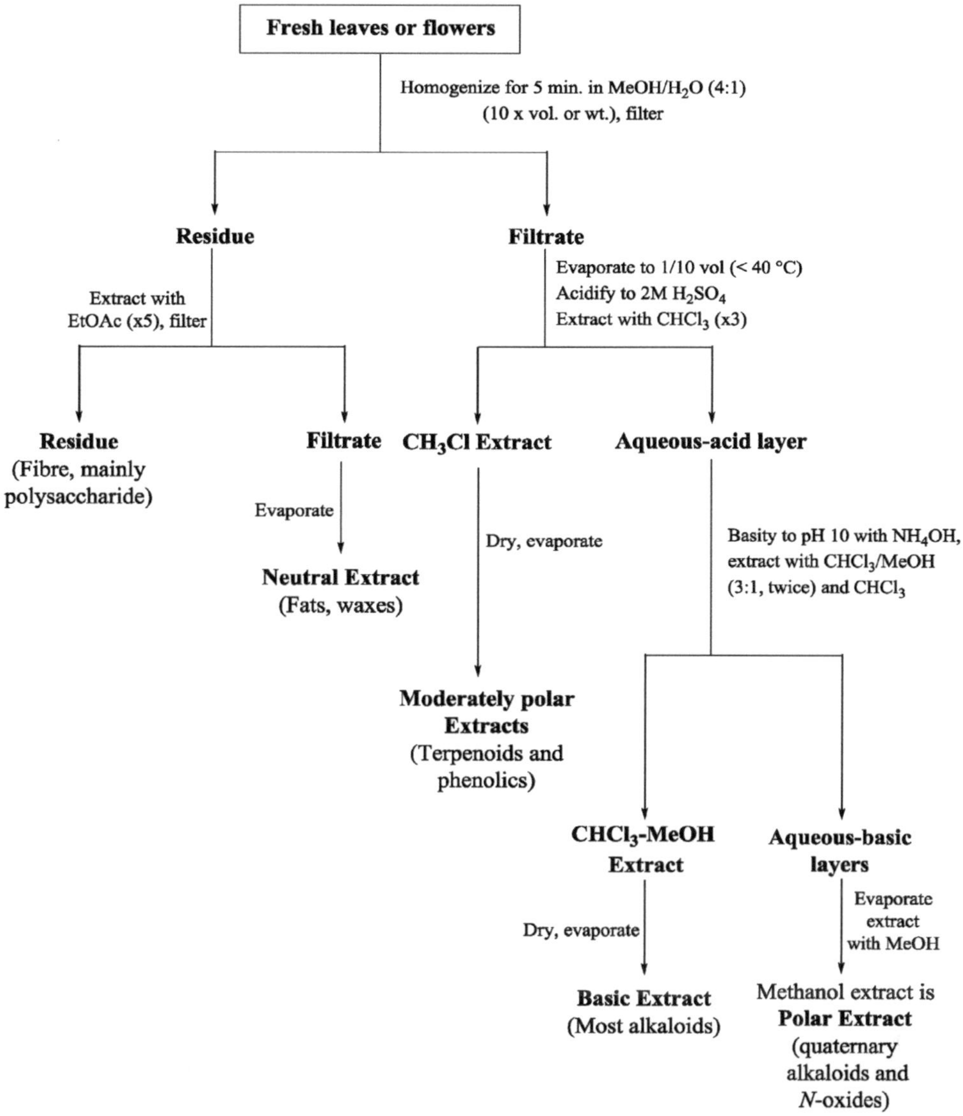

Figure 10.
A general procedure for extracting fresh plant tissues and fractionating into different classes according to polarity.

as *n*-hexane, diethyl ether, chloroform, and alcohol [98]. Fats are triglycerides that are solid or semi-solid at room temperature, while oil is also triglycerides that are liquid or clear liquid at room temperature, however, their chemistry is determined by the degree of solubility. Fats and oil may be of vegetable, animal, and marine origin [99]. Oilseeds and fats production requires several units-operations, starting with a pre-treatment stage. It is often necessary to dry the sample before oil extraction using solvents because many organic solvents are not miscible with water and cannot easily penetrate the matrix and extraction would be inefficient [100]. The processing methods used are usually neither specific to lipids, nor insure 100% recovery of the lipid material because of the nature of the matrix. Diethyl ether and petroleum ether stands as favorite solvents in the case of crude fat because they are relatively non polar, hence extract most non-polar components [98].

Extraction process of edible oils may have negative effects on taste, stability, appearance or nutritional value, preserve tocopherols, and prevent chemical changes in the triacylglycerol. Fats and oil can be extracted from plants using conventional and advanced techniques that include hot water extraction, cold pressing, solvent extraction, high-pressure solvent extraction, microwave –assisted extraction, and supercritical fluid extraction [99]. Extraction of oil involves several mechanisms for removing a liquid from a solid such as leaching, washing, diffusion and dialysis [98]. In the case of palm oil (seeds of *Elaeis oleifera*), crude oil is obtained after a digestion step followed by a pressing stage. Digestion helps the rupture or breaking down the oil-bearing cells, thus releasing the palm oil in the fruit [101–103]. Enzyme-assisted extraction (EAE) is an efficient method to improve lipid extraction from several different biomasses such as soybean, sunflower, and microalgae [104, 105].

The main side reactions reported during oil processing are (i) *trans* fatty acid formation, (ii) *cis-trans* isomerization, (iii) and physical loss [99]. Before oilseeds processing, moisture must not exceed a certain limit to prevent growth of fungi and the occurring lipase formation, resulting in a free fatty acid increase [98].

9.3 Volatile organic compounds

Volatile organic compounds (VOCs) are odorant compounds emitted from plant tissues. Plants can produce a high diversity of VOCs. They are responsible for the distinct aroma of certain dried plants, including the tea, *Camellia sinensis*. VOCs can therefore be used as an indicator of tea quality [106, 107]. Several VOCs are emitted as a natural defense mechanism against arthropods and pathogen attacks [108, 109].

Hydro-distillation (HD), steam distillation (SD), simultaneous distillation solvent extraction (SDE), microwave-assisted hydro-distillation (MWHD), supercritical fluid extraction (SFE), purge and trap, and solid phase microextraction (SPME), are used to extract VOCs [110].

Verde and collaborators conducted a work to optimize the MAE of the volatile oil terpenes from *Pterodon emarginatus* fruits and characterize the volatile compounds. According to their study, MAE proved to be feasible with a particular interest in avoiding the need of organic solvents in volatile oil extraction from plants. They proved that a minimum amount of water could be enough to bring result in extraction. That green methodology appears to be an excellent alternative to extract terpenes from aromatic plants [111].

9.4 Alkaloids

The alkaloids are low molecular weight nitrogen-containing compounds found mainly in plants and a lesser extent in microorganisms and animals. They contain one or more nitrogen atoms, typically as primary, secondary, or tertiary amines, which usually confers basicity on the alkaloids. If the free electron pair on the nitrogen atom is not involve in mesomerism, the salt formation can occur mineral acids. This fundamental property of alkaloids is used in their extraction and further clan-up. According to the nature of the nitrogen-containing structure, alkaloids are classified as pyrrolidine, piperidine, quinoline, isoquinoline, indole, etc. [27].

Two methods may be used for alkaloids extraction. One is to basify the plant material using diethylamine or ammonia and extract with an organic solvent [112, 113]. Alkaloids are substances with a basic character and their solubility is a function of pH. They are soluble in low polar organic solvents in basic medium, while in acidic medium, they are soluble in water.

Alkaloids containing basic amines can be selectively extracted using a modified version of the classical "acid–base shake-out" method (**Figure 11**).

As recommendations, mineral acids and strong bases should be avoided in extracting alkaloids (and plant material in general) because of the risk of artifact formation [3, 114, 115].

9.4.1 Extraction of caffeine

Caffeine is a natural product found in Coffee, cocoa beans, kola nuts, and tea leaves in a substantial amount. Its efficient extraction from Coffee relies heavily on the properties of caffeine and other components present in Coffee. One of the most popular species of the genus whose seeds contains caffeine is *Coffea arabica* (**Figure 12**). Several methods can be used to extract caffeine, including Ultrasonic extraction, Heat Reflux extraction, and Soxhlet extraction. Heat Reflux extraction is commonly used methods to extract caffeine from Coffee [116]. The initial solvent used in the extraction of caffeine is water. Caffeine is sparingly soluble in water at ambient temperature (2 g/100 mL) but increasing when mixed in boiled water (100°C) with a yield of 66 g/ 100 mL. Meanwhile, the solubility of caffeine in chloroform, toluene, acetone and ethyl acetate is relatively high at ambient temperature [116, 117]. Caffeine is a weakly basic, white colorless powder in its anhydrous state.

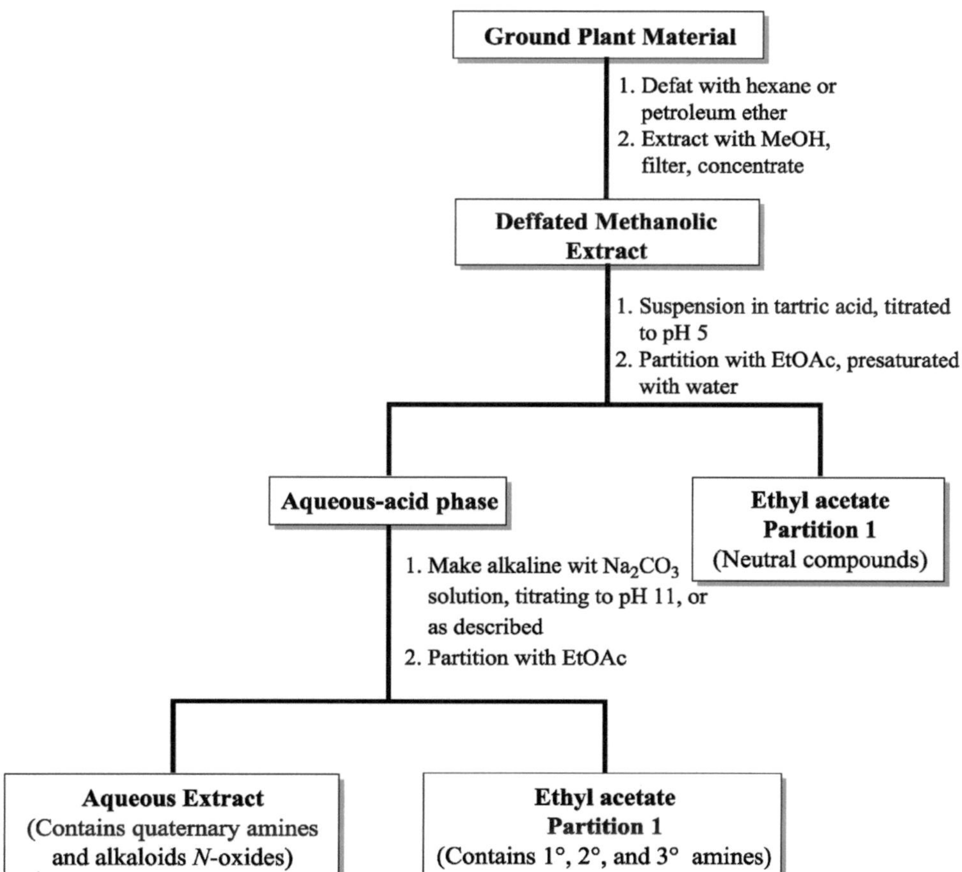

Figure 11.
General procedure to obtain alkaloidal extracts from crude plant material [114].

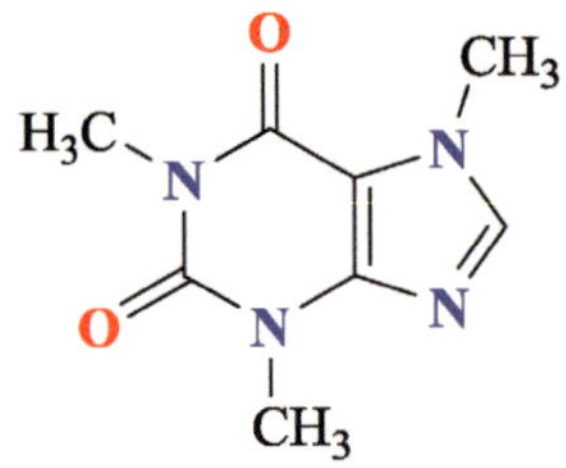

Chemical Formula: $C_8H_{10}N_4O_2$
Molecular Weight: 194.1940

Figure 12.
Chemical structure and a few data of caffeine.

There are several ways to remove caffeine from coffee. Here are few reported procedures:

9.4.1.1 Extraction procedure I: solvent extraction using dichloromethane (DCM)

Coffee seeds are firstly grounded and refluxed in an aqueous sodium carbonate solution for about 20 minutes under constant stirring. After filtration of the resulted mixture to filtrate is allowed for cooling at room temperature. The DCM is use to perform the partition of the aqueous filtrate. The process is repeated several times to extract more caffeine. The DCM fractions are then mixed with anhydrous sodium sulfate to remove water traces, the DCM-caffeine solution is filtered through reverse-phase filter paper, which will trap any water and residual matter. The DCM solution is allowed to evaporate and the white amorphous powder of caffeine is obtained [118].

The addition of sodium carbonate converts the protonated form of caffeine, which is naturally present in coffee, to its free caffeine form. During the extraction of caffeine, tannins being soluble in water and organic solvents can interfere with extraction. A weak base such as calcium carbonate or sodium sulphate can be added to break down tannins esters bonds into glucose and calcium or sodium salts of gallic acid, both of which will not be extracted into the organic solvent.

9.4.1.2 Extraction procedure II: supercritical carbon dioxide extraction

Some benefits are reported when using this method: caffeine is easily extracted from the final product after avoiding the use of flammable and toxic solvents. In this process, caffeine diffuses into supercritical CO_2 with water. Coffee beans are introduced at the top while fresh CO2 is introducing at the bottom of an extractor vessel in a continuous extraction to remove caffeine. The recovery is accomplished in a separate absorption chamber containing water. Higher temperature and pressure are mandatory to obtain great yields. A pretreatment step is needed in this process. The addition of polar cosolvents affects cosolvent solute specific chemical or physical interactions. The extraction rate is accelerated by the solvent–cosolvent interaction and makes the extraction easier. The material is humidified with ultrapure water for prewetting, this will destroy the hydrogen bonds that link the caffeine to its natural matrix. Cell membrane swelling enhances solute diffusion. Subsequently, the quality of caffeine extracted can reach a purity >94%, which is generally the standard criteria for use in the soft drink and drug companies [119].

9.4.1.3 Extraction procedure III: activated charcoal

There are some benefits to use charcoal: it is cheaper, "green," and ease to regenerate by heat and steam. The choice of active charcoal with the appropriate number of micropores and a specific area up to 1000 m^2/gram is mandatory for good absorption performance.

Cleaned green coffee beans are firstly soak in water, and the caffeine and other soluble content transferred to the aqueous phase. During the filtration through the activated charcoal, solely caffeine will continue to migrate in water. The recovered and dried coffee beans are now decaffeinated [30].

9.4.2 Extraction of morphine

The poppy straw (*Papaver somniferum* capsules) produces a white sticky latex known as opium. Usually, two weeks after the petals fall from the bud farmers harvest and collect opium. To allow the viscous latex to ooze out slowly farmers generally use sharp blade to do two to five incisions into the pod's skin. 24-hours after incisions of the pod, opium is then collected. This gummy latex, or opium (poppy tears), is a complex mixture containing at least 50 different alkaloids (**Figure 11**). Morphine is the major alkaloid, making up to 8–17% of the dry weight of opium. The chemical structure of morphine was established in 1925 despite de fact it has been used for centuries. Even if the immense majority of morphine continues to be harvested from the opium poppy, there are at least three classical processes (all old) for the extraction of morphine from simple starting material [120].

9.4.2.1 Extraction by Merck process

Cold water is used to treat the opium and the obtained aqueous solution concentrated until syrupy consistence. Powered sodium carbonate is added to precipitate hot and heated as long as ammonia given off; it is recommended that the solution remain alkaline to phenolphthalein and left aside four 24 hours at room temperature. After standing, the precipitate is filtered and cold water is use to wash several times until the wash-water become colorless. The precipitate is dissolved in alcohol at 85°C and the alcoholic solution is allowed for evaporation until dryness, and the residue is exhausted after neutralization with little amount of acetic acid. Decolorizing charcoal is used to treat the acidic solution and afterward precipitated with ammonia, avoiding excess is important. After filtration, the precipitate is washed and purified by crystallization in alcohol; concentration of the alcoholic mother-liquor yields a further quantity of morphine. This procedure was reported to be impossible to be consider for industrial scale because of the slight solubility of morphine is alcohol [120].

9.4.2.2 The Thiboumery and Mohr process

The gummy opium in divide into thin slices and treated with hot water thrice of its weight until obtain a homogeneous paste. After filtration the residue is pressed and treated again with thrice its weight in water. The resulted solutions are combined and allowed to evaporation until half their volume and poured into boiling milk of lime. One part of lime in ten parts of water should be used for four parts of opium; it is then filtered off again. The lime solutions are united and concentrated to a quantity twice the weight of the opium used. The solution is filtered, heated to boiling, and morphine is precipitated by adding ammonium chloride. The solution is filtrated after cooling at room temperature, and the precipitate is washed, then purified by solution in hydrochloric acid and crystallization of the morphine hydrochloride. It is an attractive process since there are no technical difficulties and the morphine is well separated from the secondary alkaloids. The morphine solutions are relatively clean; however, the yield might be bad. The contributory factors may be the oxidation of morphine in alkaline solution, and the fact that the lime always retains morphine [120].

9.4.2.3 The Roberson-Gregory process

Five to ten times its weight of cold distilled water is used to completely exhaust the opium. The resultant solution is evaporated to the consistency of a soft extract. The process is repeated with cold distilled water. This aqueous re-extraction causes impurities to precipitate, they are filtered off and the solution obtained is evaporated until its density is 10° Baumé. For each kilogram of opium, one hundred and twenty grams of calcium chloride are added to the boiling liquor, which is further diluted with an amount of cold water equal to its volume. A mixture of a precipitate of meconate and sulfate of calcium is thus formed and is filtered off. After filtration, the filtrate is once more concentrated to produce a new deposit which consist almost entirely of calcium meconate. After removal of the residue by filtration, the filtrate is left to stand for few days until it becomes a crystalline mass called "Gregory's salt". It is a mixture of hydrochloride and codeine hydrochloride. The crystals obtained are drained and then placed in a cloth and squeezed out in the presser. Successive crystallization is employed and each time animal charcoal is used to decolorize the solutions. To separate morphine to codeine, sufficiently pure crystals are dissolve in water and ammonia is therefore added to precipitate morphine while codeine remains in aqueous solution.

The first disadvantage of this procedure is that 20 to 25% of the morphine is left with the secondary alkaloids in the brown and viscous mother-liquids after filtration of the Gregory's salt. The second drawback is that the hydrochloride of morphine and codeine crystallize in furry needles retains the mother-liquids in which the crystallization occurred. Several successive crystallization and subsequent recoveries are required for purification, which is a time-consuming process [120].

Later in 1957, an efficient method of extraction of morphine from poppy straw was developed by Mehltretter and Weakley. Water-saturated isobutanol containing 0.23% ammonia was used to extract morphine. Almost all the alkaloid was absorbed by passing off the raw opium through a cation exchange ions resin bed. Quantitative elution of morphine from the bed was achieved with dilute aqueous alkali. After neutralization and concentration, the crude morphine is obtained, and the eluate can be converted to hydrochloride pharmaceutical grade without difficulty. The general recovery of morphine was 90% [121].

Cooper and Nicola have reported recently a straightforward process for extraction of morphine with a good overall yield (**Figures 11** and **13**). Morphine and related alkaloids can be purified from opium resin and crude extracts by extraction in the following manner: first, soaking the resin with diluted sulfuric acid, which releases the alkaloids into solution. Either ammonium hydroxide or sodium carbonate then precipitates the alkaloids. The last step separates morphine from other opium alkaloids. Today, morphine is isolated from opium in relatively large quantities: over 1000 tons per year (**Figure 14**) [30].

Till date, morphine is used as a powerful painkiller to alleviate severe pain by acting straightaway on the brain. It also possesses euphoric and hallucinatory

Figure 13.
Extraction of raw opium from poppy straw.

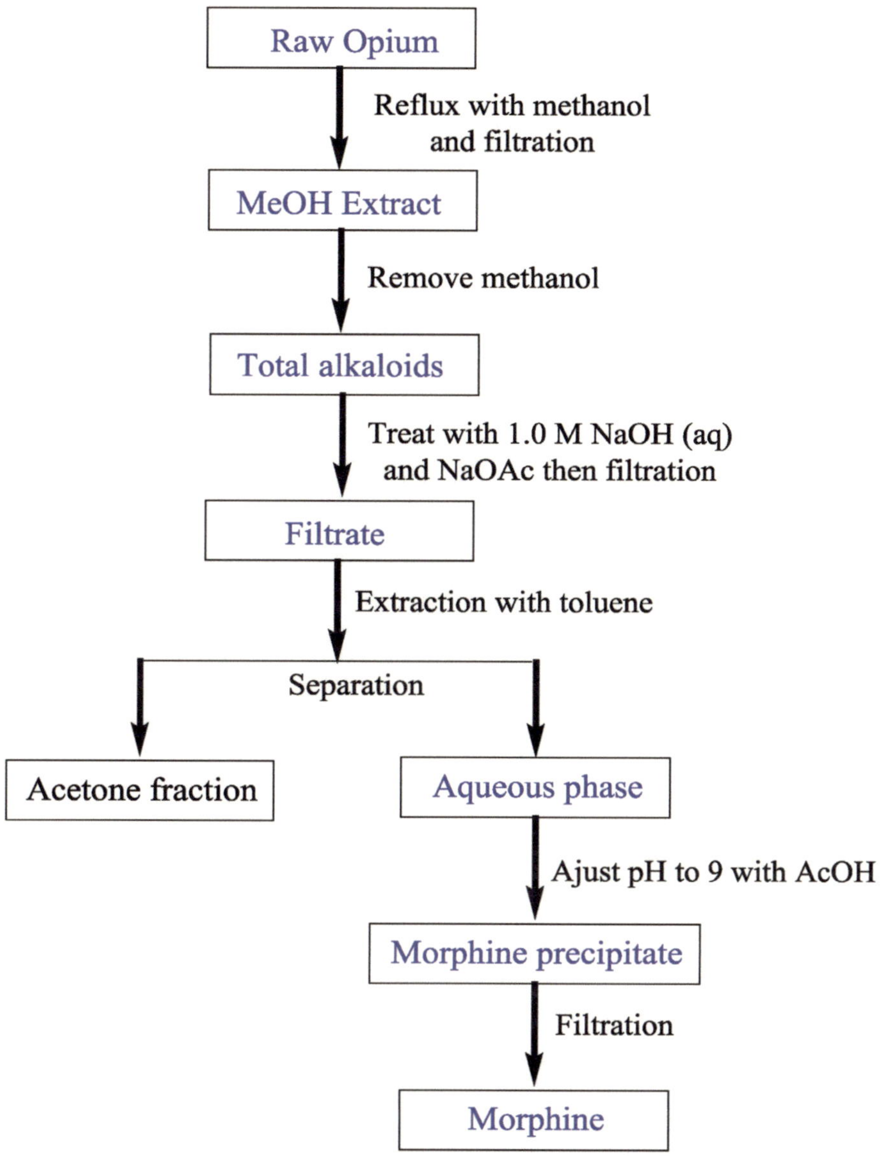

Figure 14.
Extraction protocol of morphine from raw opium by Cooper and Nicola [30].

effects. Morphine can also be chemically converted by an acetylation reaction using acetic anhydride and pyridine to create a much more potent form of the narcotic drug known as heroin [30].

9.5 Glycosides

Glycosides are relatively polar, and their polarity depends on both the number and type of sugar moieties attached to the aglycone. Cardiac glycosides have bulky steroidal aglycone, which are soluble in chloroform. However, most glycosides are extracted using polar solvents like acetone, methanol, ethanol, water or mixtures of these solvents. When extraction in done using water as solvent, enzymatic breakdown can happen. This will be avoid by using boiling water or add important proportions of alcohol or ammonium sulfate to the extract. In some cases, it may be the hydrolytic separation of the aglycone and sugar before or after extraction [122, 123].

9.6 Total phenolic and total flavonoids content

Phenolic compounds are well-known phytochemicals found in almost all plants. They can be simple phenols, benzoic and cinnamic acid derivatives, coumarins, tannins, lignins, lignans, and flavonoids [124]. Flavonoids are a group of plant constituents, the most common phenolic compound produce by plants as secondary metabolites in response to diverse biotic and abiotic factors [63, 82, 124]. They are responsible for the characteristics of flavor, color and pharmacological activities [67, 80, 125]. Because of their positive effects on human and animal health, and medical application for disease therapy and chemoprevention, interest in flavonoids increases [126, 127]. Complete extraction of phenolics is the next critical step after the sample preparation. The most common procedures of extraction of phenolics employ solvents, either organic or inorganic. Different parameters may influence the extraction yield, that includes temperature, the solvent used, time, solvent-to-sample ratio, as well as the number of repeated extractions of the plant material [124].

There is no universal extraction method and each optimized procedure is unique [82]. Due to the complex nature of the sample matrix and diverse chemical characteristics of flavonoids, it is consensual among scholars that there is no single or/and standard method to be used for every material or flavonoids to be extracted at present [67]. Maceration, water infusion, and Soxhlet extractions are generally used in research laboratories and/or in small manufacturing companies. The choice of solvent for extraction such as water, acetone, ethyl acetate, alcohols (methanol, ethanol, and propanol), and their mixtures will influence phenolics' extraction [124, 128]. The extraction of flavonoids-containing sample material are still performed by simple direct solvent extraction. It can also be extracted in a Soxhlet apparatus, first with *n*-hexane or diethyl ether to remove fats, and then with ethyl acetate or ethanol to obtain total phenols. This procedure is unsuitable for thermolabile components. A commodious and frequently used technique is sequential solvent extraction. Dichloromethane is used in the first step to remove flavonoid aglycones and non-polar components. A subsequent step using alcohol or alcohol-water mixtures will therefore extract flavonoid glycosides and other polar constituents. Cowan indicated that acetone was the most selective solvent for extracting flavonoids [129]. Anokwuru and collaborators discovered that acetone and *N*,*N*-dimethylformamide (DMF) were highly influential for removing antioxidants [130]. In most cases, flavonoids and polyphenols are coextracted [82]. Furthermore, several promising methods (Microwave-assisted extraction (MAE), Enzyme-assisted extraction (EAE), Pressurized liquid extraction (PLE), Ultrasound-assisted extraction (UAE), Matrix solid-phase dispersion (MSPD), and Supercritical fluid extraction (SFE) are nowadays used with increased yields and lower cost as main advantages [8, 82].

Due to the multiplicity of hydroxyl functions, phenols tend to be relatively polar and dissolve in aqueous alcohols. They may also be extracted or partitioned into aqueous alkali as phenolate salts as they are weak acids. A problem encountered with phenolic compounds is that they can undergo extensive polymerization reaction by polyphenol oxidation. This reaction is responsible for developing brown coloration in damaged plant material when exposed to the air and in certain extracts. The polymerization reaction is catalyzed by acid [131].

9.7 Total mixture of crude saponins

The procedure for isolating mixtures of crude saponins (i.e., steroidal or triterpene glycosides) is shown in **Figure 15**. Fats are removed from the plant material by treating with *n*-hexane and after extraction with methanol. The resultant methanol extract

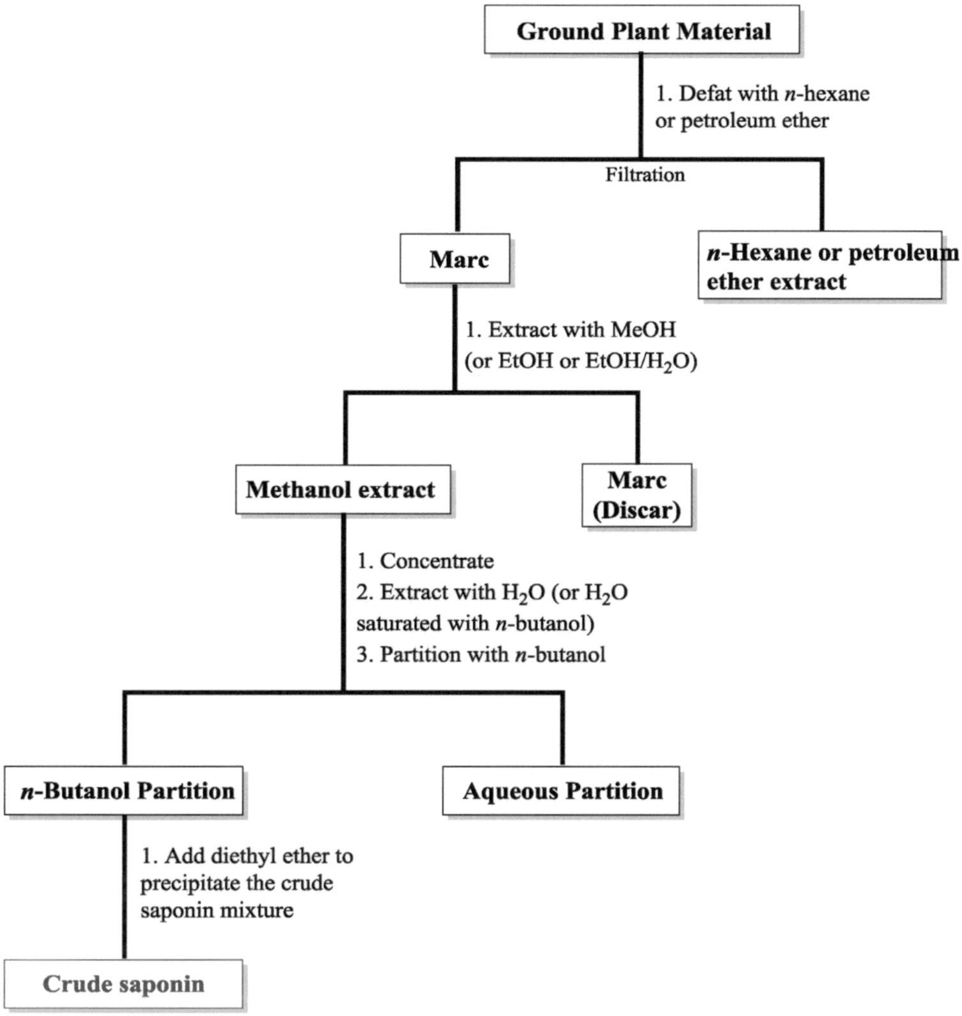

Figure 15.
General fractionation procedure to obtain a precipitate of crude saponin from plants, adapted from the literature [132].

is evaporated under vacuum and suspended in deionized water (presaturated with n-butanol), and partitioned with *n*-butanol. Diethyl ether is added to the butanol solution to precipitate the saponins [3, 132]. Selective extraction and fractionation of plant sterols (including sapogenins, bufadienolides, and cardiac glycosides) using manipulations and liquid/liquid partitioning have been described [3, 133]. Partitioning between the aqueous phthalic anhydride and organic solvent can be used to separate alcohols from non-alcohols. The alcohols partition into the aqueous layer as half-phthalates and can be regenerated by treatment with sodium methoxide in methanol. Sterols with ketone functional groups can be set-apart from non-ketones by liquid/liquid partition between organic and aqueous layers using Girard's hydrazide reagents ($H_2N.NH.CO.CH_2.NR_3^+Cl^-$), and generate ketones by acid hydrolysis [3, 133].

10. Conclusion

There is a clear and growing interest in the extraction procedure of natural products and their isolation, identification, and applications. Research innovation and safe extraction processes are of primary importance in modern analytical processes, which are economically viable and environmental friendly. In the process of plant extracting plant material, it is peremptory to reduce interference of components that may be co-extracted with the target compounds, and to bypass contamination of the extract, moreover to prevent degradation of necessary metabolites or the formation of artifact as a result of extraction conditions or solvent impurities. Regardless of the extraction procedure, the resulting solution should be filtered to remove any particulate matter. Plant extracts should be stored for short time at room temperature or in sunlight to avoid increasing risks associated with the production of artifact making and additionally degradation or isomerization of extract components. The most suitable extraction procedure depends on the matrix of the plants and the type of compost, and should follow clear selection criteria.

Conflict of interest

The authors declare no conflict of interest. The funders had no role in the design of the study; in the collection, analyses or interpretation of data; in the writing of the manuscript or in the decision to publish the results.

Author details

Fongang Fotsing Yannick Stéphane[1*], Bankeu Kezetas Jean Jules[2], Gaber El-Saber Batiha[3], Iftikhar Ali[4] and Lenta Ndjakou Bruno[5]

1 Department of Chemistry, Higher Teachers Training College, University of Maroua, Maroua, Cameroon

2 Department of Chemistry, Faculty of Science, University of Bamenda, Bambili, Cameroon

3 Department of Pharmacology and Therapeutics, Faculty of Veterinary Medicine, Damanhour University, Damanhour, AlBeheira, Egypt

4 Department of Chemistry, Karakoram International University, Gilgit, Gilgit-Baltistan, Pakistan

5 Department of Chemistry, Higher Teacher Training College, University of Yaoundé I, Yaoundé, Cameroon

*Address all correspondence to: fongangfys@yahoo.fr

References

[1] Aman D. Introductory chapter: Plant extracts. 2019. DOI:http://dx.doi.org/10.5772/intechopen.85493

[2] Swamy MK and Akhtar MS. Natural Bio-active Compounds. Volume 2: Chemistry, Pharmacology and Health Care Practices, 2019. Springer Nature Singapore Pte Ltd. (eBook) https://doi.org/10.1007/978-981-13-7205-6

[3] Jones WP and Kinghorn AD. Extraction of Plant Secondary Metabolites. In, Natural Products Isolation, 2nd Ed. Humana Press Inc. 999 Riverview Drive, Suite 208 Totowa, New Jersey 07512, 2006

[4] Cosa P, Vlietinck AJ, Berghe DV, Maes L. Anti-infective potential of natural products: How to develop a stronger *in vitro* 'proof-of-concept'. Journal of Ethnopharmacology 2006; 106: 290-302.

[5] Daily G. Nature's Services: Societal Dependence on Natural Ecosystems. Covelo, CA: Island Press. 1997; 392 pp.

[6] Botanic Gardens Conservation International, 1996. "CITES and Medicinal Plants Study: A Summary of Findings." Web resource: www.bgci.org/wellbeing/CITES_and_Med_Plants_Summary/.

[7] Wang J, Soisson S, Young C, Shoop W, Kodali S, Galgoci A, Painter R, Parthasarathy G, Tang YS, Cummings R, Ha S, Dorso K, Motyl M, Jayasuriya H, Ondeyka J, Herath K, Zhang C, Hernandez L, Alloco J, Basilio A, Tormo JR, Genilloud O, Vicente F, Palaez F, Colwell L, Lee SH, Michael B, Felcetto T, Gill C, Silver LL, Hermes JD, Bartizal K, Barrett J, Schmatz D, Becker JW, Cully D, Singh SB. Platensimycin is a selective FabF inhibitor with potent antibiotic properties. Nature 2006: 441: 358-361. Doi: 10.1038/nature04784

[8] Azwanida NN. A review on the extraction methods use in medicinal plants, principle, strength and limitation. Medicinal & Aromatic Plants 2015; 4:3. Doi: 10.4172/2167-0412.1000196

[9] Santos GS, Sinoti SBP, Cunha de Almeida FT, Silveira D, Simeoni LA, Gomes-Copeland KKP. Use of galantamine in the treatment of Alzheimer's disease and the strategies to optimize its bisynthesis unsing *in vitro* culture technique. Plant Cell Tiss Organ Cult 2020; 143: 13-29. Doi.org/10.1007/s11240-020-01911-5

[10] Scott LJ and Goa KL. Galantamine: a review of its use in Alzheimer's disease. Adis Drugs Evaluation 2000; 60(5): 1095-1122.

[11] Tajuddeen N and Van Heerden FR. Antiplasmodial natural products: an update. Malaria Journal 2019, 18; 404: 1-62. Doi.org/10.1186/s12936-019-3026-1

[12] Armstrong MJ and Okun MS. Diagnosis and treatment of Parkinson Disease: A review. F1000Research 2020; 9: 862

[13] Ho JFV, Yaakup H, Low GSH, Wong SL, Tho LM, Tan SB. Morphine use for cancer pain: a strong analgesic used only at the end of life? A qualitative study on attitudes and perceptions of morphine in patients with and advanced cancer and their caregivers. Palliative medicine 2020; 34(5): 619-629.

[14] Rathmes G, Rumisha SF, Lucas TCD, Twohig KA, Python A, Nguyen M, Nandi AK, Keddie SH, Collins EL, Rozier JA, Gibson HS, Chestnut EG, Battle KE, Humpheys GS, Amratia P, Arambepola R, Bertozzi-Villa A, Hancock P, Millar JJ, Symons TL, Bhatt S, Cameron E,

Guerin PJ, Gething PW, Weiss DJ. Global estimation of antimalarial drug effectiveness for the treatment of uncomplicated *Plasmodium falciparum* malaria 1991-2019. Malaria Journal 2020; 19: 374. Doi.org/10.1186/s12936-020-03446-8

[15] Pandey A and Tripathi S. Concept of standardization, extraction and pre phytochemical screening strategies for herbal drug. Journal of Pharmacognosy and Phytochemistry 2014; 2(5): 115-119.

[16] Fellows LE. Pharmaceuticals from traditional medicinal plants and others: Future prospects. A paper presented at the symposium "New drugs from natural sources" sponsored by I.B.C. Technical services Ltd, London, 1991. Royal Botanic Gradens, Kew.

[17] Farnsworth NR and Soejarto DD. Potential consequence of plant extinction in the United States on the current and future availability of prescription drugs. Economic Botany 1985; 39: 231-240.

[18] Doughari JH. Phytochemicals: Extraction methods, basic structures, and mode of action as potential chemotherapeutic agents, phytochemicals—a global perspective of their role in nutrition and health. In: A Global Perspective of Their Role in Nutrition and Health. Venketeshwer R. Editor. InTech; 2012. Available from: www.intechopen.com.

[19] Rungsung W, Ratha KK, Dutta S, Dixit AK, Hazra J. Secondary metabolites of plants in drugs discovery. World J Pharm Res 2015; 4:604-613

[20] Sofowora A, Ogunbodede E, Onayade A. The role and place of medicinal plants in the strategies for disease prevention. Afr J Tradit Complement Altern Med., 2013; 10(5); 210-229.

[21] Evans WC. Trease and Evans' Pharmacognosy. 16th Edition. WB Saunders Company Ltd., London, 2008.

[22] Sofowora A. The present status of knowledge of the plants used in traditional medicine in western Africa: A medical approach and a chemical evaluation. J Ethnopharmacol 1980; 2:109-118.

[23] Astutik S, Pretzsche J, Kimengsi JN. Asian medicinal plants' production and utilization potentials: A review. Sustainability, 2019; 11: 5483. Doi:10.3390/su11195483

[24] Smith-Hall, C.; Larsen, H.O.; Pouliot, M. People, plants and health: A conceptual framework for assessing changes in medicinal plant consumption. J. Ethnobiol. Ethnomed. 2012; 8, 43.

[25] Schippmann, U.; Leaman, D.; Cunningham, A.B. A comparison of cultivation and wild collection of medicinal and aromatic plants under sustainability aspects. In Bogers J, Craker LE, Lange D. Medicinal and Aromatic Plants: Agricultural, Commercial, Ecological, Legal, Pharmacological and Social Aspects; Eds.; Springer: Dordrecht, The Netherlands, 2006; pp. 75-95. ISBN 978-1-4020-5447-1.

[26] Slikkerveer, L.J. The challenge of non-experimental validation of MAC plants. In Medicinal and Aromatic Plants: Agricultural, Commercial, Ecological, Legal, Pharmacological and Social Aspects; Bogers, R.J., Craker, L.E., Lange, D., Eds.; Springer: Dordrecht, The Netherlands, 2006; pp. 1-28.

[27] Peter KV. Handbook of herbs and spices. Woodhead Publishing Limited Abington Hall, Abington Cambridge CB1 6AH England, 2001.

[28] Akhtar I, Javad S, Yousaf Z, Iqbal S, Jabeen K. Microwave assisted extraction

of phytochemicals an efficient and modern approach for botanicals and pharmaceuticals. Pak. J. Pharm. Sci. 2019; 32(1):223-230.

[29] Omeroglu PY, Acoglu B, Özdal T, Tamer CE, Çopur OU. Extraction techniques for plant-based bio-active compounds. In Mallappa KS and Mohd SA. Natural Bio-active Compounds. Volume 2: Chemistry, Pharmacology and Health Care Practices. Springer, Springer Nature Singapore, 2019. (eBook) https://doi.org/10.1007/978-981-13-7205-6.

[30] Cooper R and Nicola G. Natural Products Chemistry: Sources, Separations, and Structures. CRC Press Taylor & Francis Group 6000 Broken Sound Parkway NW, Suite 300 Boca Raton, 2015.

[31] Sasidharan S, Chen Y, Saravanan D, Sundram KM, Yoga Latha L. Extraction, isolation and characterization of bioactive compounds from plants' extracts. Afr J Tradit Complement Altern Med 2011; 8:1-10.

[32] Bendary E, Francis RR, Ali HMG, Sarwat MI, El Hady S. Antioxidant and structure-activity relationships (SARs) of some phenolic and anilines compounds. Annals of Agricultural Science 2013, 58(2): 173-181. http:/dx.doi.ogr/10.1016/j.aoas.2013.07.002

[33] Nath R, Roy S, De B, Choudhury MD. Anticancer and antioxidant activity of *Croton*: a review. International Journal of Pharmacy and Pharmaceutical Sciences 2013; 2(2): 1-8.

[34] Huang W, Zhang X, Wang Y, Ooi V, Chung HY, Li Y. Chinese Medecine 2010; 5(23): 1-6.

[35] Chouna JR, Tamokou JDD, Nkeng-Efouet-Elango P, Lenta BN, Sewald N. Antimicrobial triterpenes from the stem bark of *Crossopteryx febrifuga*. Z. Naturforsch. 2015; 70(7-8c): 169-173.

[36] Sidjui LS, Nganso YOD, Toghueo RMK, WAKEU BNK, Dameue JT, Mkoumga P, Adhikari A, Lateef M, Folefoc G, Ali MS. Kostchyienones A and B, new antiplasmodial and cytotoxicity of limonoids from the roots of Pseudocedrela kotschyi (Schweinf.) Harms. Z. Naturforsch. 2015: 1-8. https.//doi.org/10.1515/znc-2017-0102

[37] Online Etymology Dictionary (website): http://www.etymonline.com; 2013.

[38] Bernard G and Dromard A. Book of etymology and medical terminology: Lexicon etymology (in French). Livret d'étymologie et de terminologie médicale: Lexique d'étymologie, 2011: 1-4.

[39] Alain R. Le Robert micro-poche (2nd Ed) (in french). 1994: 14 and 126.

[40] Cammack R, Atwoot TK, Campbell PN, Parish JH, Smith AD, Stirling JL, Vella F. Oxford Dictionary of Biochemistry and Molecular Biology (2nd Ed). Oxford University Press, Oxford New York 2006: 74-75.

[41] Guaadaoui A, Benaicha S, Elmajdoub N, Bellaoui M, Hamal A. What is a bioactive compound? A combined definition for a preliminary consensus. International Journal of Nutrition and Food Sciences 2014; 3(3): 174-179. Doi: 10.11648/j.ijnfs.20140303.16

[42] Dictionary of Food Science and Technology (2nd Ed). International Food Information Service (IFIS Editor), 2009: 47-48.

[43] Neeraj Varma. Phytoconstituents and Their Mode of Extractions: An Overview. Res. J. Chem. Env. Sci., 2016; 4(2): 08-15.

[44] Biesalski H-K, Dragsted LO, Elmadfa I, Grossklaus R, Muller M,

Schrenk D, Walter P, Weber P. Bioactive compounds: Definition and assessment of activity. Nutrition. 2009; 25: 1202-1205.

[45] Patel K, Panchal N, Ingle P. Techniques Adopted for Extraction of Natural Products Extraction Methods: Maceration, Percolation, Soxhlet Extraction, Turbo distillation, Supercritical Fluid Extraction. International Journal of Advanced Research in Chemical Science 2019; 6(4): 1-12. DOI: http://dx.doi.org/10.20431/2349-0403.0604001.

[46] Zhang Z and Li G. A review of advances and new developments in the analysis of biological volatile organic compounds. Microchemical Journal 2010; 95(2):127-139.

[47] Harborne JB. Phytochemical methods. Chapmann and Hall 1998, 3rd edition, London. 302 pages.

[48] Sing CC, Hasmida MN, Siti H M-S, Sarajul FM, Akil A, Waseem AW, Mohd M, Abdullah A. A Glimpse into the Extraction Methods of Active Compounds from Plants. Critical Reviews in Analytical Chemistry. 2020; DOI: 10.1080/10408347.2020.1820851

[49] Hussain MK, Saquib M, Khan MF. Techniques for Extraction, Isolation, and Standardization of Bio-active Compounds from Medicinal Plants. In Swamy MK and Akhtar MS. Natural Bio-active Compounds. Volume 2: Chemistry, Pharmacology and Health Care Practices, 2019. Springer Nature Singapore Pte Ltd. (eBook) https://doi.org/10.1007/978-981-13-7205-6.

[50] Handa SS. An Overview of extraction techniques for medicinal and aromatic plants. In Handa SS, Khanuja SPS, Longo G, Rakesh DD. Extraction technologies for medicinal and aromatic plants. ICS-UNIDO, Trieste, Italy. 21-52. 2006.

[51] Chuo S C, Ahmad A, Mohd-Setapar SH, Ripin A. Reverse micelle extraction - an alternative for recovering antibiotics. Der Pharma Chemica 2014 ; 6 : 37.

[52] Jurinjak TA, Benkovic M, Valinger D, Jurina T, Belscak-Cvitanovic A, Gajdos KJ. Optimizing Bioactive Compounds Extraction from Different Medicinal Plants and Prediction through Nonlinear and Linear Models. Ind. Crops Prod. 2018; 126: 449-458. DOI: 10.1016/j.indcrop. 2018.10.040

[53] Visht S and Chaturvedi S. Isolation of natural products. Current Pharma Research, 2012, 2(3), 584-599.

[54] Handa SS, Khanuja SPS, Longo G, Rakesh DD. Extraction Technologies for Medicinal and Aromatic Plants, (1stedn), no. 66. Italy: United Nations Industrial Development Organization and the International Centre for Science and High Technology. 2008.

[55] Rasul MG. Extraction, Isolation and Characterization of Natural Products from Medicinal Plants. International Journal of Basic Sciences and Applied Computing 2018; 2(6): 1-6.

[56] Rassem HA, Nour AH, Yunus MR. Techniques for extraction of essential oils from plants: a review. Aust J Basic Appl Sci., 2016; 10: 117-127.

[57] Wang L and Weller CL. Recent advances in extraction of nutraceuticals from plants. Trends in Food Science & Technology 2006, 17; 300-312.

[58] Soxhlet F. Die gewichtsanalytische Bestimmung des Milchfettes. Dingler's Polytechnisches Journal (in German) 1879; 232: 461-465.

[59] Luque de Castro MD and Garcia-Ayuso LE. Soxhlet extraction of solid materials: An outdated technique with a promising innovative future. Analytica Chimica Acta 1998; 369: 1-10.

[60] Mandal V, Mohan Y and Hemalath S. Microwave assisted extraction-an innovative and promising extraction tool for medicinal plant research. Pharmcognostic Reviews 2007; 1(1): 7-18.

[61] Alupului A, Călinescu I, Lavric V. Microwave extraction of active principles from medicinal plants. U.P.B. Sci. Bull., Series B, 2012; 74(2): 129-142.

[62] Castro-López C, Rojas R, Sánchez-Alejo EJ, Niño-Medina G, Martínez-Ávila GCG. Phenolic compounds recovery from grapefruit and by products: an overview of extraction methods. In Morata A and Loira I. Grape and Wine Biotechnology. IntechOpen: London, UK, 2016. DOI: 10.5772/64821. Available from: https://www.intechopen.com/books/grape-and-wine-biotechnology/phenolic-compounds-recovery-from-grape-fruit-and-by-products-an-overview-of-extraction-methods.

[63] Chavez-Gonzales ML, Sepulveda L, Verma DK, Luna-Garcia HA, Rodriguez-Duran LV, Ilina A, Aguilar CN. Conventional and emerging extraction processes of flavonoids. Processes 2020; 8, 434. 1-30. Doi:10.3390/pr8040434

[64] Jaitak V, Bikram SB, Kaul VK. An efficient microwave-assisted extraction process of stevioside and rebaudioside-A from *Stevia rebaudiana* (Bertoni), Phytochem Anal., 2009; 20(3): 240-245.

[65] Bandar H, Hijazi A, Rammal H, Hachem A, Saad Z, Badran B. Techniques for the extraction of bioactive compounds from Lebanese *Urtica dioica*. American Journal of Phytomedicine and Clinical Therapeutics 2013; 1(6): 507-513.

[66] Rodsamran, P.; Sothornvit, R. Extraction of Phenolic Compounds from Lime Peel Waste Using Ultrasonic-Assisted and Microwave-Assisted Extractions. Food Biosci. 2019; 28 : 66-73. DOI : 10.1016/j.fbio.2019.01.017

[67] Chaves JO, De Souza MC, Da Silva LC, Lachos-Perez D, Torres-Mayanga PC, Machado APF, Forster-Carneiro T, Vasquez-Espinoza M, Gonzalez-de-Peredo AV, Barbero GF, Rostagno MA. Extraction of flavonoids from natural sources using modern techniques. Frontiers in Chemistry 2020; 8(507887): 1-20. Doi: 10.3389/fchem.2020.507887.

[68] Richter BE, Jones BA, Ezzell JL, Porter NL, Avdalovic N, Pohl C. Accelerated solvent extraction: a technique for sample preparation. Analytical Chemistry 1996; 68(6): 1033-1039

[69] Wu K, Ju T, Deng Y, Xi J. Mechanochemical assisted extraction: a novel, efficient, eco-friendly technology. Trends in Food Science & Technology2017, 66: 166-175.

[70] Suan L. A review on plant-based rutin extraction methods and its pharmacological activities. Journal of Ethnopharmacology 2013; 150(3): 805-817.

[71] Wijngaard, H., Hossain, M. B., Rai, D. K., and Brunton, N. (2012). Techniques to extract bioactive compounds from food by-products of plant origin. Food Res. Int. 46, 505-513. doi: 10.1016/j.foodres.2011. 09.027

[72] Evstafev SN and Chechikova EV. Transformation of wheat straw polysaccharides under dynamic conditions of subcritical autohydrolysis. Russ. J. Bioorg. Chem. 2016; 42: 700-706. Doi: 10.1134/S1068162016070050

[73] Lachos-Perez D, Brown AB, Mudhoo A, Martinez J, Timko MT, Rostagno MA, Forster-Carneiro T. Applications of subcritical and supercritical water conditions for

extraction, hydrolysis, gasification, and carbonization of biomass: a critical review. Biofuel Research Journal 2017; 4:611-626. Doi: 10.18331/BRJ2017.4.2.6

[74] Plaza, M., and Turner, C. Trends in analytical chemistry pressurized hot water extraction of bioactives. Trends Anal. Chem., 2015; 71, 39-54. Doi: 10.1016/j.trac.2015.02.022

[75] Lv GP, Huang WH, Yang FQ, Li J, Li SP. Pressurized liquid extraction and GC–MS analysis for simultaneous determination of seven components in *Cinnamomum cassia* and the effect of sample preparation. Journal of Separation Science 2010; 33(15): 2341-2348. https://doi.org/ 10.1002/ jssc.201000208

[76] Verma, D.K.; Dhakane, J.P.; Mahato, D.K.; Billoria, S.; Bhattacharjee, P.; Srivastav, P.P. Supercritical Fluid Extraction (SCFE) for Rice Aroma Chemicals: Recent and Advance Extraction Method. In Science and Technology of Aroma, Flavour and Fragrance in Rice; Verma, D.K., Srivastav, P.P., Eds.; Apple Academic Press: Waretown, NJ, USA, 2018; pp. 179-198.

[77] Bubalo MC, Vidovic S, Redovnikovic IR, Jokic S. New perspective in extraction of plant biologically active compounds by green solvents. Food and Bioproducts Processing 2018. DOI: https://doi. org/10.1016/j.fbp.2018.03.001

[78] Conde-Hernández LA, Espinosa-Victoria JR, Trejo A, Guerrero-Beltrán JÁ. CO_2-supercritical extraction, hydrodistillation and steam distillation of essential oil of rosemary (*Rosmarinus officinalis*). J Food Eng. 2017; 200: 81-86.

[79] Khaw KY, Parat MO, Shaw PN, Falconer JR. Solvent supercritical fluid technologies to extract bioactive compounds from Natural sources: A review. Molecules 2017; 22: 1186: 1-22. Doi: 10.3390/molecules22071186

[80] De Luna SL, Ramirez-Garza RE, Salvidar SOS. Environmentally freiendly methods for flavonoids extraction from plant material: Impact of their operating conditions on yield and antioxidant properties. The Scientific World Journal 2020, 2020(6792069): 1-38. Doi. org/10.1155/2020/6792069

[81] Bonizou E, Karageorgou I, Batra G, Dourtoglou VG, Lalas SI. Pulse electric field extraction of antioxidant activity determination of *Moringa oleifera* dry leaves: a comparative study with other extraction technique. Beverages. 2019; 5(8): 1-13. Doi:10.3390/ beverages5010008

[82] Tzanova M, Atanasov V, Yaneva Z, Ivanova D, Dinev T. Selectivity of current extraction techniques for flavonoids from plant materials. Processes 2020, 8, 1222:1-30. Doi:10.3390/pr8101222.

[83] Martel JP. Device for High Speed Production of Aromatic Essential Oils from Perfume-Generating Plants or Parts Thereof. U.S. Patent 4,406,745, 27 September 1983.

[84] Périno S, Chemat-Djenni Z, Petitcolas Emmanuel, Giniès C, Chemat F. Downscaling of Industrial Turbo-Distillation to Laboratory Turbo-Clevenger for Extraction of Essential Oils. Application of Concepts of Green Analytical Chemistry. Molecules 2019; 24: 2734. Doi:10.3390/ molecules24152734w.

[85] Sonaglio D, Ortega GG, Petrovick PR, Bassani VL. Desenvolvimento tecnológico de produção de fitoterápicos. In C.M.O. Simões, *et al.* (Eds.), Farmacognosia: Da planta ao medicamento, UFRGS/UFSC, Porto Alegre/Florianópolis (2003), pp. 289-326.

[86] Martins PM, Lanchote AD, Thorat BN, Freitas LAP. Turbo-extraction of glycosides *Stevia rebaudiana* using a fractional factorial design. Revista Brasileira de Farmacognosia 2017; 27(4): 510-518. Doi: 10.1016/J.Bjp.2017.02.007

[87] Arsenault JC. Beginner's guide to SPE Solid-Phase Extraction. Waters Corporation. Water, Sep-Pak, Oasis, UPLC, Nova-Pak, 34 Maple street, Milford, MA 01757. 2012.

[88] Sridhar A, Ponnuchamy M, Kumar PS, Kapoor A, Vo D-VN. Techniques and modeling of polyphenol extraction from food: a review. Environmental Chemistry Letters 2021; 1-35. Hptps://doi.org/10.1007/s10311-021-01217-8

[89] Martinez-Correa HA, Bitencourt RG, Kayano AC, Magalhaes PM, Costa FTM, Cabral FA. Integrated extraction process to obtain bioactive extracts of *Artemisia Annua* L. Leaves using Supercritical CO2, Ethanol and Water. Ind. Crops Prod. 2017; 95: 535-542. DOI: 10.1016/j.indcrop.2016.11.007

[90] Tamborrino, A.; Romaniello, R.; Caponio, F.; Squeo, G.; Leone, A. Combined Industrial Olive Oil Extraction Plant Using Ultrasounds, Microwave, and Heat Exchange: Impact on Olive Oil Quality and Yield. J. Food Eng. 2019, 245, 124-130. DOI: 10.1016/j.jfoodeng.2018.10.019

[91] Fongang FYS and Bankeu KJJ. Terpenoids as Important Bioactive Constituents of Essential Oils. [Online First], IntechOpen 2020, DOI: 10.5772/intechopen.91426. Available from: https://www.intechopen.com/online-first/terpenoids-as-important-bioactive-constituents-of-essential-oils.

[92] Benmoussa H, Elfalleh W, He S, Romdhane M, Benhamou A, Chawech R. Microwave hydrodiffusion and gravity for rapid extraction of essential oil from Tunisian cumin (*Cuminum cyminum* L.) seeds: optimization by response surface methodology. Industrial Crops & Products 2018; 124: 633-642. Doi.org/10.1016/j.indcrop.2018.08.036

[93] Dawidowicz AL, Rado E, Wianowska D, Mardarowicz M, Gawdzik J. Application of PLE for the determination of essential oil components from *Thymus Vulgaris* L. Talanta 2008; 76: 878-884.

[94] Fahlbusch KG, Hammerschmidt FJ, Panten J, Pickenhagen W, Schatkowski D, Bauer K, Garbe D, Surburg H. Flavors and Fragrances. Ullmann's Encyclopedia of Industrial Chemistry 2003. Doi:10.1002/14356007.a11_141

[95] Meyer-Warnod B. Natural essential oils: extraction processes and application to some major oils. Perfume. Flavorist 1984, 9: 93-104.

[96] Arnould TWE. Aromatherapy for the Whole Person. UK: Stanley Thornes 1981, pp: 22-26.

[97] Yousefi M, Rahimi-Nasrabadi M, Pourmortazavi SM, Wysokowski M, Wysokowski T, Ehrlich H, Mirsadeghi S. Supercritical fluid extraction of essential oils. Trends in Analytical Chemistry 2019; 118: 182-193. Doi: 10.1016/j.trac.2019.05.038

[98] Ghouila Z, Sehailia M, Chemat F. Vegetable oils and fats: extraction, composition and applications. In Li Y and Chemat F. plant based "Green chemistry 2.0". moving from evolutionary to revolutionary. Springer Nature Singapore Ltd. 2019. Ebook ISBN 978-981-13-3809-0. Doi.org/10.1007/978-981-13-3810-6.

[99] Ogori AF. Source, extraction and constituents of fats and oils. Journal of

Food Science and Nutrition 2020, 6; 060: 1-8. Doi: 10.24966/FSN-1076/100060

[100] Chemat F, Fabiano-Tixier AS, Vian MA, Allaf T, Vorobiev E. Solvent-free extraction of food and natural products. TrAC Trends Anal. Chem 2015: 71: 157-168.

[101] Cakaloglu B, Ozyurt VH, Otles S. Cold press in oil extraction. A review. The Ukrainian Food Journal 2018; 7(4): 640-654.

[102] Virgilio P, Tuomas H, Marjukka K, Risto S, Jussi H, Linnanen L. Sustainability of palm oil production and opportunities for finnish technology and know-how transfer. Lappeenranta University of Technology Press 2009.

[103] Gunstone F. Palm oil supplying much of world demand for fats and oils. INFORM 2001, 12:141.

[104] Gerken HG, Donohoe B, Knoshaug EP. Enzymatic cell wall degradation of *Chlorella vulgaris* and other microalgae for biofuels production. Planta 2013, 237:239-253. https://doi.org/10. 1007/s00425-012-1765-0

[105] Shankar D, Agrawal YC, Sarkar BC, Singh BPN. Enzymatic hydrolysis in conjunction with conventional pretreatments to soybean for enhanced oil availability and recovery. J Am Oil Chem Soc 1997; 74:1543-1547.

[106] Zheng X-Q, Li Q.-S, Xiang L-P, Liang Y-R. Recent advances in volatiles of teas. Molecules 2016, 21, 338.

[107] Yang Z, Baldermann S, Watanabe N. Recent studies of the volatile compounds in tea. Food Res. Int. 2013, 53, 585–599.

[108] Villamar-Torres R, Jazayeri SM, Luiba-Delfini G, Cruzaty LCG, Viot C-R. Volatile organic compounds: plant natural defense mechanisms against herbivorous arthropods and an opportunity for plant breeding of cotton. Scientia Agropecuaria 2018, 9(2): 287-297. Doi: 10.17268/sci. agropecu.2018.01.14.

[109] War AR, Paulraj MG, Ahmad T, Buhroo AA, Hussain B, Ignacimuthu S, Sharma HC. Mechanisms of plant defense against insect herbivores. Plant Signaling and Behavior 2012, 7(10): 1306-1320.

[110] Adamova T, Hradecky J, Panek M. Volatile organic compounds (VOCs) from wood and wood-based panels: methods for evaluation, potential health risks, and mitigation. Polymers 2020; 12(2289): 1-21. Doi: 10.3390/polym12102289

[111] Verde GMV, Barros DA, Oliviera MS, Aquino GLB, Santos DM, De Paula JR, Dias LD, Pinero M, Pereira MM. A green protocol for microwave-assisted extraction of volatile oil terpenes from *Pterodon emarginatus* Vogel. (Fabaceae). Molecules 2018, 23,651: 1-12. Doi: 10.3390/molecules23030651.

[112] Mukherjee PK. Quality control of herbal drugs: an approach to evaluation of botanicals, 1st edition. Business Horizons Pharmaceutical Publishers; India: 2002.

[113] Dermarderosian A. The Review of Natural Product, 1st edition. Fact and comparisons; St. Louis, Missouri: 2001.

[114] Cordell, G. A. Introduction to the Alkaloids: A Biogenetic Approach. Wiley-Interscience, New York, 1981.

[115] Hesse, M. Alkaloids: Nature's Curse or Blessing? Wiley-VCH, Weinheim, Germany, 2002.

[116] Shinde RR and Shinde NH. Extraction of caffeine from Coffee and

preparation of anacin drug. International Journal of Engineering research and Technology 2017; 10(1): 236-239.

[117] Subila S and Sirley N. Determination of Caffeine in different tea samples. IOSR, Journal of Applied Chemistry 2016; 9(1): 75-78.

[118] Chaugule A, Patil H, Pagariya S, Ingle P. Extraction of Caffeine. International Journal of Advanced Research in Chemical Sciences 2019; 6(9): 11-19. DOI : http://dx.doi.org/10.20431/2349-0403.0609002

[119] Tello J, Viguera M, Calvo L. Extraction of caffeine from Robusta coffee (*Coffea canephora* var. Robusta) husks using supercritical carbon dioxide. Journal of Supercritical Fluids 2011; 59: 53-60.

[120] Barbier A. The extraction of Opium alkaloids. United Nations Office on Drugs and Crime 1950; 3: 22-29.

[121] Mehtretter CL and Weakley FB. Extyraction of morphine fron Poppy capsules and its recovery by ion exchange. Journal of the American Pharmaceutical Association (Scientific ed.) 1957; 46(3):193-196.

[122] Kalia AN. Textbook of industrial pharmacognosy, 1st edition. CBS Publishers and Distributors Pvt. Ltd; Noida, India: 2005.

[123] Mohammed A. Pharmacognosy and Phytochemistry, 1st edition. CBS Publishers and Distributors Pvt. Ltd; Noida, India: 2008.

[124] Khoddami A, Wilkes MA, Roberts TH. Techniques for analysis of plant phenolic compounds. Molecules 2013; 18: 2328-2375. Doi:10.3390/molecules18022328

[125] Qing-Wen Z, Li-Gen L, Wen-Cai Y. Techniques for extraction and isolation of natural products: a comprehensive review. Chinese Medecine 2018; 13(20): 1-26. https://doi.org/10.1186/s13020-018-0177-x

[126] Panche AN, Diwan AD, Chandra SR. Flavonoids: An overview. J. Nutr. Sci. 2016; 5: e47.

[127] Rana AC and Gulliya B. Chemistry and Pharmacology of Flavonoids—A Review. Indian J. Pharm. Educ. Res. 2019; 53: 8-20.

[128] Garcia-Salas, P.; Morales-Soto, A.; Segura-Carretero, A.; Fernández-Gutiérrez, A. Phenolic compound-extraction systems for fruit and vegetable samples. Molecules 2010; 15: 8813-8826.

[129] Cowan MM. Plant products as antimicrobial agents. Clin. Microbiol. Rev. 1999; 12: 564-582.

[130] Anokwuru CP, Anyasor GN, Ajibaye O, Fakoya O, Okebugwu P. Effect of extraction solvents on phenolic, flavonoid and antioxidant activities of three nigerian medicinal plants. Nat. Sci. 2011; 9: 53-61.

[131] Bohlin L. Natural Products Isolation, Drug Discovery Today. 1998; 3(12): 536-537.

[132] Hostettmann K, Hostettmann M, Marston A. Saponins: in Terpenoids (Charlwood B. V., Banthorpe D. V., eds.), Methods in Plant Biochemistry 1991. Dey PM and Harborne JB eds., 7, Academic Press, San Diego, CA, pp. 435-471.

[133] Klyne W. The Chemistry of the Steroids. Wiley, New York, 1957.

Chapter 10

Controversy, Adulteration and Substitution: Burning Problems in Ayurveda Practices

Puneshwar Keshari

Abstract

Ayurveda is an Indian traditional system of medicine. In present era, world is looking towards herbal medicine because of acceptability and safety. Medicinal plants constitute an effective source of Ayurvedic and other traditional system of medicines as well as modern medicine. In India, about 80% of the rural population depends on herbal medicines in primary health care level. A large percentage of plants used in herbal industries are subject of controversy. Non-availability of plants, poor understanding and parallel evolved knowledge systems are some of the reasons attributed to it. The existing practices of polynomial nomenclature system of Sanskrit, different perceptions in various communities, vernacular equivalents, all are cumulative factors for controversy, adulteration and substitution. " Sandigdha Dravaya " is a term used for that type of medicinal plants which are mentioned in Ayurvedic classics but their exact botanical source is not known. Adulterants and substitutes are the common practices in herbal raw material trade. Adulteration is a debasement of an article. The motives for intentional adulteration are normally commercial that which involves deterioration, admixture, sophistication, inferiority, spoilage and other unknown reasons. Substitution is a replacement of equivalent drugs in place of original drugs. The principles to select substitute drugs are based on similar Rasa, Guna, Virya, Vipaka and mainly the Karma. At present the adulteration and Substitution of the herbal drugs is the burning problem in herbal industry and in Ayurvedic practices. So it is necessary to develop reliable methodologies for correct identification, standardization and quality assurance of Ayurvedic drugs.

Keywords: Controversy, Substitution, Adulteration, Ayurveda, Pratinidhi dravya

1. Introduction

Ayurveda is an Indian traditional system of medicine. It is a science of life and believed to be prevalent for last 5000 years in Indian Subcontinent. It is one of the most noted systems of medicine in the world [1]. In Ayurvedic system of medicine, treatment is based on *Chikitsa Chatuspad* (Tetra-pod of treatment) and for success of treatment, all these pods most contain special qualities [2]. *Aushadh* (Drugs) is one of the major pod, and for success of treatment potent drugs are the primary requirement. Medicinal plants are the major source of drugs in Ayurveda.

India is one of the world's top 12 mega diversity countries. It has more than one fourth (8000) of the world's known medicinal plant species (30,000), which provides an important source of medicinal raw materials for traditional medicine systems as well as for pharmaceutical industries [3]. Medicinal plants are globally valuable sources of new drugs. There are over 1300 medicinal plants used in Europe, of which 90% are harvested from wild resources; similar figure in India also. Furthermore, up to 80% of people in developing countries are totally dependent on herbal drugs for their primary healthcare, and over 25% of prescribed medicines in developed countries are derived from wild plant species [4]. Due to an increasing demand for medicinal plants and a loss and fragmentation of natural habitats, close to 300 species of Indian medicinal plants have been so far assessed as under threat in the wild. Around 1,000 species are estimated to be facing various degrees of threat across different biogeographic regions in the country [5]. Due to such a high demand and less availability of natural sources and unavailability of crude genuine drugs, practices of substitution and adulteration are increasing day by day. Similarly a large percentage of plants used in herbal industries are subject of controversy. Non-availability of plants, poor understanding and parallel evolved knowledge systems i.e. knowledge of naming of plants by identifying species with partly similar or fully similar properties, inherent qualities of accent and dialects, nonmedical literature describing flora etc. are some of the reasons attributed to it [6].

At present the adulteration and Substitution of the herbal drugs is the burning problem in herbal industry and in Ayurvedic practices. Due to adulteration, faith in herbal drugs has declined and led to one of the greatest drawbacks in promotion of Ayurveda and Herbal products. Adulterants are also creating health hazards or adverse events. Similarly controversy is creating problem for uniformity in standardization and reliability of Ayurvedic products and due to use of substitutions, it is difficult to get the appropriate effects as the genuine drugs could give.

2. Enumeration of controversy, adulteration and substitution

2.1 Controversy and controversial drugs

Controversial drugs or *Sandigdha Dravyas* are those plants which are mentioned in Ayurveda classics but their botanical identification is not clear. The Ayurvedic and Sanskrit literature has described a herb with many synonyms, which do not precisely indicate the botanical source but many a times attribute to therapeutic utility of the plant [7]. For a single herb various synonyms are mentioned in Ayurvedic lexicons on the basis of morphology, habitat, origin, therapeutic uses etc. by using different similes which are leading causes of controversy. Quantum of information gained from Ayurvedic and other Sanskrit literature revealed various incidences where on common vernacular name is used for two or more entirely different plant species in Ayurvedic and other traditional system of medicines [7] e.g. *Amrita* is used both for *Tinospora cordifolia,* and *Terminalia chebula* which are totally different drugs. Synonyms of herbs are also given according to the local languages. India is a country having a variety of languages and population dependent on different tribal and folklore medicine. Sometimes this is also responsible for confusion in the nomenclature of different plants having similar name.

2.2 Causes of controversy

2.2.1 Mistake done during copying of manuscripts

In past there was no printing machine, Acharyas had written the manuscript-manually in *Bhurja-Patra* or *Taalpatra* or other substances. During copying of these manuscripts by editors or translators, mistakes might have occurred, which ultimately created controversy.

Single synonym given for multiple plants- In Ayurvedic lexicons single synonym is used for two or more than two herbs which are totally different in morphology which creates controversy. These types of practices come in existence mainly during *Nighantu* periods e.g. *Amrita* is used for both *Tinospora cordifolia* (willd.) Miers ex Hook & Thoms and *Terminalia chebula* Retz [8–10].

Geographical variation- India is a countryof multi diversity having high Himalayas to sea level area and world highest rainy area to Thar Desert. Every area has its own types of plant diversity, the plant which found in northern India mayn't found in southern part. So due to unavailability of those species another species are used for the same purpose, which ultimately creates controversy. For example *Convulvulus microphyllus* Sieb. ex Spreng is used by the name of *Sankhpushpi* in north India but due to geographical variation, it is not available in southern part and there *Clitoria ternatea* Linn. is used [8–10].

S. No.	Sanskrit Name of the drugs	Botanical Sources
1.	Brahmi	1. *Bacopa monnieri* (L.) Pennel (Scrophulariaceae) 2. *Centella asiatica* (L) urban (Apiaceae)
2.	Jeevanti	1. *Leptadenia reticulata* Wight and Arn. (Asclepiadaceae) 2. *Desmotrichum fimbriatum* Bl. Bidr (Orchiaceae) 3. *Cimicifuga foetida* Linn (Ranunculaceae)
3.	Shankhapushpi	1. *Convolvulus pluricaulis* Choisy (Convulvulaceae), 2. *Evolvulus alsinoides* (Convulvulaceae), 3. *Canscora decussate* Schult (Gentianaceae), 4. *Clitorea ternatea* Linn.(Papilonaceae).
4.	Daruharidra	1. *Berberis aristata* DC (Berberidaceae), 2. *Coscinium fenestratum* (Gaertn.) Colebr. (Menispermaceae),
5.	Rasana	1. *Vanda tessellata* Loud and Loud (Orchidaceae), 2. *Alpinia galanga* (L.) Willd (Scitaminaceae), 3. *Pleuchea lanceolata* C.B.Clarke. (Compositae) 4. *Viscum album* (Loranthaceae), 5. *Withania coagulens* (Stocks) Dunal (Solanaceae), 6. *Aristolochia indica* L.(Aristolochiaceae) 7. *Inula racemosa* Hook.f. (Asteraceae) 8. *Rauwolfia serpentine* (L.) Benth. ex Kurz (Apocynaceae), 9. *Lochnera rosea* (Apocynaceae) 10. Enicostemma littorale Blume (*E. littorale*) (Gentianaceae)
6.	Talishpatra	1. *Abies webbiana* Lindl.(Pinaceae) 2. *Taxus baccata* Linn.(Pinaceae) 3. *Rhododendron anthopogon* D. Don.(Ericaceae)
7.	Pashanabheda	1. *Aerva javanica* Juss. (Amarantaceae) 2. *Ammania baccifera* Linn. (Lythraceae) 3. *Bergenia ligulata* Wall (Saxifragaceae) 4. *Bryophyllum pinnatum* (Lam.)Kurz. (Crassulaceae) 5. *Coleus aromaticus* Benth. (Lamiaceae) 6. *Rotula aquatica* Lour.(Boraginaceae) 7. *Bridelia montana* (Roxb.) Willd. (Euphorbiaceae) 8. *Homania riparia* (Euphorbiaceae) 9. *Ocimum basillicum* L.(Lamiaceae)

Table 1.
List of some controversial drugs.

Poor understanding of Sanskrit word in different context- Ayurvedic classics are mainly written in Sanskrit language [8–10]. Same word in different contexts give different meaning, and due to poor understanding of this type of words by commentator further creates controversy, for example *Pippala* denotes *Bodhivriksha* when used in malegender and the same in female gender denotes long pepper.

Substitute leading controversy- Due to non-availability or high cost in the market, there are chances of substitution of drugs. If this practice continues for long time the original identity of a plant may become obscure and the substitute will be considered as the original, which ultimately creates controversy later on. For example-*Pashanbheda* is used as urolithiasis (*Ashmaribhedana*) as the name indicates, so drugs like *Bryophyllum pinnata* (Patharchuda), *Aerva lanata Juss* etc. are used by name of *Pashanbheda*. But originally *Bergenia ligualata* (Wall.) Engle is identified as the source of *Pashanbheda* [8–10].

Parallel evolving knowledge system- Identifying species and naming them with partly similar or fully similar properties, inherent qualities of accent and dialects may create controversy. For example *Brahmi* is mentioned in Ayurveda calssics as brain tonic. *Mandukparni* is another drug mentioned as *Medhya Rasayan* (braintonic) in *Charaka Samhita*. *Bacopa monnieri* [8–10].

(L.) Pennel. is source for *Brahm*i but in North India *Centella asiatica* (L) urban (*Mandookparni*) is called as Brahmi due to similarity in therapeutic effects.

Vernacular names- Somewhere same name is used in different languages but having different meaning and identity which is also a cause of controversy, e.g. M*atala* in Tamil refers to *Punica granatum L*inn. Where as in Kannada it pertains to *Citrus medica* [8–10].

Non Ayurvedic literature also creates controversy- In poetry *Kamala, Utpala, Kumuda, Kalhara* all are referred as same plant lotus but botanically they are different species [8–10].

Polynomial nomenclature- Multiple names for single plant are given in Ayurvedic lexicons. This type of trends aroused during *Nighantu* Period. Different *Nighantu* written by different authors gave multiple names for a single drug especially for better understanding about the drug but they created controversy later on [8–10] (**Table 1**).

3. Adulteration

Adulteration is a practice of substituting original crude drug partially or wholly with other similar looking substances but later is either free from or inferior in chemical and therapeutic properties. In simple word, it is debasement of an article [1]. On the basis of motive; adulteration is intentional or direct and accidental or indirect adulteration. Direct or intentional adulteration is mainly done for commercial benefits [11]. Deterioration, Admixture, Sophistication, Substitution, Inferiority and Spoilage are methods of adulteration. Intentional impairment in the quality of drug is Deterioration. Addition or mixing one substance to another accidentally or carelessly or due to ignorance is Admixture. It is a type of unintentional adulteration. Sophistication is the intentional or deliberate type of adulteration in which some totally different substance is added in place of genuine drug while Inferiority refers to adding of any substandard drug, and Spoilage is due to the attack of microorganisms or parasitic infestation [1].

Deterioration, Admixture, Sophistication, Substitution, Inferiority and Spoilage are methods of adulteration. Intentional impairment in the quality of drug is Deterioration. Addition or mixing one substance to another accidentally or carelessly or due to ignorance is admixture.

It is a type of unintentional adulteration. Sophistication is the intentional or deliberate type of adulteration in which some totally different substance is added in place of genuine drug while Inferiority refers to adding of any substandard drug, and Spoilage is due to the attack of microorganisms or parasitic infestation [1].

3.1 Major intentional types of adulteration

- Substitution with substandard commercial varieties [1, 12, 13]

- This is the most common type of adulteration in which low standard drugs are mixed which are morphologically, chemically and therapeutically resembles to the original crude drugs, for example Arabian senna is used instead of Indian senna.

- Using superficially similar inferior drugs – In this type of adulteration adulterants are superficially similar in appearance but may or may not having any chemical or therapeutic value as the original crude drugs. For example papaya seeds are adulterated with *Piper nigrum*, saffron is admixed with dried flowers of *Carthamus tinctorious*.

- Using artificially manufactured substance – In this type of adulteration artificially manufactured substances resemble to original crude drugs, are adulterated. This type of adulteration is done for costlier drugs. For example Calcium carbonate compounds are used by name of *Vansha lochan*.

- Using exhausted drug - This type of adulteration is usually done for those drugs which contain volatile oil, for examples fennel, clove, coriander, caraway etc. In this type same crude drug is adulterated but after extracting major chemical constituents, e.g. Volatile oil is extracted from bud of *Lavanga* (clove) and exhausted buds are adulterated. In this case sometimes extra additives are used to make the exhausted drugs attractive.

- Using synthetic chemicals to enhance natural character- Synthetic chemicals are used as adulterant which enhances the natural characteristics of original drug, for example, Citral is added in citrus oil like oil of lemon or orange oil.

- Presence of vegetative matter of same plant- Instead of proper used parts of crude drugs other parts of same species or miniature species grown around the large species are mixed with genuine crude drugs. For example instead of *Moola* (root) of

- *Bala* (*Sida cardifolia*) stem or whole parts of plant is used. This type of adulteration occurs in both intentional and unintentional adulteration.

- Harmful adulterants - For increasing weight of crude drugs for commercial profit, some harmful substances are added with genuine crude drugs, for example stone pieces and sand particles mixed in *Guggulu* (gum of *Commiphora mukul*).

- Adulteration of powders- The drugs which are commonly found in powder forms are adulterated with powder of other substances resembling the same, examples are dextrin in ipecacuanha and *Kampillak* (*Malotous phillipinensis*) powder is adulterated with Annatto dye (*Bixa orellana* Linn.) [1, 12, 13].

3.2 Reason for adulteration

1. Intentional adulteration is done mainly due to commercial benefits, when there is high demand but less availability of drugs.

2. Unintentional adulteration is done due to following reasons [14, 15]

    - Confusion in vernacular names- e.g. *Aerva lanata* (source of *Pashanbheda* in south) adulterated as *Bergenia ligualata.*

    - Lack of knowledge about authentic source e.g. *Calophyllum inophyllum* is adultered with *Mesua ferra.*

    - Similarity in color and morphology – For example *Mucuna utilis* and *Mucuna deeringiana* are used for *Mucuna pruriens*

    - Careless collection/improper collection – Definite part of herb should be collected in particular season, particular place and particular part of plant should be collected but ignorance of these things during collection and drugs collected carelessly may cause adulteration.

    - Improper storage- Due to improper storage physical factors such as air (oxygen), humidity, light, and temperature can bring about deterioration directly or indirectly and use of such type of drug acts as adulterant.

    - Imperfect preparation- Some of crude drugs should be processed before marketing, during such processing improper technique may destroy active constituents e.g. over drying of crude drugs, removal of cork from zinger etc. (**Table 2**).

S.No.	Genuine drugs	Adulternats
1.	Mussabar(*Aloe barbadensis*)	Black catechu (*Acacia catechu*)
2.	Nagkeshara (*Mesua ferrea*)	Buds of *Mammea suriga* and *Calophyllum inophyllum*
3.	Punarnawa (*Boerhavia diffusa*)	*Trianthema portulachastrum*
4.	Ashoka (*Saraca indica*)	*Polyalthia longifoia*
5.	Kutaja (*Holarrhena antidysenterica*)	*Wrightia tinctoria, Wrightia tomentosa*
6.	Guggulu (Gum of *Commiphora wightii*)	Gum resin of *Boswellia serrata, Hymenodictyon excelsura*
7.	Bol or Myrrh (*Commiphora myrrha*)	Gum of *Commiphora wightii*

Table 2.
List of few commonly used adulterants in Ayurveda.

4. Substitution

Substitution is a replacement of equivalent drugs in place of original drugs on the basis of similar pharmacological actions and therapeutic uses. In Ayurveda, substitution is described by the name of *Abhava Pratinidhi Dravya*. During *Samhita* Period concept of adulteration and substitution was not existed but later on this practices come in existence. But *Vagbhatta* has mentioned that the *dravya* having similar *Ras* (taste), *Guna*(characteristic), *Veerya* (potency) and *Vipaka* should be used in absence of each other. So *Abhava Pratinidhi Dravya* is a replacement of original drug basically having similar *Rasa, Guna, Veerya, Vipak* and mostly on *Karma*. Description of *Abahva Pratinidhi Dravyas* are mentioned in *Bhavaprakash Nighantu*, *Yogratnakar* and *Bhaishajya-Ratnawali* [16–18]. There are 47 drugs of plant origin (*Sthavar Dravya*), 2 drugs of animal origin (*Jangam Dravya*), 7 drugs Minerals- Metals origin (*Bhoumya Dravya*) and 5 food materials (*Ahara Dravya*) mentioned for *Abhava Pratinidhi Dravya* in *Bhavaprakash Nighantu* [9, 16].

5. Need for substitution

- **Non-availability of the drug**: Some drugs mentioned in Ayurvedic lexicon are not available nowadays, so those drugs are substituted by other drugs having similar therapeutic value [9, 10, 12, 13, 15]. For example most of drugs from *Astavarga* are not easily available so those drugs are substituted by other ones e.g. *Meda* and *Mahameda* are substituted by *Shatavari*.

    - **Uncertain identity of the drug**: The drugs which are mentioned in Ayurvedic classics but their botanical identity is not clear those are substituted by known one e.g. for the herb *Lakshmana,* different species such as *Arlia quinquefolia, Ipomea sepiaria* etc. are considered.

- **Cost of the drug**: *Kumkuma* (*Crocus sativus*) is more costly so it is substituted by less expensive *Kusumbha* (*Carthamus tinctorius* Linn.).

- **Geographical distribution of the drug**: *Rasna (Pluchea lanceolata*) is used in Northern India while in southern parts *Alpinia galanga* is used as Rasana and *Vanda roxburghii* is considered as source in Bengal.

- **The adverse reaction of the drug**: *Vasa* (*Adhatoda vasica*) is good *Rakta-Pittahara* (antihaemorrhagic) drug, but having abortificiant activity, so instead of this drug *Laksha* (*Lacifer lacca*), *Ashoka* (*Saraca asoka*) etc. are used in pregnant women for the same purpose.

- **Seasonal availability of drug**- *Punarnawa* (*Boerhaavia diffusa*) is commonly not found throughout the year so for that *Trianthema portulacastrum* (*Varshabhu*) can be used as substitute, which is found throughout the year [9, 10, 12, 13, 15].

6. Types of substitution

- **Substitution with totally different drug** - Use of *Danti* (*Baliospermum montanum)* as a substitute of *Chitraka* (*Plumbgo zelenycum*) [9, 10, 12, 13, 15, 19].

- **Substitution of species belonging in same family** – *Datura metal* is substituted by *Datura stamonium*.

- **Using different species having common Sanskrit name** - Two types of *Gokshura* are used, they are *Tribulus terrestris* (Laghu Gokshur) and *Padalium murex* (*Brihat Gokshura*).

S.No.	Main drugs	Substitutes
1.	*Plumbago zeylanicum* Linn	*Baliospermum montanum* Muell
2.	*Valeriana wallichii* D C	*Saussurea lappa* C B Clarke
3.	Punarnawa (*Boerhavia diffusa*)	*Trianthema portulachastrum*
4.	Ashoka (*Saraca indica*)	*Polyalthia longifoia*
5.	*Marsedenia tenacissima* W	*Odina woodier* Roxb.
6.	*Clerodendrum serratum* Spreng	*Solanum xanthocarpum* Schrad & Wendl
7.	*Piper cubeba* Linn.f.	*Cyperus rotundus* Linn.

Table 3.
List of few examples of substitute drugs (herbs) [20, 21].

- **Using different parts of same plant** – Instead of root of *Sida cordifolia* whole plants of *Sida cordifolia* is used.
- **Due to similar action**- Aamalki (Embelica officinalis) is taken instead of Bhallatak (Semicarpus anacardium) for Rasayan *karma* (rejunaive action) (**Table 3**) [20, 21].

7. A case of substitution in Nepal

Rohitaka is mentioned in almost all classics of Ayurveda such as Brihattrayee (Charaka Samhita, Sushruta Samhita, and Ashtanga Hridya) and other lexicons such as Sharangdhara Samhita, Bhavaprakasha Nighantu, and Yoga Ratnakar. It is also mentioned in maximum numbers of Nighantu (Ayurvedic lexicons). It is mentioned as Yakritpleehagulmodar Roga-hara (useful in liver diseases, spleen disorders, and abdominal lumps). Tecomella undulata (Sm.) Seem. from Bignoniaceae family is a genuine source of Rohitaka. It is commonly known as "Rohida" or "Desert Teak" and an important deciduous, ornamental, and medicinal tree [19].

In crude drug market of Nepal *Rhododendron arboreum* Sm. is sold by the name of Rohitaka and considered as substitute of *Tecomella undulata* (Sm.) Seem. The Nepali name of Rohitaka is given as Guransa in Chandra Nighantu which is a hand written famous manuscript in Nepal and kept in Singha Darbar Vaidyakhana Vikas Samiti (SDVKVS) in Kathmandu. In this manuscript manually drawn picture of Rohitaka is given which is Guransa and is botanically identified as *Rhododendron arboretum* Sm. Traditional practitioners use this drug for liver disorders like Jaundice (Kamala), hepatitis, hepatomegaly etc. Leaves, flowers, bark are used for various purposes traditionally and in Ayurvedic practices [19].

The genuine source of Rohitaka is identified as *T. undulata* (Sm.) Seem. However, due to non-availability of the genuine source, various other drugs are used as substitute. Stem bark of *R. arboreum* Sm. is commonly found by the name of Rohitaka in herbal raw drugs trade in Nepal. The rationality behind selection of Pratinidhi Dravya (substitute drug) isbased on similarity in Rasa, Guna, Veerya, and Vipaka with that of original drug. On organoleptic evaluation, both drugs are bitter in taste substantiated the similarity in Rasa. The pharmacognostic and analytical studies have confirmed the genuinity and purity of both the drugs (*R. arboreum* Sm. and *T. undulata* [Sm.] Seem.). Both the drugs have few common phytochemicals such as carbohydrates, alkaloids, tannins, and phenols which are responsible for their pharmacological actions. Chromatographic study showed the presence of 18 phytoconstituents in *R. arboreum* and 24 phytoconstituents in *T. undulata* extracts, and among them, three phytoconstituents having Rf of 0.30. 0.45, 0.66 are common [22].

8. Discussion

Controversy, Adulteration and Substitution are interrelated with each other. Substitution practices if exists for long time the original identity of a plant may become obscure and the substitute will be considered as the original, leading to create controversy. Nonavailability and high market price of crude drugs led to adulteration. Similarly controversy about authentic botanical source of medicinal plants dealt in classical Ayurveda texts led a cause of substitution because of lack of proper authentication, the drugs having similar morphology or similar therapeutic effects might be practiced. Controversy, adulteration and substitution create problems for standardization of Ayurvedic practices and herbal products. Substitution of genuine drug is need of time because of unavailability of genuine drugs due to deforestation, global warming, lack of adequate cultivation practices etc. Although substitution should be only done for endangered and red listed plants and the major constituent of a preparation should not be substituted. The rational substitution in Ayurveda is based on similarities in *Guna* of both the drugs and not on inferior qualities. It should be properly validated in contemporary context using both Ayurvedic principles and Modern Scientific tools. World Health Organization (WHO), in its publication on quality standards for medicinal plant materials, recommends rejecting any batch of raw material, which has more than 5% of any other plant part of the same plant (e.g. stem in leaf drugs), never the less if they are derived from the authentic plant. Based on these standards, adulteration whether, intentional or unintentional, should be rejected. Collectors, suppliers and traders should be educated for authentic sources of drugs. Intentional adulteration should be discouraged by strictly implementing the regulatory laws. Due to adulteration faith in Ayurvedic practices and drugs has declined and adulteration in market samples is the greatest drawback in promotion of herbal drugs. So for quality, safety and standardization purpose of Ayurvedic products and practices the problem related with controversy, substitution and adulteration of drugs should be resolved for its worldwide acceptance.

The prime factor for resolution of controversy is the proper authentication of botanical source of medicinal herbs mentioned in classics, for this, literature review, ethno-botanical survey, medicinal plants survey and drug evaluation (morphological, microscopic, chemical, physical and biological evaluation) should be done. Similarly for determination and detection of adulteration, various steps of drug evaluation should be applied. Substitution of drugs should only suggested when therapeutic efficacy of substituted drug is similar to original one.

The uniformity in selection of crude drugs for pharmaceutical preparations and practices should maintain the standardization of Ayurvedic products and for this Ayurvedic pharmacopeia of India (API) and Ayurvedic formulary of India (AFI) is playing a vital role, so maximum number of drugs mentioned in classics and practiced traditionally should be incorporated in API and AFI.

9. Conclusion

- Controversy about drugs is mainly due to polynomial system of nomenclature in classical texts.

- *Naama-Roopa* (nomenclature and morphology) of drugs are clear in *Samhitas*, controversy aroused mainly due to *Nirukti* (basonyms)and *Paryaya* (synonyms) are given by different *Nighantus*.

- Proper identification of original botanical source is even a great problem till date.

- Adulteration and Substitution are different. The most essential criteria for substitution is the pharmacological activity rather than morphology or phytoconstituents.

- Substitution of the herbs is the need of the hour with more than 300 medicinal plants becoming red listed.

- Adulteration is a malpractice not only done intentionally but accidentally due to involvement of untrained personnel in collection and trade.

- Controversy about authentic botanical source of medicinal plants dealt in classical Ayurvedic texts and problem regarding substitution and adulteration should be resolved by integrated research and those sources should be validated which have more potency for described pharmacological activities.

Author details

Puneshwar Keshari
Department of Ayurveda and Alternative Medicine, Ministry of Health and Population, Government of Nepal, Teku, Kathmandu, Nepal

*Address all correspondence to: dr.puneshwarkeshari@gmail.com

References

[1] Kokate C K, Purohit A P and Gokhele S B: Pharmacognosy. Chapter-7, 15th edition. Pune:Nirali Prakashan, 2014.

[2] Agnivesha, Charaka samhita, Sutra Sthan, chapter 9th, Khuddakchatush-padiyam Adhyaya, Shloka no.2, Jaydev Vidyalankara, editor. 9th edition. Vol. 1. Varanasi: Chaukhamba Surbharati Prakashan; Reprint1999. p.74

[3] Bhattacharyya R, Bhattacharya S, Chaudhuri S. Bhattacharyya R, Bhattacharya S, Chaudhuri S. Conservation and documentation of the medicinal plant resources of India. Biodiversity and Conservation 2006;15:2705-2717

[4] Shi-Lin Chen, Hua Yu, Hong-Mei Luo, Qiong Wu, Chun-Fang Li, André Steinmetz. Conservation and sustainable use of medicinal plants: problems, progress, and prospects. Chin Med 2016; 11(37):1- 10

[5] WWF. Sustainable Harvest of Medicinal Plants in India. FRLHT. Februar 2007 (www.wwf.de/fileadmin/.../HG_Sustainable_Harvest_of_Medicinal_Plants_in_India.pd.)

[6] www.shodhganga.inflibnet.ac.in/bitstrea m/10603/6813/11/11_chapter%203.pdf Dixit VK. Controversial ayurvedic herbs.J Adv Pharm Tech Res 2011;2:78-80.

[7] Sharma. P.V. Dravya-Guna Vigyan. Vol.5. Varanasi: Chaukhambha Bharati Academy, Reprint 2014.

[8] Vaghela B, Soni H, Shukla L. A Concept of Herbal Pratinidhi Dravyas (Substitute drugs) In Ayurved. Pharmagene 2013;1(3):85-88

[9] Pravin R. Joshi, Bhupesh R. Patel, Vinay J. Shukla. An overview on the substitution of drugs in Ayurveda and their evaluation methods. AYU. Oct-Dec. 2012; 33(4): 481-484

[10] Poornima B, Adulteration and substitution in herbal drugs a critical analysis. IJRAP 2010; 1(1): 8-12.

[11] Pawan Kumar Sagar. Adulteration and Substitution in Endangered, ASU Herbal Medicinal Plants of India, their Legal Status, Scientific Screening of Active Phytochemical Constituents. IJPSR, 2014; Vol. 5(9): 4023-4039

[12] Dr. Poonam. Adulteration of crude drugs burning problem. International Journal of Applied Research 2016; 2(2): 99-101

[13] Mitra S K, Kannan R. A Note on Unintentional Adulterations in Ayurvedic Herbs. Ethnobotanical Leaflets 2007; 11:11-15. (https://www.researchgate.net/publication/27654557)

[14] Om Prakash, Jyoti, Amit Kuma, Pavan Kuma, Niranjan Kumar Manna. Adulteration and Substitution in Indian Medicinal Plants: An Overview. Journal of Medicinal Plants 2013;1(4): 127-132

[15] Sharma PC, Yelne MB, Dennis TJ. Editors. Database on Medicinal Plants used in Ayurveda, Vol 1. New Delhi: CentralCouncil for Research in Ayurveda and Siddha, Reprint 2002

[16] Bhava Mishra. Bhavaprakash Nighantu, Vidyotini Hindi Commentary, Mishraka Varga, 6th Chapter, shloka no.138- 168. Mishra B, Shankar. editor. 10th edition. Vol.1, Varanasi: Chaukhamba Sanskrit sansthan 2002. pp. 181-183

[17] Anonymous. Yogratnakar, Vidyotini Hindi commentary, Shri Lakshmipati Shastri Commentator. Brahmashankar Shastri. editor.Reprint, 2015, pp. 171

[18] Shastri Ambikadatta "Baishajya Ratnavali" 18th Edition. Varanasi: Chaukhambha Sanskrit Sansthan 2005.

[19] Keshari P, Pradeep. Pharmacognostical and chromatographic evaluation of market sample of Rhododendron arboreum stem bark as a source plant for Rohitaka in Nepal. Journal of Pharmacognosy and Phytochemistry. 2017; 6(5): 296-306.

[20] Sharma PC, Yelne MB, Dennis TJ. Editors. Database on Medicinal Plants used in Ayurveda, Vol 2. New Delhi: Central Council for Research in Ayurveda and Siddha, Reprint 2005.

[21] Prachi A. Khaire, Prashant B. Nandwate. Substitutes for Ayurvedic medicinal herbs: A Review. International Journal of Recent and Futuristic Ayurveda Science 2017; 2 (1): 107-114

[22] Keshari P, Pradeep. Comparative pharmacognostic evaluation of Tecomella undulata and Rhododendron arboreum as two different sources of Rohitaka. International Journal of Green Pharmacy. 2018; 12(04): 242-252

Chapter 11

Phytochemical Profile and Antiobesity Potential of *Momordica charantia* Linn.

Pushpa Anantrao Karale, Shashikant Dhawale and Mahesh Karale

Abstract

Momordica charantia L. is growing in many tropical and subtropical regions; the fruits of bitter melon are also gradually becoming popular for treating diabetes and associated diseases. Over 248 compounds belonging to the lipids, phenolics and terpenoids class are reported by diverse studies. However, *M. charantia* L. appears to be an inimitable species that synthesizes a diverse range of natural products in the fruits, leaves, stems and roots. The cucurbitane types of triterpenes exist in the various tissues of the plant in their aglycone as well as glycosylated forms. The bitter melon seems to exert their lipid lowering and antiobesity effects via several mechanisms like PPARs, LXRs, SREBPs, and Sirts mediated fat metabolism in various tissues, prevent adipocyte hypertrophy and visceral fat accumulation. *M. charantia* L. has been comprehensively studied worldwide for its therapeutic properties to treat a number of diseases like diabetes, dyslipidaemia, obesity, and certain cancers. This chapter apparently displays an encompassing literature review on vast potential of bitter melon as antiobesity agent and assembles data on complex phytochemistry.

Keywords: obesity, phytochemicals, bitter melon, cucurbitane type terpenoids

1. Introduction

Momordica charantia L. or Bitter Melon, also known as balsam pear or Karela, is a vegetable and common food in Indian cuisine and has been used comprehensively in folk medicine. *Momordica charantia* L. is tropical or subtropical creeping belonging to family *Cucurbitaceae* and widely used as medicinal herb from ancient time (**Figure 1**). The Latin name *Momordica* means "to bite" referring to the serrated edges of the leaf, which appear as if they have been bitten. The major regions of *M. charantia* L. cultivation are Asia including China, India, Sri Lanka and Thailand, central and South America and North America [1]. In Ayurveda, the fruit is considered as tonic, stomachic, stimulant, emetic, antibilous, laxative and alterative. Bitter melon has been used in various Asian traditional medicine systems for a long time. It is well recognized, the plant is extensively in use in the Chinese, Indian Ayurvedic, and Indonesian systems of medicines as well as in Japan [2].

The therapeutic significance of the plant is symbolized by the fruits which contain about half a dozen seeds per gram of the fresh fruit. As the name implies, the fruits are bitter and bitterness enhance with the level of maturity and hence

Figure 1.
The image describes plant parts of M. charantia *L. unripe fruit; ripe fruit and seeds of plant.*

earlier harvesting required to battle bitter taste. The leaves and young shoots of bitter melon recognized to be used in traditional medicine as an herbal tea. The range of pharmacological activities reported for bitter melon is rapidly increasing in recent years and its claimed uses and potential applications for cancer and other diseases have been extensively reviewed. Likewise, the range of medicinal claims range from diabetes, hypertension, obesity, cancer, as well as AIDS. *M. charantia* L. possesses various beneficial effects, including anti-cancer, anti-viral, antioxidant, antiulcer, anti-obesity, anti-HIV, cytotoxic, anti-inflammatory, reduction of cholesterol, inhibition of protein tyrosine phosphatase 1B, and anti-osteoporosis [3]. This chapter aims to highlight the complex phytochemistry and extensive review on antiobesity potential of bitter melon with possible targets.

2. Plant profile

2.1 Botanical description

Cucurbitaceae family is known to comprise some 101 accepted genera and the genus Momordica L. itself comprises of some 50 accepted species within the family [4]. *M. charantia* L. was known previously by several synonyms including the *Cucumis argyi* H. Lev., *Cucumis intermedius* M. Roem., *Momordica anthelmintica* Schumach. & Thonn., *Momordica charantia subsp. abbreviata* (Ser.) Greb., *Momordica charantia* var. *abbreviata* Ser., *Momordica charantia* f. *muricata* (Willd.)., *Momordica charantia* var. *muricata* (Willd.) Chakrav., *Momordica charantia* var. *pseudobalsamina* Griseb., *Momordica charantia* var. *zeylanica* Hitchc., *Momordica elegans* Salisb., *Momordica indica* L., *Momordica jagorana* K.Koch., *Momordica muricata* Willd., *Momordica papillosa* Peckolt ex Rosenthal., *Momordica roxburghiana* G. Don., *Momordica senegalensis* Lam., *Momordica thollonii* Cogn., *Momordica zeylanica* Mill. The botanical description of different parts of the plant demonstrated in **Table 1**. The taxonomic hierarchy of bitter melon within the plant kingdom is as follow:

Kingdom: Plantae
Subkingdom: Viridiplantae
Superdivision: Embryophyta
Division: Tracheophyta
Subdivision: Spermatophytina
Class: Magnoliopsida
Super order: Rosanae
Order: Cucurbitales

Plant parts	Description
Leaves	Broadly ovate to orbicular in outline, cordate, narrowly decurrent on to petiole, sparsely pubescent to densely villous on veins beneath, deeply palmately 3–7-lobed, lobes variously sinuate-dentate or lobulate. Leaf lamina 10 × 12.5 cm.
Flowers	Flowers are monoecious and solitary. **Male flowers:** peduncle 0.3–5 cm long; bract 2–17 mm long, broadly ovate or reniform, sessile, cordate, amplexicaul, pedicel 2–9.5 cm long. Receptacle-tube 1–5 mm long; lobes 3–7 mm long, ovate-lanceolate. Petals 1.0–2.5 cm long, pale to deep yellow, ovate to obovate. **Female flowers:** peduncle 0.2–5 cm long; bract 1–12 mm long; pedicel 1–10 cm long; ovary 8–11 × 2–4 mm, ovoid-rostrate to fusi-form, ridged, pilose on ridges, tuberculate; receptacle-tube 1–3 mm long, lobes 2–5 mm long, lanceolate; petals 0.7–1.2 cm long.
Fruits	Fruit 2.5–4.8 × 1.5–2.3 cm, ovoid-rostrate or ellipsoid, longitudinally ribbed and tuberculate, bright orange-red, dehiscent into 3 valves; fruit-stalk 3.4–15 cm long.
Seeds	Seeds 8–11, 4.5–8 × 2–3.5 mm, enveloped in sticky red pulp, ovate-elliptic to oblong in outline; faces flattened, sculptured, with sinuate edges; margins grooved.
Petiole	Petiole 0.5–7 cm long.

Table 1.
Botanical description of Momordica charantia *Linn.*

Family: Cucurbitaceae
Genus: Momordica Linn.
Species: *Momordica charantia* L.-balsampear

2.2 Phytochemistry

The main constituents of bitter melon are triterpenoids, saponins, protein, polysaccharides, steroid, alkaloid, lipid, and phenolic compounds. Several bioactive compounds of *M. charantia* L. have been recorded and the literature shown that these were responsible for various pharmacological effects as depicted in **Table 2** [5].

2.2.1 Triterpenoids

The most abundant phytochemical components of bitter melon fruits are the triterpenoids class of secondary metabolites, and are well-known for their bitterness

Sr. No.	Bioactive compounds	Distribution	Pharmacological effects
1	Triterpenoids	Leaves, stem, fruits	Chemo protective, anticancer, antioxidant, antidiabetic
2	Peptides and proteins	Seed	Antiviral, anti-tumor, immune suppressant, antimicrobial
3	Phenolics	Fruit and seed	Antioxidant, anti-inflammatory, immunostimulant
4	Saponin	Fruit, root, seed	Antihyperglycemic, hypolipidemic, antiviral, bacteriostatic
5	Polysaccharide	All parts of plant	Antioxidant, antidiabetic, immune enhancement, neuroprotective, antitumor
6	Lipid	Seed	Anti-tumor, antioxidant
7	Steroids	Fruit and pericarp	Antimicrobial

Table 2.
Bioactive components of Momordica charantia *Linn. And their pharmacological effects.*

and toxicity. These were divided in two types primarily the cucurbitane-type and to a less extent oleane type which may occur either in their glycosylated or aglycone forms. The sugar monomers as β-D-glucopyranosyl, β-D-allopyranosyl, β-D-xylopyranosyl occur in cucurbitane-type triterpenes either by their own as O-linked glycosides, or in different combinations as disaccharides or polysaccharides. The rare glycoside in these compounds was the 3-keto-glucoside [6]. An extremely large and certainly exhibited 193 number of cucurbitane-type triterpenes isolated from bitter melon having various pharmacological effects (**Table 3**) [7]. The fruits are predominance source of terpenoids with great deal of structural diversity but the leaves, stems and roots have also been shown to be good sources of these compounds [8].

Momordicosides A and B were isolated firstly from the seeds of bitter melon fruits; While, Momordicosides C, D and E were isolated as minor components of the seeds [9]. The study on the fruits of *M. charantia* L. further added a Momordicosides F1, F2, G, I, K and L as novel compounds in previous one [10]. The Momordicosides I and Momordicosides M along with other compounds have also been isolated from the fruits of bitter melon [11]. Further additions of Momordicosides M-O as a new compound were confirmed by chemical examination of the fresh fruits [12]. The Vietnamese origin dried fruits of *M. charantia* had shown existence of an auxiliary three pioneering Momordicosides U, V and W. The cucurbitane-type triterpenes which are known to be responsible for the bitterness of the leaves and vines of the *M. charantia* L. are Momordicines. The Momordicine IV and the malonyl derivative of Momordicine II, Momordicine V was readily available in the leaves of bitter melon. The isolation of Momordicines VI, VII, and VIII were first confirmed from the stems and leaves of bitter melon [13].

The Goyaglycosides-a, −b, −c, −d, −e, −f, −g, and -h were isolated from the fresh fruits of Japanese *M. charantia* L. A novel compound Goyaglycoside I was an additional triterpene isolated from the immature fruit of bitter melon [14]. The novel Cucurbitane-type triterpene called karavilagenins A-C and five new triterpene glycosides called karavilosides I-V were isolated from the dried fruit of *M. charantia* L. [15]. The three novel compounds that they names charantosides A, B and C were obtained from fruits of bitter melon. The methanol extracts of the fruits of Japanese *M. charantia* L. shown charantosides I, II, III-VI, VII and VIII, however charantoside IX and X were other novel compounds reported by the studies [16].

Another group of interesting triterpenoids are those known by their trivial names kuguacins. Kuguacins A-E and Kuguacins F-S were isolated from the roots and the leaves of bitter melon plant respectively [17]. Kuguacins II-VI was novel compounds isolated together with various other known compounds from the fruit of *M. charantia* L. From the aqueous ethanolic extracts of fresh fruits isolated eight novel cucurbitane-type glycosides that they named kuguasaponins A-H. The ethanolic extract of fruits of *M. charantia* L. identified 15 cucurbitane-type triterpene glycosides including 4 new compounds, kuguaosides A-D [18].

2.2.2 Flavonoids and phenolic compounds

A number of phenolic compounds with many biological activities have been isolated from bitter melon including, coumaric, caffeic, and ferulic acids as well as the caffeic acid ester, chlorogenic acid, Benzoic, gallic and gentisic acid. The major flavonoids and phenolic acids in the dried leaves of bitter melon were also analyzed and found to be rutin, gentisic acid and coumaric acid [19]. While the phenolic acids and flavonoids as well as their glycosides can be readily extracted by water, their non-polar derivatives may be present in the oil components of the plant.

Phytochemical Name	Plant parts	Chemical structure	Pharmacological effects
Momordicoside A & B	Fruits and seeds	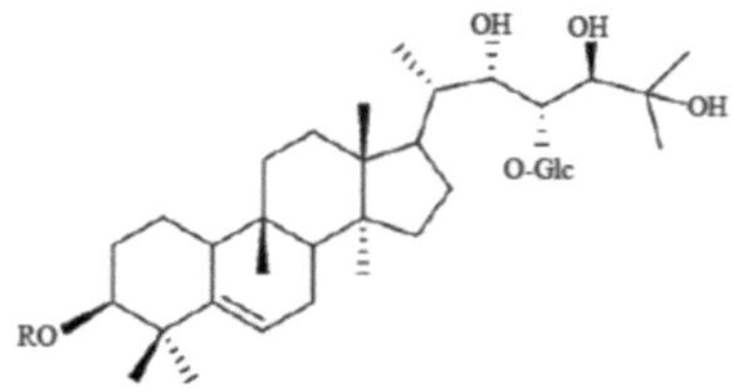	Antidiabetic, anti-obesity, Anticancer
Momordicoside K	Leaves, fruits and roots	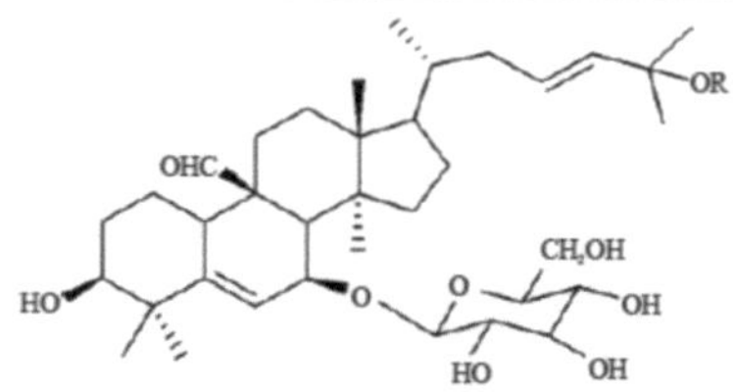	Antiproliferative, Hypoglycemic, Anti-obesity, Antioxidant
Momordicoside I & F1	Fruits	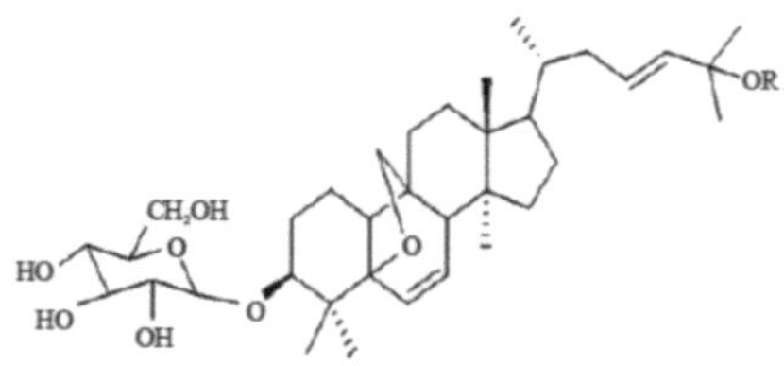	Antiproliferative, Hypoglycemic, anti-obesity, Disaccharidase
Momordicoside G & F2	Leaves and fruits	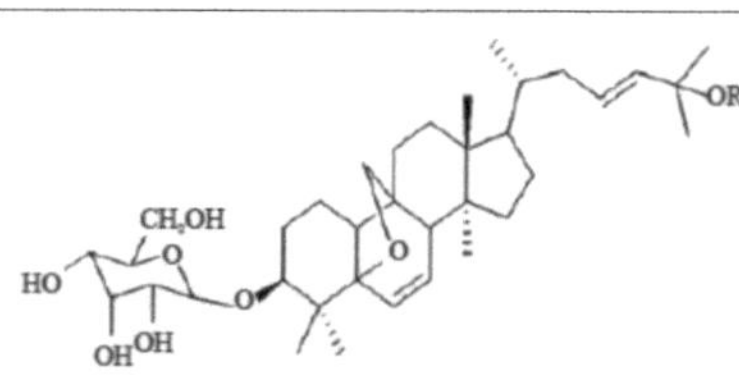	Antiproliferative, Hypoglycemic, Anti-obesity
Momordicine I & II	Leaves, vines and fruits	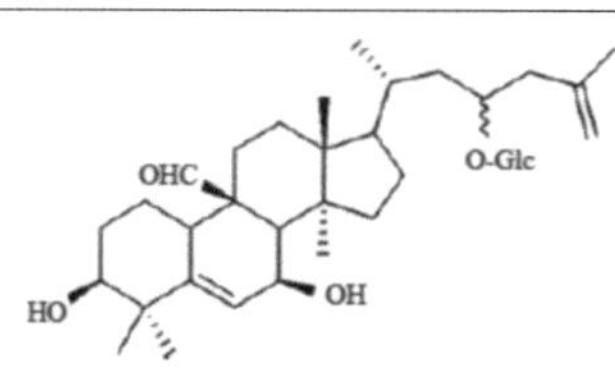	Cytotoxic, Anti-inflammatory, Antiviral, Immunomodulatory, Anti-obesity
Goyaglycoside-a & b	Leaves and vines	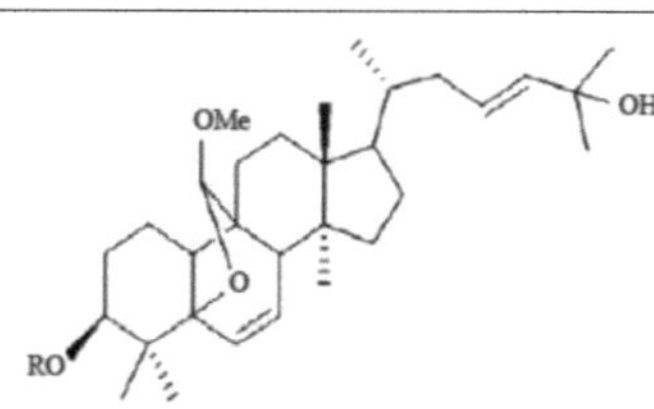	Antiproliferative, Hypoglycemic, Anti-obesity, Anticancer
Goyaglycoside-c & d	Fruits	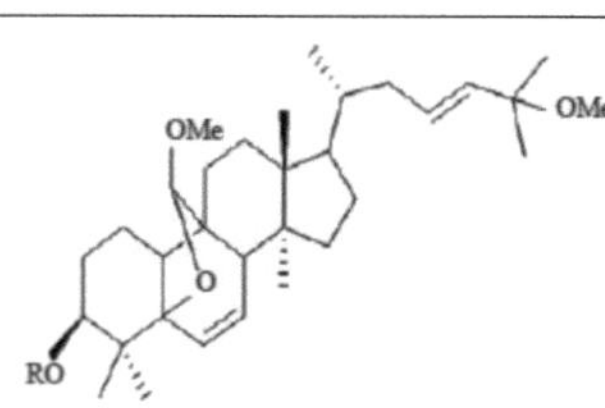	Antiproliferative, Hypoglycemic, Anti-obesity, Anticancer
Goyaglycoside-e	Fruits	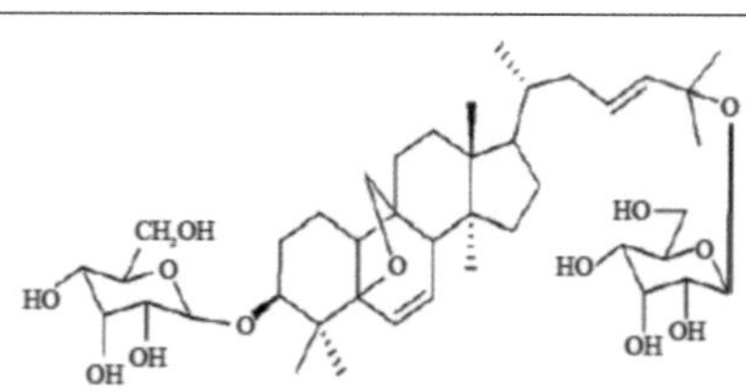	Cytotoxic, Anti-obesity, Antidiabetic

Phytochemical Name	Plant parts	Chemical structure	Pharmacological effects
Goyaglycoside-f	Fruits	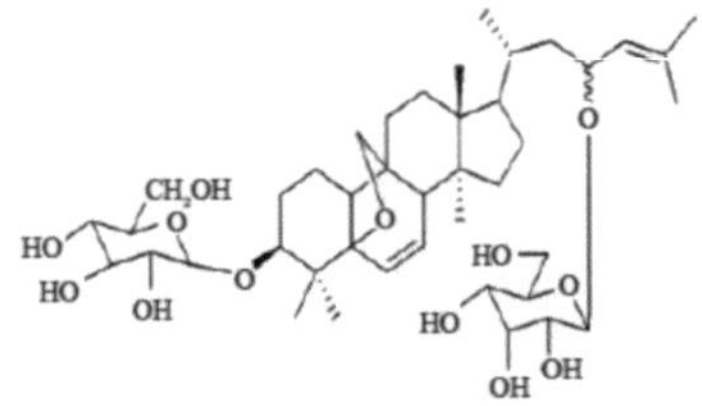	Anti-obesity, Antidiabetic
Goyaglycoside-g	Fruits	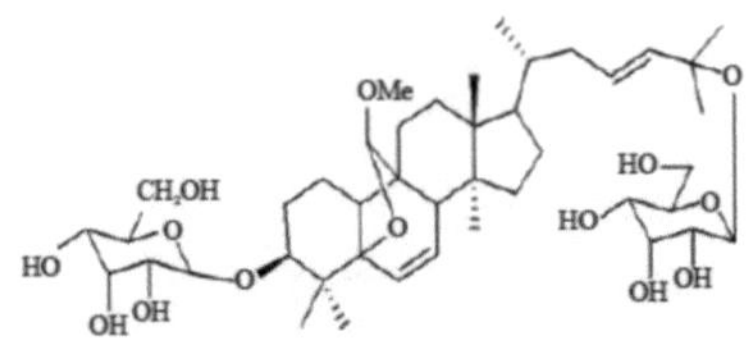	Hypoglycemic, Anti-obesity
Goyaglycoside-h	Fruits	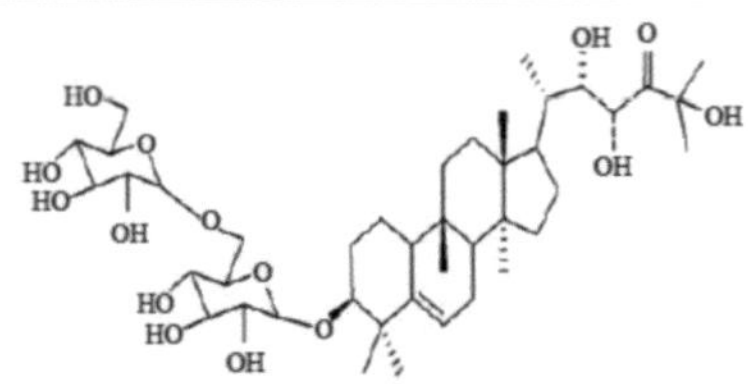	Hypoglycemic, Anti-obesity
Kuguacin B	Roots	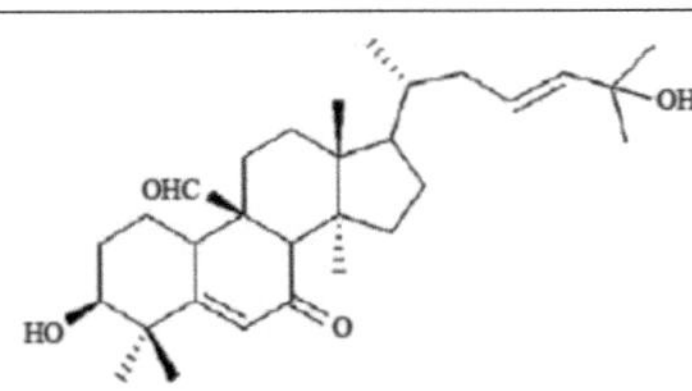	Anticancer, Antidiabetic
Kuguacin C & D	Roots	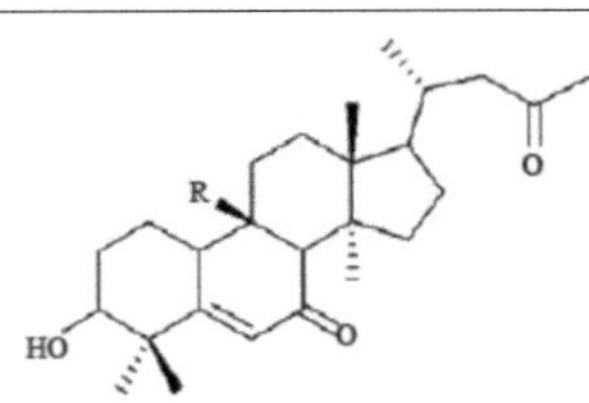	Anti-HIV-1, Antidiabetic
Kuguacin E	Roots	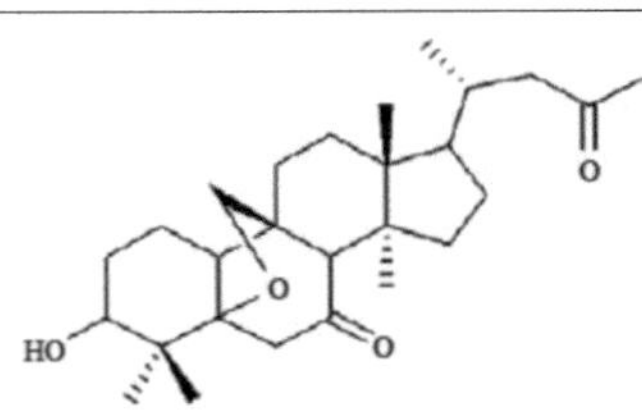	Anti-HIV-1, Hypoglycemic
Kuguacin F	Leaves and vines	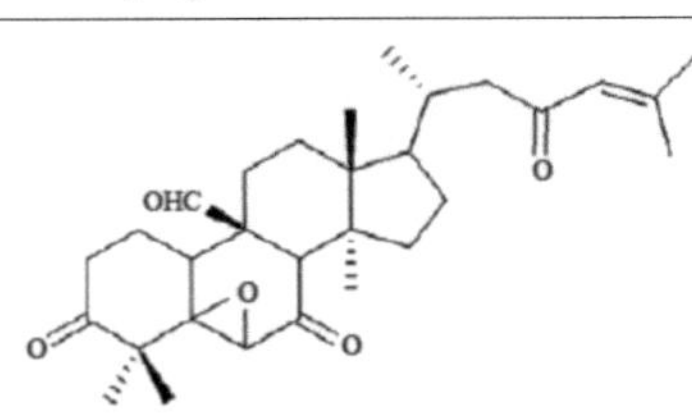	Hypoglycemic, Lipid lowering
Kuguacin G	Leaves and vines	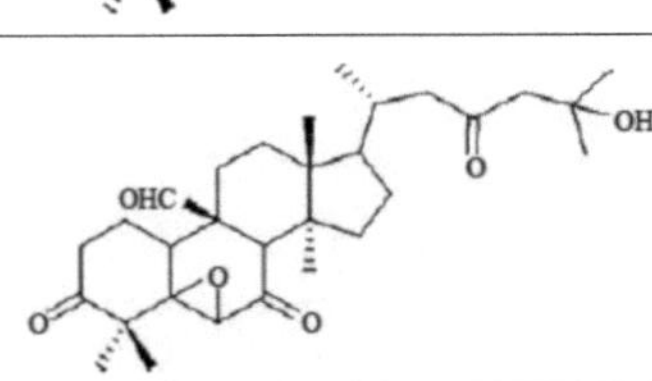	Antidiabetic, Anti-obesity, Anticancer

Phytochemical Name	Plant parts	Chemical structure	Pharmacological effects
Kuguacin K	Leaves and vines	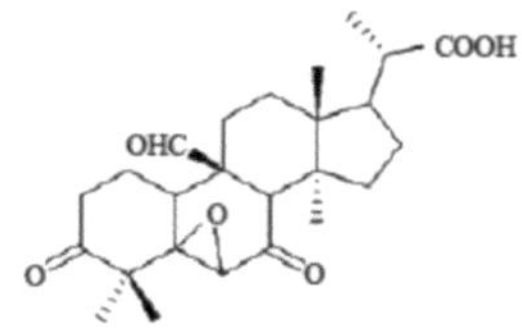	Anticancer, Antiproliferative, Hypoglycemic
Kuguacin H	Leaves and vines	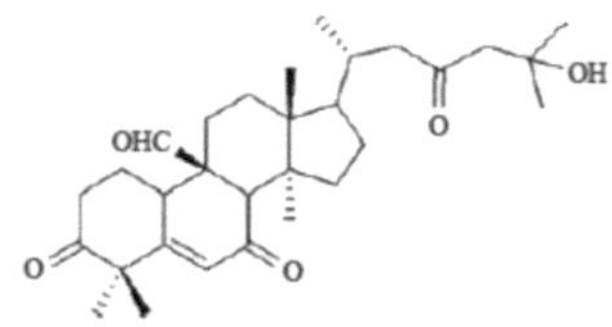	Hypoglycemic, Antiproliferative
Kuguacin I	Leaves and vines	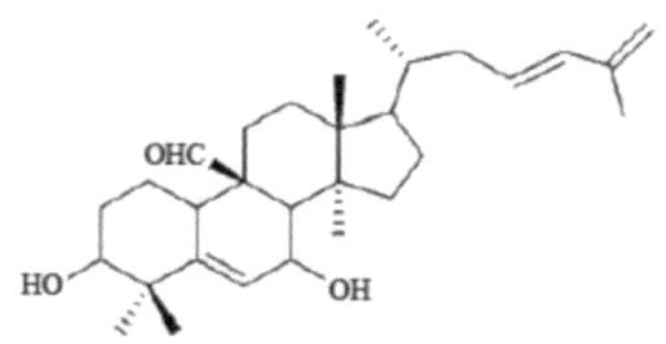	Anticancer, lipid lowering
Kuguacin J	Leaves and vines	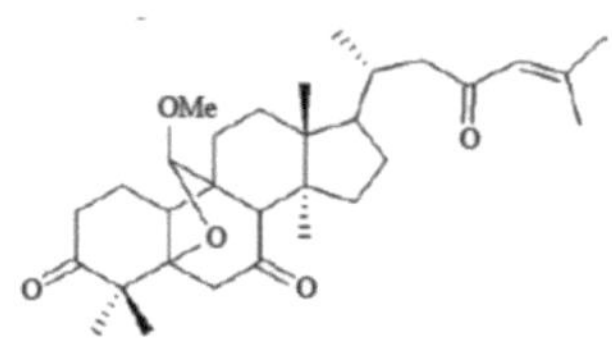	Anticancer, Hypoglycemic
Kuguacin L	Leaves and vines	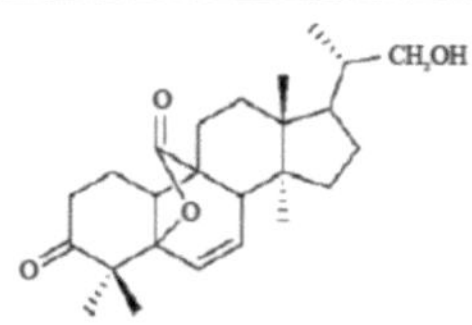	Antiproliferative, Hypoglycemic
Kuguacin M	Leaves and vines	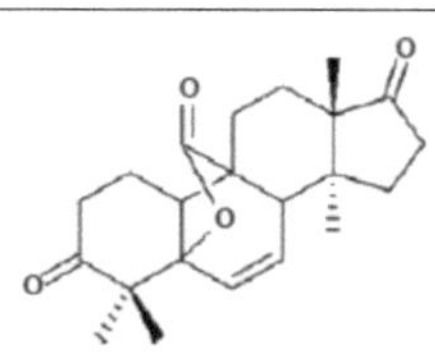	Hypoglycemic, Anticancer
Kuguacin P & Q	Leaves and vines	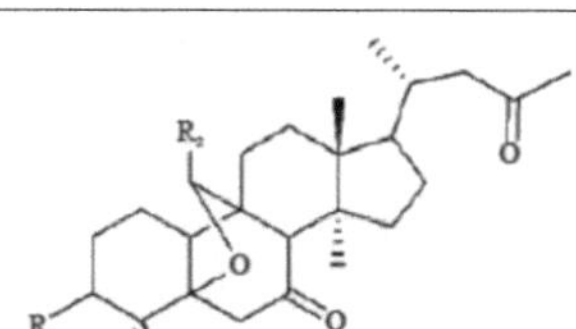	Hypoglycemic, Antiproliferative
Kuguacin R	Leaves, stems and fruits	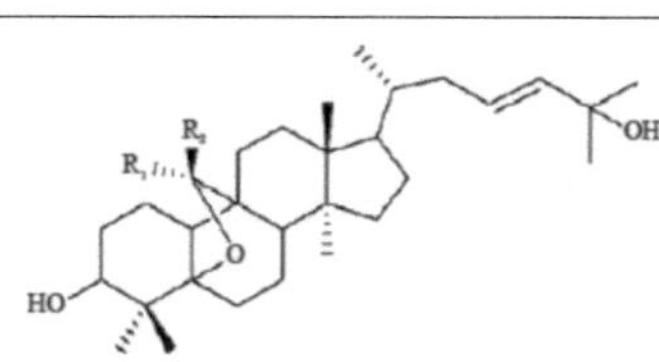	Antioxidant, Hypoglycemic, lipid lowering
Kuguacin S	Leaves and vines	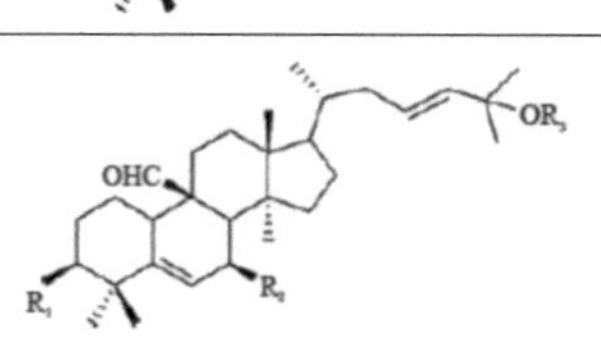	Lipid lowering, hypoglycemic

Phytochemical Name	Plant parts	Chemical structure	Pharmacological effects
Charantoside I	Fruits	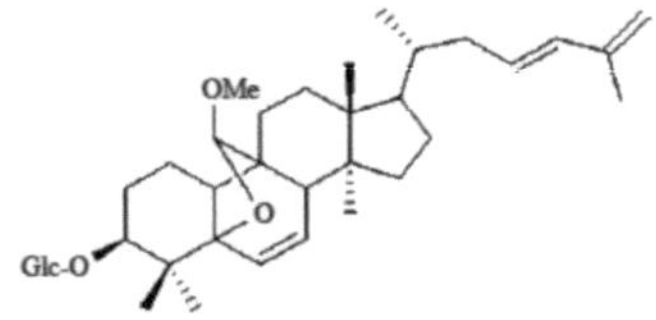	Hypoglycemic, Anti-obesity
Charantoside II	Fruits	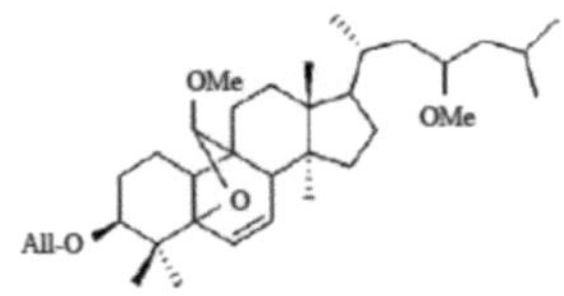	Hypoglycemic, Anti-obesity
Charantoside III	Fruits	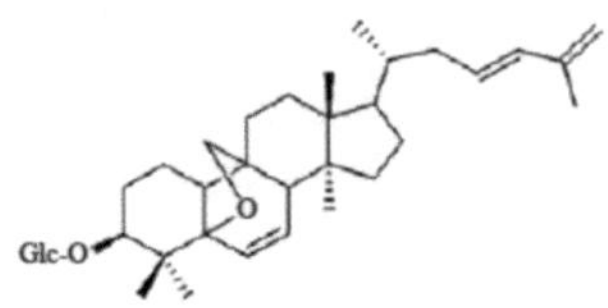	Hypoglycemic, Anti-obesity
Charantoside IV	Fruits	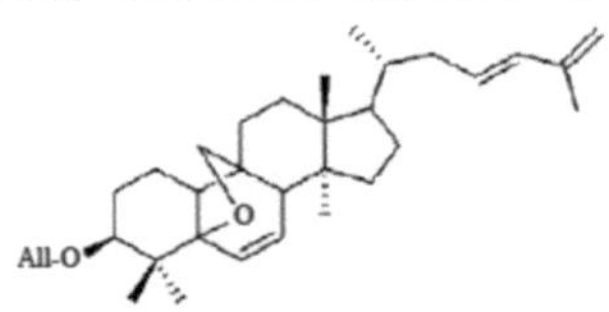	Hypoglycemic, Anti-obesity
Charantoside V	Fruits	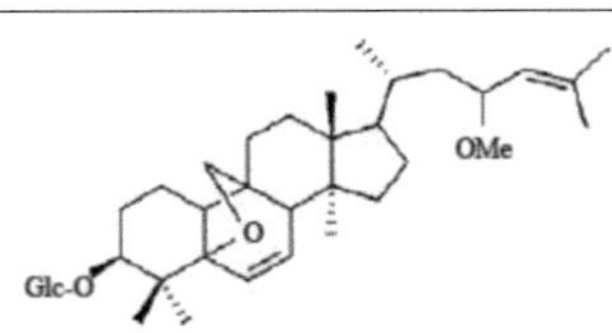	Hypoglycemic, Anti-obesity
Charantoside VII	Fruits	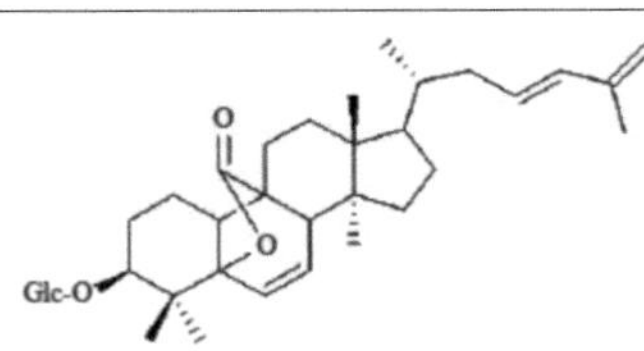	Hypoglycemic, Anti-obesity
Charantoside VIII	Fruits	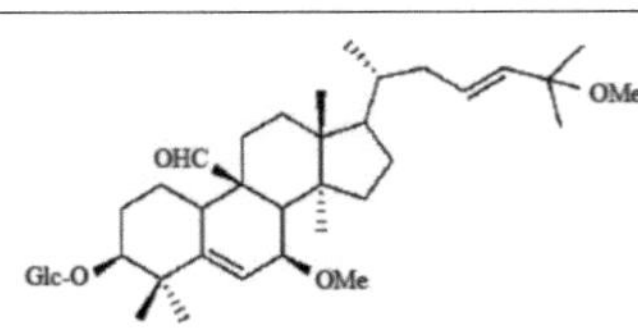	Hypoglycemic, Anti-obesity
Goyasaponin I	Fruits	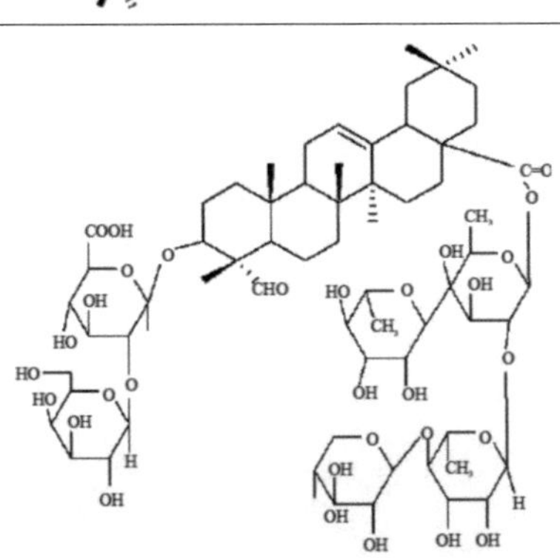	Hypoglycemic, Anti-obesity

Phytochemical Name	Plant parts	Chemical structure	Pharmacological effects
Goyasaponin II	Fruits	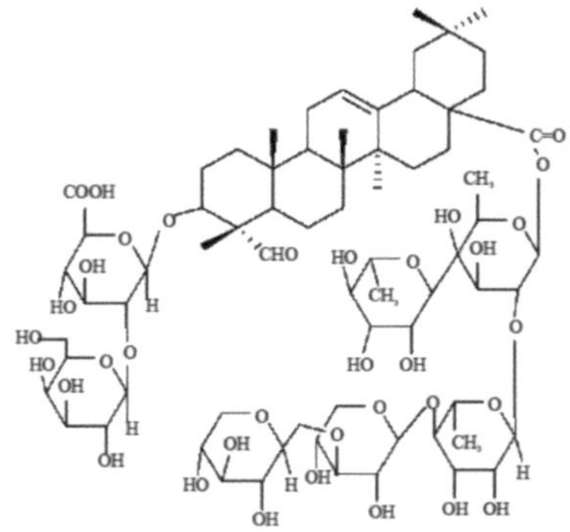	Hypoglycemic, Anti-obesity
Goyasaponin III	Fruits	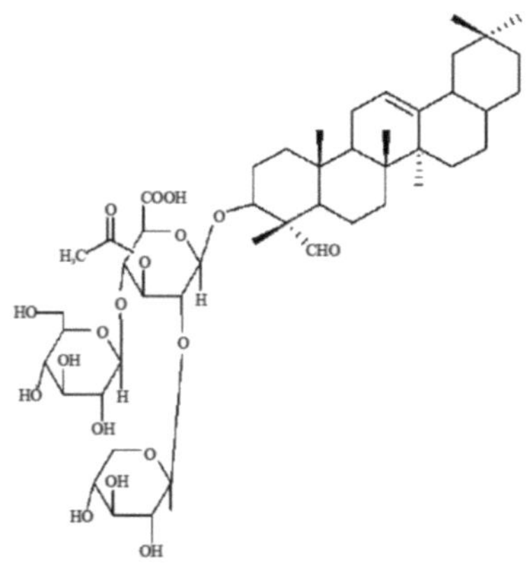	Hypoglycemic, Anti-obesity
Karavilagenin C	Fruits	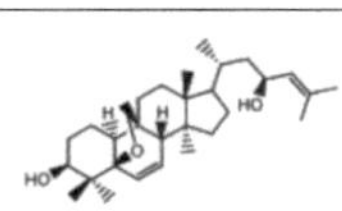	Antidiabetic
Karaviloside I, II & III	Fruits	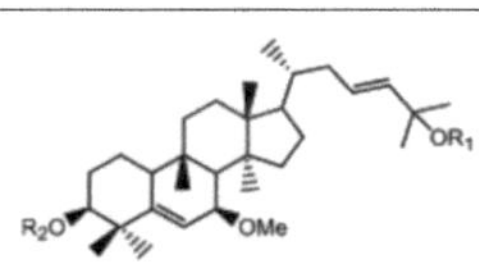	Antiproliferative, Antidiabetic, Hypolipidemic
Kuguaglycoside G	Roots	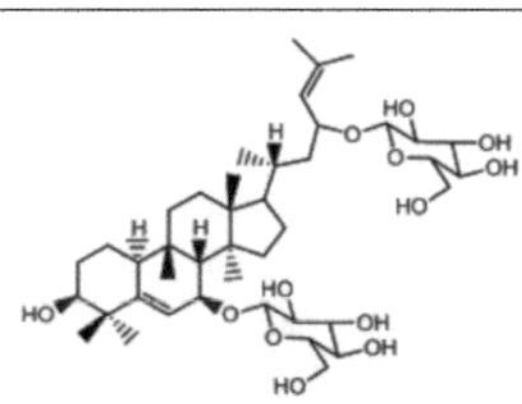	Cytotoxic, Antiproliferative, Hypoglycemic, Anticancer
Momordicinin	Fruits	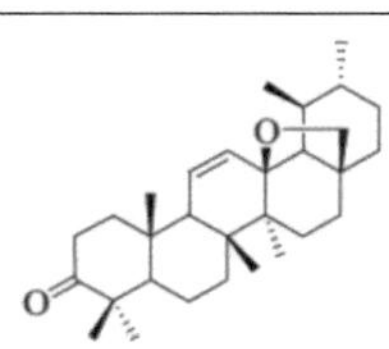	Antidiabetic, Anti-obesity, lipid lowering

Table 3.
Pharmacological effects of cucurbitane type of triterpenoids of Momordica charantia.

2.2.3 Other components

Other than the bioactive compounds, unsaturated fatty acids, alkaloids, amino acids minerals and vitamins are also present in bitter melon. The extracts of bitter melon shows nine kinds of unsaturated fatty. It has also been demonstrated that 12, 13 and 12 carbon fatty acids are found in young, mature, and senescent leaves of *M. charantia* L. representing 87.3%, 95.25%, and 83.11% of the total fatty acids respectively. The contents of total amino acids and the free amino acids of *M. charantia* L. were 11.99% and 2.36% as determined by acid hydrolysis and amino acid analysis. In addition, bitter melon is a natural source of vitamins; ascorbic acid was detected in the range of 440–780 mg in the fruit fraction. Vicine is an alkaloidal agent that has been isolated from the seeds of bitter melon, which is responsible for hypoglycaemic activity [20].

3. Antiobesity and lipid lowering effects

The various experimental studies reported decrease in serum TC, TG and LDL-C concentrations and an increase in serum high density lipoprotein-cholesterol (HDL-C) for bitter melon by different authors. The action of bitter melon in lowering fat has been supported by plentiful studies, its effect on the level of serum FFAs have been contradictory with some authors showing reduction, some shown same level, and others reported an increased levels. For example, the serum FFAs concentration increased in obese rats treated with bitter melon shown by Chen et al. [21]. An increased level of TG and LDL-C in the serum that may arises due to either overproduction by the liver or defective removal from the circulation or the overall dyslipidaemia in diabetes. It is not clear why bitter melon increase the serum level of FFAs. It may facilitate fat mobilization due to the suppression of lipogenesis or lipid deposition. While other revealed *M. charantia* L. could lower the serum and liver TG levels [22].

The findings of experimental research conclude that, *M. charantia* L. may reduce the fasting insulin, TG, cholesterol and epididymal fat, which were increased by HFD. The dwindling of insulin resistance, improves glucose tolerance, and increases insulin signaling under HFD-induced insulin-resistance and elevated serum lipids may also shown by bitter melon. The administration of an aqueous extract of unripe fruits of bitter melon improved glucose and insulin tolerance together with inhibition of plasma apoB-100 and apoB-48. The animal study had shown that, the evidence of a potent inhibitor of apoB secretion and TG synthesis as well as the plasma lipid and VLDL effects of bitter melon juice [23]. Overall the studies on the fruits, seeds, and aerial parts of *M. charantia* Linn have been shown to reduce adiposity, lower serum insulin and normalize glucose tolerance in rats fed with a HFD. The body weight and visceral fat mass of bitter melon treated obese rats were shown to be lowered [24].

While further study revealed that, bitter melon supplementation into HFD notably suppressed the levels of fatty acid synthase (FAS), acetyl-CoA carboxylase-1 (ACC-1), lipoprotein lipase (LPL) and adipocyte fatty acid binding protein [25]. Water extract of *M. charantia* L. fruits at a dose 1 g/Kg body weight revealed to be effectual in improving the obesity-induced hyperglycaemia and hyperleptinemia [26]. This in progression propose that bitter melon can reduce insulin resistance, visceral fat accumulation and adipocyte hypertrophy probably by down-regulating the expression of key lipogenic genes or proteins in adipose tissues. Aqueous fruit extract of *M. charantia* L. significantly reduce the level of serum TG, TC, LDL and VLDL at a dose of 350 mg/Kg body weight in experimental rats [27]. Numerous animal studies have been designated the efficacy of bitter melon in the amelioration of weight gain and regulation of lipid metabolism [28].

The methanolic extract of fruit of bitter melon showed antidiabetic and antihyperlipidemic action during different seasons of the year, this suggests that antidiabetic and hypolipidemic activity of *M. charantia* L. may fluctuate on quantity and quality of active constituents during different seasons of the year and reach the peak during spring [29]. The bitter melon seed oil had shown significantly decreased in body-weight, Lee's index, fat index and adipose size in the HFD mice. Meanwhile, the serum FFAs levels returned to normal at the dosage of 10 g/kg [30]. *Momordica charantia* L. extracts have anti-obesity effects and the ability to modulate lipid prolife of mice fed a HFD by suppressing body weight gain, visceral tissue weight, plasma and hepatic lipid concentrations, and lipid peroxidation along with increasing lipid metabolism. The plasma TG, TC, and LDL-C levels along with hepatic TG and TC concentrations considerably lowered in mice fed a HFD by *Momordica charantia* L. extracts. Also elevated plasma HDL-C levels and fecal TG concentration shown in animals treated with the extracts. The extracts comprise anti-obesity effects in mice fed a HFD by inhibiting lipid peroxidation whereas increasing lipid metabolism [31]. Bitter melon extract showed useful benefit on body weight gain and fat deposition.

Moreover, bitter melon reduced the lipid accumulation during differentiation from a pre-adipocyte to adipocyte and down-regulated PPAR [32]. PPAR is considered the master regulator of adipogenesis during differentiation of pre-adipocyte to adipocyte [33]. Bitter melon juice inhibited adipocyte differentiation by reducing PPAR, SREBP, and perilipin mRNA gene expression and by increasing lipolysis in primary human adipocyte [34]. Several transcriptional regulatory factors like AMPK, PPAR, and PGC-1 regulate the mitochondrial biogenesis, which would be a possible way of increasing lipid metabolism and utilization in energy demanding cells and tissues [35]. PGC-1 stimulates mitochondrial biogenesis and respiration in multiple cell types and modulates biological programs normally associated with increased oxidative metabolism. Also decreased plasma level of TGs, cholesterol, and FFA in plasma of rats fed a HF diet revealed by bitter melon supplementation due to up regulation and activation of PGC-1 [36].

Bitter melon affects on various body organs to treat obesity and diabetes as [37]:

1. Liver

    - Increased β-oxidation
    - Increased PPAR-α and PPAR-gamma expression
    - Increased expression of the transcription coactivator PGC-1α
    - Decreased fatty acid synthesis
    - Decreased fat

2. Pancreas

    - Increased insulin secretion
    - Prevents cell damage
    - Increased PPAR-α and PPAR- gamma expression in skeletal muscle

3. Fat cells

- Inhibited adipocyte hypertrophy
- Inhibited adipocyte differentiation
- Increased PPAR-gamma expression
- Increased expression of the transcription co-activator PGC-1α
- Decreased visceral fat mass

AMPK synchronized PPAR and PGC-1 activation encouraged most of the transcriptional signal to augment fatty acid oxidation and mitochondrial function [38]. Recent investigation also reported that increased hepatic AMPK p, AMPK, AMPK-2, and Sirt1 content in HF diet fed mice with supplementation of 1.2% bitter melon extract [39]. LXRs were first recognized as orphan members of the nuclear receptor plays an important role in lipid and cholesterol metabolism and oxidized derivatives of cholesterol act as ligands for the LXRs. A high cholesterol diet fed mice develop enlarged fatty livers, degeneration of liver cells, high cholesterol levels in liver, and impaired liver function by LXR knockout [40]. The *M. charantia* L. extract supplementation decreased hepatic LXR which was responsible for decreased serum TC and LDL-C, HDL-C in high cholesterol diet Wistar rats [41]. Bitter melon extract was a potent inhibitor of lipogenesis and stimulator of lipolysis in 3 T3-L1 pre-adipocyte shown by researcher [42].

4. Toxic effects

The severe adverse reactions were not reported during the short term studies while extensive data on the potential toxic effect of bitter melon are not available. Bitter melon fruits are edible and assumed to be well tolerated, at the same time toxicological evidences were reported to discover its therapeutic potential for diabetes. The two cases of acute intoxication reported after taking bitter melon tea [43]. The fruit and seeds demonstrated greater toxicity than the leaf or aerial parts of the plant. Abdominal pain as a side effect has also been reported in some studies [44]. The antifertility and abortifacient effects of the *M. charantia* L. reported in animals also value advance investigation. An acute disease favism characterized by hemolytic anemia, in individuals with a hereditary loss of the enzyme glucose-6-phosphatase has been shown by vicine found in fava bean. Consequently, the presence of vicine in bitter melon seeds was also suggested to put patients with glucose-6-phosphatase deficiency at risk [45]. Although there have been no reports on favism induced by bitter melon, individuals susceptible to the disease should avoid eating the fruit.

Several studies have been directed to reduce the bitterness of *M. charantia* L. preparations attributes to the triterpene compounds and increasing tolerability by the general public through various formulation approaches. Some recent studies used β-cyclodextrin at 0.25–2% concentrations to improve sensory quality, total phenolic content, antioxidant activity and antidiabetic potential of *M. charantia* L. juice [46]. Various encapsulation methods of bitter melon extracts along with optimized spray-drying techniques were also scrutinized to obtain the powder [47].

5. Conclusion

M. charantia L. has been broadly studied globally for its medicinal properties and to treat a number of diseases like diabetes, dyslipidaemia, obesity, and certain cancers. The extracts of fruits and different compounds seem to exert their beneficial effects via several mechanisms like PPARs, LXRs and SREBPs mediated fat metabolism in various tissues, reduces visceral fat accumulation and adipose hypertrophy. Isolated compounds from this plant like cucurbitane type of terpenoids, flavonoids and phenolic acids possesses antiobesity potential. These mechanisms will be directly related to controlling and treating diabetes mellitus, dyslipidaemia, obesity and related cardiovascular complications. Thus, further studies are required to conduct more double blind randomized trials with bitter melon extracts in obese population. In this chapter, we summarized phytoconstituents of bitter melon and its antiobesity potential. This compilation of phytochemicals and antiobesity activity of *Momordica charantia* L. will help the researchers in designing new untried strategies.

Author details

Pushpa Anantrao Karale[1*], Shashikant Dhawale[1] and Mahesh Karale[2]

1 Department of Pharmacology, School of Pharmacy, Swami Ramanand Teerth Marathwada University, Nanded, Maharashtra, India

2 Department of Microbiology, Dayanand Science College, Latur, Maharashtra, India

*Address all correspondence to: pushpakarale7711@gmail.com

References

[1] Subratty AH, Gurib-Fakim A, Mahomoodally F. Bitter melon: An exotic vegetable with medicinal values. Nutr Food Sci 2005;35:143–147.

[2] Grover JK, Yadav SP. Pharmacological actions and potential uses of *Momordica charantia*: A review. J Ethnopharmacol 2004;93:123–132.

[3] Fang EF, Ng TB. Bitter gourd (*Momordica charantia*) is a cornucopia of health: a review of its credited antidiabetic, anti-HIV, and anti-tumor properties. Curr Mol Med 2011;11(5):417–436.

[4] Joseph B, Jini D. Antidiabetic effects of Momordica charantia (bitter melon) and its medicinal potency. Asian Pac J Trop Dis 2013;3(2):93–102

[5] Svobodova BB, Calhelha L, Heleno RC, Alves S, Walcott MJ. Bioactive properties and phenolic profile of *Momordica charantia* L. medicinal plant growing wild in Trinidad and Tobago. Ind Crop Prod 2017;95:365–373.

[6] Akihisa T, Higo N, Tokuda H, Ukiya M, Akazawa H, Tochigi Y. Cucurbitane-type triterpenoids from the fruits of *Momordica charantia* and their cancer chemo-preventive effects. J Nat Prod 2007;70:1233–1239.

[7] Karale P., Dhawale S. C., Karale M. A. Antiobesity potential and complex phytochemistry of Momodica charantia Linn. With promising molecular targets. Indian J Pharm Sci. 2020; 82(4): 548–561.

[8] Chang CI, Chen CR, Liao YW, Cheng YW, Chen YC, Chou CH. Cucurbitane-type triterpenoids from *Momordica charantia*. J Nat Prod 2006; 69(8):1168–1171.

[9] Miyahara Y, Okasbe H, Yamauchi T. Studies on the constituents of *Momordica charantia* L. II. Isolation and characterization of minor seed glycosides, momordicosides C, D and E. Chem Pharm Bull 1981;29:1561–1566.

[10] Okabe H, Miyahara Y, Yamauchi T. Structures of momordicosides F1, F2, G, I, K and L, novel cucurbitacins in the fruits of *Momordica charantia* L. Tetrahedron Lett 1982a; 23(1):77–80.

[11] Li QY, Liang H, Chen HB, Wang B, Zhao YY.A new cucurbitane triterpenoid from *Momordica charantia*. Chin Chem Lett 2007; 18(7):843–845.

[12] Nguyen XN, Phan VK, Chau VM, Ninh KB, Nguyen XC, Le MH. Cucurbitane-type triterpene glycosides from the fruits of *Momordica charantia*. Magn Reson Chem 2010; 48:392–396.

[13] Okabe H, Miyahara Y, Yamauchi T. Structures of momordicine I, II and III. The bitter principles in the leaves and vines of *Momordica charantia* L. Chem. Pharm. Bull 1982b;30:4334–4340.

[14] Murakami T, Emoto A, Matsuda H, Yoshikawa M. Medicinal foodstuffs XXI. Structures of new cucurbitane-type triterpene glycosides, goyaglycosides-a,-b, −c, −d, −e, −f, −g, and -h, and new oleanane-type triterpene saponins, goyasaponins I, II, and III, from the fresh fruit of Japanese *Momordica charantia* L. Chem. Pharm. Bull 2001;49:54–63.

[15] Nakamura S, Murakami T, Nakamura J, Kobayashi H, Matsuda H, Yoshikawa M. Structures of new cucurbitane-type triterpenes and glycosides, karavilagenins and karavilosides, from the dried fruit of *Momordica charantia* L. in Sri Lanka. Chem Pharm Bull 2006;54:1545–1550.

[16] Zhao GT, Liu JQ, Deng YY, Li HZ, Qiu MH. Cucurbitane-type triterpenoids from the stems and leaves of *Momordica charantia*. Fitoterapia 2014;95:75–82.

[17] Yue J, Sun Y, Xu J, Cao J, Zhao Y. Cucurbitane triterpenoids from the fruit of *Momordica charantia* L. and their anti-hepatic fibrosis and anti-hepatoma activities. Phytochemistry 2019;157: 21–27.

[18] Yue J, Xu J, Cao J, Zhang X, Zhao Y. Cucurbitane triterpenoids from *Momordica charantia* L. and their inhibitory activity against α-glucosidase, α-amylase and protein tyrosine phosphatase 1B (PTP1B). J Funct Foods 2017;37:624–631.

[19] Minh NP. Extraction of polyphenol in bitter melon (*Momordica charantia*). IJMRD 2014;1(4):115–125.

[20] Krawinkel MB, Keding GB. Bitter gourd (*Momordica charantia*): A dietary approach to hyperglycemia. Nutr Rev 2006;64:331–337.

[21] Chen Q, Chan LL, Li ET. Bitter melon (*Momordica charantia*) reduces adiposity, lowers serum insulin and normalizes glucose tolerance in rats fed a high fat diet. J Nutr 2003;133:1088–1093.

[22] Wehash FE, Abpo-Ghanema II, Saleh RM. Some physiological effects of *Momordica charantia* and Trigonella foenum-graecum extracts in diabetic rats as compared with cidophage®. World Acad Sci Eng Technol 2012;64: 1206–1214.

[23] Nerurkar PV, Lee YK, Motosue M, Adeli K, Nerurkar VR. *Momordica charantia* (bitter melon) reduces plasma apolipoprotein B-100 and increases hepatic insulin receptor substrate and phosphoinositide-3 kinase interactions. Br J Nutr 2008;100:751–759.

[24] Chen Q, Li ET. Reduced adiposity in bitter melon (*Momordica charanita*) fed rats is associated with lower tissue triglyceride and higher plasma catecholamines. Br J Nutr 2005;93: 747–754

[25] Huang HL, Hong YW, Wong YH, Chen YN, Chyuan JH, Huang CJ. Bitter melon (*Momordica charantia* L.) inhibits adipocyte hypertrophy an down regulates lipogenic gene expression in adipose tissue of dietinduced obese rats. Br J Nutr 2008;99:230–239.

[26] Shih CC, Lin CH, Lin WL. Effects of *Momordica charantia* on insulin resistance and visceral obesity in mice on high-fat diet. Diabetes Res Clin Pract 2008;81:134–143.

[27] Rajalakshmi A, Senthikumar B, Devi K. Antihyperglycemic and antihyperlipidemic effect of aqueous fruit extract of *Momordica charantia* against alloxan induced diabetic rats. Int J Pharma Res Sch 2013;2(4):54–60.

[28] Fernandes NP, Lagishett CV, Panda VS, Suresh RN. An experimental evaluation of the antidiabetic and antilipidemic properties of a standardized *Momordica charantia* fruit extract. BMC Complement Altern Med 2007;7:29.

[29] Kolawole OT and Ayankunle AA. Seasonal variation in the anti-Diabetic and hypolipidemic effects of *Momordica charantia* fruit extract in rats. European Journal of Medicinal Plants 2012;2(2): 177–185.

[30] Li X, Yi X, Shuang W, Qianchun D, Chun-Yan W, Xiang-Tao C et al. Novel bitter melon extracts highly yielded from supercritical extraction reduce the adiposity through the enhanced lipid metabolism in mice fed a high fat diet. Journal of Nutrition & Intermediary Metabolism 2016;6:26–32.

[31] Wang J and Ho KR. The effects of *Momordica charantia* on obesity and lipid profiles of mice fed a high-fat diet. Nutrition Research and Practice 2015;9 (5):489–495.

[32] Popovich DG, Li L and Zhang W. Bitter melon (*Momordica charantia*)

triterpenoid extract reduces preadipocyte viability, lipid accumulation and adiponectin expression in 3T3-L1 cells. Food and Chemical Toxicology 2010;48(6):1619–1626.

[33] Wakabayashi KI, Okamura M, Tsutsumi S. The peroxisome proliferator-activated receptor gamm/ retinoid X receptor alpha/heterodimer targets the histone modification enzyme PRSet7/Setd8 gene and regulates adipogenesis through a positive feedback loop. Molecular and Cellular Biology 2009;29(13):3544–3555.

[34] Nerurkar PV, Lee YK, and Nerurkar VR. *Momordica charantia* (bitter melon) inhibits primary human adipocyte differentiation bymodulating adipogenic genes. BMC Complementary and Alternative Medicine 2010;10:34.

[35] Puigserver P. and Spiegelman BM. Peroxisome proliferator activated receptor-alpha coactivator 1alpha (PGC-1α): transcriptional coactivator and metabolic regulator. Endocrine Reviews 2003;24(1):78–90.

[36] Ching RHH, Yeung LOY, Tse IMY, Sit WH, Li ETS. Supplementation of bitter melon to rats fed a high-fructose diet during gestation and lactation ameliorates fructose-induced dyslipidemia and hepatic oxidative stress in male offspring. Journal of Nutrition 2011;141(9):1664–1672.

[37] Md Ashraful Alam, Riaz Uddin, Nusrat Subhan, Md Mahbubur Rahman, Preeti Jain, and Hasan Mahmud Reza. Beneficial Role of Bitter Melon Supplementation in Obesity and Related Complications in Metabolic Syndrome. Journal of Lipids, 2015 ; 1–18.

[38] Canto C. and Auwerx J. PGC-1α, SIRT1 and AMPK, an energy sensing network that controls energy expenditure. Current Opinion in Lipidology 2009;20(2):98–105.

[39] Yu Y, Zhang XH, Ebersole B, Ribnicky D, Wang ZQ. Bitter melon extract attenuating hepatic steatosis may be mediated by FGF21 and AMPK/ Sirt1 signaling in mice. Scientific Reports 2013;3:3142.

[40] Peet DJ, Turley SD, Ma W. Cholesterol and bile acid metabolism are impaired in mice lacking the nuclear oxysterol receptor LXR-α Cell 1998;93 (5):693–704.

[41] Matsui S, Yamane T, Takita T, Oishi Y, Kobayashi-Hattori K. The hypocholesterolemic activity of *Momordica charantia* fruit ismediated by the altered cholesterol- and bile acidregulating gene expression in rat liver. Nutrition Research 2013;33(7):580–585.

[42] Chikkavadaragudi RS, Vishwanath P, Prashant A, Rangaswamy C, Maduvanahalli NS, Hattur B. Fifty percent ethanolic extract of *Momordica charantia* inhibits adipogenesis and promotes adipolysis in 3T3-L1 pre-adipocyte cells. Rep Biochem Mol Biol 2017;6(1):23–32.

[43] Hulin A. Intoxication aigue par *Momordica charantia* (sorrossi). A propos de deux cas (acute intoxication due to *Momordica charantia* (sorrossi). Study of two cases). Sem. Hop 1988;64: 2847–2848.

[44] Dans AM, Villarruz MV, Jimeno CA, Javelosa MA, Chua J, Bautista R. The effect of *Momordica charantia* capsule preparation on glycemic control in type 2 diabetes mellitus needs further studies. J Clin Epidemiol 2007;60:554–559.

[45] Dutta PK, Chakravarty AK, Chowdhury US, Pakrashi SC. Vicine, a Favism-inducing toxin from *Momordica charantia* Linn. seeds. Indian J Chem 1981;20:669–671.

[46] Deshaware S, Gupta S, Singhal RS, Joshi M, Variyar PS. Debittering of

bitter gourd juice using β-cyclodextrin: Mechanism and effect on antidiabetic potential. Food Chem 2018;262:78–85.

[47] Tan SP, Kha TC, Parks SE, Stathopoulos CE, Roach PD. Effects of the spray-drying temperatures on the physiochemical properties of an encapsulated bitter melon aqueous extract powder. Powder Technol 2015; 281:65–75.

Chapter 12

Medicinal Plants and Its Pharmacological Values

Smita G. Bhat

Abstract

Plants have been used as a source of medicine for the treatment of different diseases from thousands of years ago. There is numerous evidences are available for use of plants as a medicine in the treatment of diseases in Indian, Egyptian, Chinese, Greek and Roman system of medicine. Pharmacognosy is the study of medicines derived from natural sources, mainly from plants which may further lead to development of new drug. The exploration, extraction and screening of biological diversity such as herbs, spices, microbes and other natural resources is the worldwide activity in recent years. Phytochemicals are the naturally available bioactive compounds which are derived from different plant parts and are primarily responsible for biological activities. The most important chemical compounds which are present in the plants are alkaloids, phenols, saponins, carbohydrates, terpenoids, steroids, flavonoids and tannins etc.

Keywords: Medicinal plants, pharmacognosy, phytochemicals, biological activities

1. Introduction

Since from ancient period man depended on nature for their survival and lives strictly connected with nature. Man depends upon surrounding environment for their livelihood, healthcare, and sustenance and also for basic needs (food, fibers, shelter, clothing and gum). Besides providing basic necessities, plants also provided his requirement of medicine. Along with the plant man has been started using animal products and other bio-resources available in nature for preparation of medicine. As a result, different traditional medicine systems have evolved based on environmental condition, social and cultural background with respect to the ethnic group in different countries [1, 2].

Plants are served as major natural resources for traditional as well as modern medicinal system all over the world. The therapeutic potential of plants and plant products can be traced back to thousands of years ago. The information with respect to medicinal benefits of plants with other therapies has been preserved in several documentations in Babylonia, Egypt, China, Greece and Rome etc. Previous works of Theophrastus (370–287 B.C.), Aristotle (384–322 B.C.), Hippocrates (460–370 B.C.) and Dioscorides (50–100 A.D) are providing evidence that Greeks and Romans are familiar with many of today's plant drugs. The "Ebers Papyrus", the best known Egyptian pharmaceutical record documented over 700 drugs represents the history of Egyptian medicine (1500 BC). Erh-ya (300B.C), Svu-ching (1000B.C) and Ben-tsao (1250A.D) are the early herbal documentations available in China, describes more than 600 medicinal plants [3]. In Asia, the earliest records of

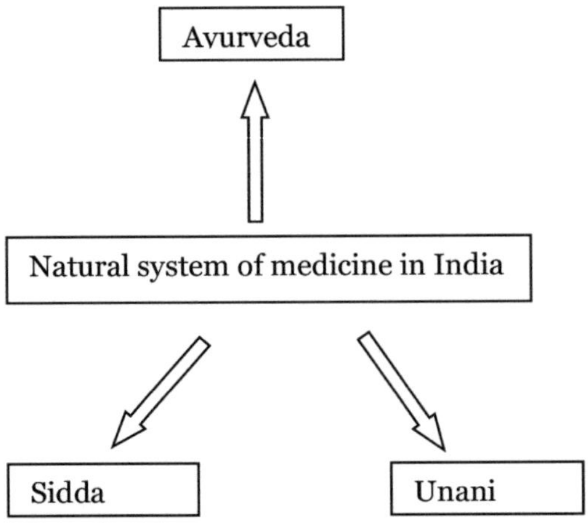

Figure 1.
Natural system of herbal medicine used in India.

plants usage are found in the clay tablets in Mesapotamia (1700 BC). In India, herbal remedies and health care preparations are also described in ancient texts like 'Charka Samhita' (100–800 B.C), 'Sushruta Samhita' (800–700 B.C), 'Rigveda' (1400–1800 B.C) and 'Atharva-veda' (4500–2500 B.C). Ayurveda is the fundamental source of Vedic knowledge for understanding remedial properties of plants (1000 BC). Ayurveda is considered as ancient medicinal system and it is the compilation of 'Charka Samhita', 'Sushruta Samhita' and 'Ashtanga Hridaya Samhita' [4]. In addition to Ayurveda, Siddha and Unani are other conventional systems of medicine providing additional information of plant based drugs used in India. 'Unani' system of medicine originated in Greece and introduced to India by Arabs and Persians after the discovery of sea route to India. During 10th to 15th century, 'Sidda' system of medicine originated in southern parts of India and is parallel to Ayurvedic system of medicine (**Figure 1**) [5].

Now a day the herbal medicine has renewed attention and hopeful both practical and scientific view points. Herbal remedies are complex mixture different parts of single herb or many herbs which may sometime produce synergistic effect with each other ensuing in the increased therapeutic potential of drug. The identification of biologically active compound responsible for its medicinal property and there is a crucial requirement for quality control. So the correct identification and quality assessment is important to ensure quality of herbal medicine, which contributes to its safety and efficacy. Therapeutic action of herbal formulation depends on its photochemical constituents. The photochemical investigation of the medicinally important plants should be carried out, as this would be beneficial in standardization, quality assessment and efficacy of herbal drugs. Thus pharmacognosy is considered as important tool to study medicinal plants for their identification, validation and standardization [6, 7].

2. Bioprospecting of medicinal plants

Biodiversity prospecting or bioprospecting of the medicinal plants is the worldwide activity in the current years. Biodiversity prospecting is the exploration, extraction and screening of biological diversity and indigenous knowledge for commercially valuable genetic and biochemical resources. In early stages, bioprospecting mainly focused on the plants from the forest ecosystem. But in recent years, various other forms of biodiversity such as insects, algae and microorganisms have been explored with substantial success [8]. During recent years this activity involves the use of advanced technologies to develop new pharmaceuticals, agrochemicals, cosmetics and other bi-products from biological diversity [9]. With the help of advanced technology and sophisticated techniques and tools it has become effective way to conduct research on metabolic response of living system, genetic manipulation and novel drug discovery through bioprospecting. Various bio-active molecules have been isolated and tested for their pharmacological activities [10].

3. Pharmocognosy of medicinal plants

The term pharmocognosy was first time coined by the Austrian physician Schmidt in 1811. A "crude drug" means a dried unprepared natural material of plant, animal or mineral origin, which is used for medicine. The word pharmacognosy is derived from the Greek word *pharmakon*-drug and *gnosis*- knowledge. Pharmacognosy is the study of medicines derived from natural sources, mainly from plants which may further lead to development of new drug. Phytochemicals ('Phyto' means plant) are biologically active natural chemical constituent of plants such as sugar, amino acids, protein, chlorophyll, alkaloids, flavonoids, steroids, tannins etc. Phytochemicals are active ingredients which possess therapeutic properties and are considered as a medicine or drug. More than 4000 phytochemicals have been obtained cataloged and are classified by protective function, physical and chemical characteristics of which 150 phytochemicals have been studied in details [11]. Latest outcome suggest that majority of phytochemicals have beneficial activities like anti- microbial, anti-malarial, anti- diabetic, anti-arthritic and anti-cancerous etc. The medicinal, biological and pharmaceutical value of phytoconstituents helps in the utilization and exploration of plant resources in recent years. The chemical information of plant coupled with medicinal properties and supported by other biological activities will add additional value for development of valuable herbal drugs [12].

During pharmacognostic investigations, physico-chemical analysis also considered as important parameter in evaluation and identification of crude drug. Macroscopic and microscopic analysis is necessary for the detection of adulterants, contaminants of herbal drug and for assessing quality before going for further study. The extractive value and solubility value is useful to evaluate specific chemical constituent dry yield in different solvents. Ash value analysis is useful in determination of unrelated matter (sand and soil) adhering to the surface of plant [13]. Moisture content is essential for evaluation of stability of crude drug. Fluorescence analysis is a reliable tool for standardization of crude drug. The different chemical constituent present in the plant extract showed characteristic fluorescence when illuminated suitably. Certain chemical substances that are not naturally fluorescence themselves are treated with different reagent to attain fluorescence [14].

4. Phytochemicals of medicinal plants

The curative properties of medicinal plants are due to presence of major group of active components which are mainly alkaloids, triterpenoids, essential oils and phenolic compounds etc. Alkaloids are the secondary metabolites of plants having noticeable pharmacological activity. Roots, leaves, bark and seeds are common parts of plants which contain alkaliods. In general the alkaloids occur as salts of citric acid, oxalic acid, acetic acid and tartaric acid. These are mostly colorless, water insoluble and non-polar solvents soluble in nature. Pharmacologically, alkaloids act as cardiac depressants, antihypertensive, anti-leukemic, analgesic, nerve stimulants and local anesthetic. Triterpenoids are made up of six isoprene units. Saponins, sterols and cardiac glycosides are chief triterpenes. The medicinal plants which have saponins are roots of *Glycyrrhiza glabra*, tuberous roots of *Asparagus racemosus* and roots of *Smilax glabra*. Typically sterols are animal substances but recently detected from plants also. Ergosterol, stigmasterol, campesterol and β- sitosterol are chief sterols derived from plants. In several plants characteristic odor is due to presence of essential oils or volatile oils which occur in lysigenous or schizogenous cavities, in glandular hairs or in specialized tubes. A variety of plant parts such as leaves of lemongrass, bark of cinnamon, flower buds of clove, nutmeg seeds and camphorwood contains volatile oils. The phenolic compound are soluble in water and includes phenols, phenolic acids, phenyl propanoids, coumarins, phenyl propenes, flavonoid pigments, anthocyanins, flavonols, flavones and tannins [15].

5. Biological activity of medicinal plants

Due to the presence of bioactive molecules plants are used as phytomedicine to cure many complaints. *Catharanthus roseus* has 'vinblastine' and 'vincristine' and used in cancer. *Rauwolfia serpentina* is hypotensive due to presence of 'serpentine', 'reserpine' and 'ajmalicine'. *Papaver somniferum* contains 'morphine' and 'codeine' and is analgesic and sedative. 'Artemisinin' is effective against malaria derived from *A. annua*. Similarly, the bioactive components of plants 'Withanaloides' reported from *Withania somnifera* useful in treatment of arthritis. 'Charantin' a steroidal saponin, isolated from *Momordica charantia* reported for anti-diabetic activity. 'Diospyrin' reported from *Diosyros species* acts as anti-leishmanial agent. 'Tephdidoside' is a flavanol glycoside derived from *Tephorsea candida* found to be active against human cancer [16]. 'Berberine' derived from *Berberis vulgaris* reported for antidiabetic, hepatoprotective, antimicrobial activity. 'Digoxin'obtained from *Digitalis lanata* used in heart diseases. Similarly, 'Quinine' isolated from *Cinchona robusta* acts as antimalerial, antiparasitic agent. Another compound 'Allicin' isolated from *A. sativum* reported for its cardioprotective, anti-inflammatory activity [17].

More than 35,000 plant species have been investigated and resulted in the discovery of anticancer drugs such as 'Vincristine', 'Vinblastine', 'Taxol', 'Etoposide analogs', 'Camptothecin' etc. Many number of effective drugs derived from higher plants were alkaloid 'Paclitaxol', isolated from *Taxus brevifolia* used in treatment of ovarian and breast cancers [18]. 'Andrographolide', active diterpine derived from *A. paniculata* acts as a noble anticancer agent against cancers of breast, ovary, stomach, colon, prostate, kidney, nasopharynx malignant melanoma and leukemia. 'Thymoquinone' and 'dithymoquinone' of *Nigella sativa* shows anticancer activity against different types of cancers such as colon, prostate, pancreas, uterus, malignant ascites, malignant lymphoma, malignant melanoma, sarcomas and leukemia. However, 'Plumbagin' isolated from *Plumbago zeylanica* hinders growth and spread of breast cancer, liver cancer, fibro-sarcoma, malignant ascites and leukemia by cell proliferation [19].

A wide range of reports are available on phytochemicals and pharmacological activity of medicinal plants. Several workers have reported biological activity of medicinal plants. Pharmacognostic and preliminary phytochemical analysis of *Colocasia esculenta* dried tubers were calculated. The tubers are good source of carbohydrate, protein and starch. Nutritional analysis showed moisture content (56.8%), ash content (1.22%), carbohydrate (3000 mg/100gm), protein (824 mg/100gm) and starch (2700 mg/100gm) in dry tubers. Phytochemical analysis revealed the presence of alkaloids, glycosides, flavonoids, terpenoids, saponins and phenols [20]. *A. aspera* important medicinal plants reported in Ayurvedic literature with number of medicinal property. Phytochemical investigation of plant extracts were subjected to qualitative screening test for various constituents. This revealed the presence protein, glycosides, alkaloids, tannins and phenolic compound, steroid reducing sugars and saponin glycosides [21]. Pharmacognostic and phytochemical evaluation of *Tridax procumbens* were studied. The quantitative microscopical and histological study is done revealed the presence of Tricomes, palisade tissue, trachieds and vessels in powder microscopy. Phytochemical analysis of the whole plant is done and the presence of carbohydrates, sterioids, phenols and tannins were reported and quercitin is confirmed using HPTLC [22].

In parallel, phytochemical screening of *Cinnamomum zeylanicum* shows the presence of carbohydrate, glycoside, protein, tannins, saponins, flavonoids and terpenoids. The proximate analysis revealed that water soluble extractive values of leaves was 29.75, total ash value was 9.75, acid insoluble ash was found to be 2.50 and sulphated ash was 37.35. Anti- microbial activity of the disk diffusion method showed that chloroform and hydro-ethanol extracts of leaf were more effective against Gram–positive bacteria in vitro [23].

Antimicrobial activity of different extract of *Cynodon dactylon* was tested against disease causing bacterial pathogens using the agar well diffusion method. Areal parts of ethanol extract show more activity against *Pseudomonas aeruginosa* and *Staphylococcus aureus* with zone of inhibition 13.83 ± 0.29 mm and 2.0 ± 0.10 mm respectively. A total 20 compounds were identified from the hydroalcoholic extract of the whole parts through GC–MS analysis. Among all, hexadecanoic acid, ethyl ester inolenic acid, ethyl ester was the major components of the hydroalcoholic extract and hexadecanoic acid ethyl ester was abundant. The antioxidant activity of the hydro-alcoholic extract of aerial parts was studied *in vitro* by different methods. Of this superoxide radical scavenging assay revealed a maximum inhibition of 93.33%. Total antioxidant capacity equivalent of ascorbic acid was 172.39 mg/g of extract. Similaly, anticancer activity of methanolic extracts of leaves was studied in ascitic lymphoma (ELA) in Swiss albino mice. Results demonstrate that methanolic extract was found to be antiproliferative at lower concentrations and induced apoptotic cell death in COLO 320 DM cells [24].

Pothos scandens another medicinally important plant screened for its various biological activity. At lower concentration of ethanol extract was found to have more antimicrobial activity compare to other. The ethanol extract of root showed significant free radical scavenging activity with DPPH and superoxide radical scavenging activity (IC50 0.284 mg/mL and 70.84%). The active compounds of ethanol extract of aerial parts investigated through GC–MS analysis. This depicts that Dodecanoic acid, tetra decanoicacid and n- hexadecanoic acid acts as antioxidant. Similarly cytotoxicity of same plant was evaluated against MCF-7 (breast cancer) cell lines by MTT assay and results revealed that the extract has significant cytotoxic activity with an IC50 of 90.18 ± 5.20 μg/ml and also cell death of MCF-7 treated with the extract was due to the induction of apoptosis, which was confirmed by comet assay [25].

The in vitro cytotoxic activity of *Colocasia gigantea* extract on cervical cancer (Hela) and human white blood cells (WBC) was conducted. Bioassay-guided fractionation method showed that not all parts promote cytotoxic activity. The leaf fraction of dichloromethane showed significant cell proliferation effect on Hela cells, but not on WBCs. The n-hexane tuber fraction only exhibited significant cytotoxicity on Hela cells (IC_{50} 585 μg/ml) and encouraged WBC cell proliferation. From the GC–MS spectrometry it was found the 4, 22-Stigmastadiene-3-one,

Diazoprogesterone, 9-Octadecenoic acid (Z)-, hexyl ester, and Oleic Acid were the components of n- hexane tuber fraction which had cytotoxic potential. Tuber fraction of n- hexane shows potential for cervical cancer treatment [26].

6. Current status of herbal medicine

World Health Organization estimated that 80% of the populations of developing countries still depend on plant drugs for their primary health care needs. According to survey of World Health Organization, the practitioners of traditional medicinal system treat about 8% of patients of India, 85% in Burma and 90% in Bangladesh. India comprises of 2.4% of the total geographical area of the world. The country accounts for an average of 8% of the total global biodiversity with approximately 49,000 species of plants of which 4,900 are endemic [27]. Approximately 2,65,000 species of seed plants exists on earth and less than half of these have been studied systematically for their chemical composition and medicinal value [28].

Greater part of drugs now available in the market is simple semi-synthetic derived from naturally occurring substances. Up to 50% the approved herbal drugs used today are from either directly or indirectly synthesized from natural products including plants, microorganisms, fungi and animals. According to an estimate, about 25% of the world pharmaceutical products find a significant degree of origin in indigenous communities, which represents more than a 2000 billion dollar share market [29].

In many developed countries, the percentage of the population which has used herbal medicines at least once is 48% in Australia, 70% in Canada, 42% in USA, 38% in Belgium and 75% in France. Malaysia, spent US$ 500 million annually on herbal health care, compared to about US$ 300 million on allopathic medicine. In USA, annual spending on conventional medicines was estimated at US$ 2700 million. In Australia, Canada and the United Kingdom, annual expenditure for herbal medicine is estimated US$ 80 million, US$ 2400 million and US$ 2300 million respectively. In several parts of the world, outflow on herbal medicinal products is not only significant, but also growing rapidly [30].

The allopathic medicine has side effects and hazardous to human beings. After realizing toxicity and adverse effects of allopathic medicines, a shift in universal trend from synthetic to herbal medicines has been observed both in developed and developing countries [27]. The most important facts about herbal medicine are that these medicines have no side effect and available in low of cost. Therefore high dose of herbal medicine or wrong medicine consumed by patient mistakenly does not cause any adverse effect on the body.

7. Conservation of medicinal plants

Medicinal plants are the basic raw materials of pharmaceutical industries and is highly depends on medicinal plant for extraction of medicinally important compounds. During this time with the advancement of science and technology, over growing demand of pharmaceutical industries, the useful medicinal plants were over exploited by the men. So there is a need of conservation and propagation of valuable, rare and endangered medicinal plants by using advanced biotechnology methods [31].

Plants occupy a major sector of health care system and represent a most important natural resource. Therefore conservation of species is most effectively achieved through the management of the wild population and natural habitats. In most of cases medicinal plants either do not produce seeds or too small seeds. In order to overcome these barriers *ex-situ* techniques can be used to complement *in-situ* methods and for some instance it may be appropriate for some species. So conservation of medicinal plants can be accomplished by the *ex-situ* that is outside the natural habitat by cultivating and maintaining plants through long term conservation of plant propagules in plant tissue culture repositories [32]. *In vitro* techniques have been increasingly applied for mass propagation and conservation of germplasm as it has superiority over alternative strategies. Hence there is a need for conservation of medicinal plant biodiversity for the present and forth coming generation by adapting the appropriate strategy with proper conservation method [33, 34].

Recently in India, several institutes and organizations involved in different aspects of drug discovery and conservation medicinal plant from natural resources. Initiative work designed for finding novel bioactive compound from plant, fungi, microbes etc. are set up by Council of Scientific and Industrial Research (CSIR), Central Drug Research Institute (CDRI), Regional Research Laboratory (RRL), Jammu and Kashmir. Golden Triangle Partnership (GTP) in collaboration with Department of Ayush, CSIR and ICMR involved in the validation of traditional ayurvedic medicine for effective drug discovery. During last few decades, the Department of Biotechnology and Government of India has set up two Micropropagation Technology Parks at National Chemical Laboratory (NCL), Pune and Tata Energy Research Institute (TERI), New Delhi [16].

8. Conclusion

Since time immemorial plants are utilized as chief source of therapeutic agents. The medicinal plants are not only the source of healthcare but also an important product of world trade. In last few years the trade of medicinal plant is increase rapidly because herbal drugs are easily available at lowest prices and less side effects [35]. The remedial properties of plants due to presence of intricate chemical components with different compositions and biological function.

Herbal medicine is widely practiced in worldwide and is considered to effective and affordable. Recently significant attention has been made to utilize eco-friendly and bio-friendly plant based product for cure of different human diseases [28]. This increases global pharmaceutical demand from last few years. Due to increasing demand of supply of plants as a raw material in pharmaceutical industries, biological diversity of plants is in danger. Therefore there is a need to advance research for the development and characterization of natural drugs with the help of better screening methods from plants and other natural sources. However, medicinal plants often being subjected to scientific validation and for discovery of safe and potential natural drug to fight against diseases [36, 37].

Conflict of interest

The authors declare no conflict of interest.

Author details

Smita G. Bhat
P.G. Department of Studies in Botany, Karnatak University,
Dharwad, Karnataka, India

*Address all correspondence to: smitabhat999@gmail.com

References

[1] Daniel EV. The pulse as an icon in the Siddha medicine. In: E.V. Daniel and J.F. Pugh, Contribution to Asian studies, E.J. Brill-Leiden, The Netherlands. 1984; 115-119.

[2] Thomas B and Rajendran A. Less Known Ethnomedicinal Plants Used by Kurichar Tribe of Wayanad District, Southern Western Ghats Kerala, India. Bot Res Int. 2013; 6 (2): 32-35.

[3] Adhikari PP and Paul SB. History of Indian traditional medicine: a medical inheritance. Asian J Pharm Clin Res. 2018; 11(1): 421-426.

[4] Katiyar CK. Safety aspects of Ayurveda: International conclave on traditional medicine, Department of AYUSH, Ministry of Health and Family Welfare, Government of India, New Delhi, and NISCAIR, CSIR, New Delhi. 2006;299-306.

[5] Raju YR, Yugandhar P and Savithramma N. Documentation of ethnomedicinal knowledgeof hilly tract areas of East Godavri District of Andhra Pradesh, India. Int J Pharm Pharm Sci. 2014; 6(4): 369-374.

[6] Kokate CK, Purohit AP and Gokhale SB. Drugs containing glycoside. In: Pharmacognosy, 21st edition, Pune, Nirali Prakashan. 2009; 158-239.

[7] Dinesh Kumar, Zulfiqar AB, Vijender Kumar I A, Chashoo NA, Khan MY. Pharmacognostical and phytochemical evaluation of *Angelica Archangelica* Linn., Int J Drug Dev & Res. 2011; 3(3): 173-188.

[8] Kumar P and Tarui N. Identifyng the contribution of indigenous knowledge in bioprospecting for conservation stategy. In Bridging scales and Epistemologioes Conference, March 2004. Alexandrea, Egypt. 2004.

[9] Artuso A. Bioprospecting, Benefit Sharing and Biotechnological Capacity Building. World Development.2002; 30(8): 1355-1368.

[10] Juan B. Bioprospecting and Drug Development, Parameters for a Rational Search and Validation of Biodiversity. J Microbial Biochem Tech. 2017; 9:1 1000 e128.

[11] Sharma M and Kumar A. Pharmacognostial Characterization of Some Selected Medicinal Plants of Semi-Arid Regions. J Pharmacogn Phytochem.2013; 1(6); 216-228.

[12] Vivekraj PS, Vinotha A, Vijayan and Anand GV. Preliminary Phytochemical Screening and GC-MS Analysis of Methanolic Extract of *Turnera subulata* Smith (Passifloraceae). J Phytopharmacol. 2017; 6(3): 174-177.

[13] Kumar D, Bhat ZA, Kumar V, Chashoo IA, Khan NA, Shah MY. Pharmacognostical and phytochemical evaluation of *Angelica Archangelica* Linn., Int J Drug Dev &Res.2011; 3(3): 173-188

[14] Roy P, Mandal P, Panda S, Mitra S, Subba A. Pharmacognosy and Phytochemical Screening of some Plant Derived Medicine to Treat Dysmenorrheal Pain by the Rajbanshi Community. Pharmacog J. 2018; 10(4):38-46

[15] Divya R. Ethnobotanical survey of traditional medicinal plants of Jhunjhunu (Rajasthan) and its neighbourhood. Thesis submitted to Shri Jagdish Prasad Jhabarmal Tibrewala University, Rajasthan, 2017.

[16] Bhutani KK and Gohil VM. Natural products drug discovery in India: Status and appraisal. Indian J Exp Biol. 2010; 48:199-207.

[17] Shakya AK. Medicinal plants: Future source of new drugs. Int J Herbal Med. 2016; 4(4): 59-64.

[18] Shaikh AM, Shrivastava B, Apte KG and Navale SD. Medicinal Plants as Potential Source of Anticancer Agents: A Review. J Pharmacogn Phytochem. 2016; 5(2): 291-295.

[19] Debjit B, Umadevi M, Sampath KP and Duraivel S. Traditionally Used Anticancer Herbs In India. J Med Plants. 2013; 1(3): 56-74.

[20] Krishnapriya TV and Suganthi A. Biochemical and phytochemical analysis of *Colocasia esculenta* (L.) Schott tubers. Int J Res Pharm Pharmaceut Sci. 2017; 2(3): 21-25.

[21] Dhale DA and Bhai S. Pharmacognostic characterization and phytochemical screening of *Achyranthus aspera* Linn. Curr Agri Res J. 2013; 1(1):51-57.

[22] Kuladeep G and Pathak AK. Pharmacognosy and phtytochemical evaluation of *Tridax procumbens* Linn. J Pharmaco Phytochem. 2013; 1(5):42-46.

[23] Paliwal R, Madungurum MA and Naziru D. Phytochemical analysis, physiochemical activity and antibacterial effects of Cinnamon zeylanicum (dalchini) extracts. Int J Engineer Sci Res Technolo. 2018; 7(4):162-170.

[24] Rawal JR and Priya SR. Determination of Bioactive Components of *Cynodon dactylon* by GC-MS Analysis & it's In Vitro Antimicrobial Activity. Int J Pharm Life Sci. 2016; 7(1): 4880-4885.

[25] Singh TG, Gupta S, Singh S and Gupta R. Pharmacological and Phytochemical updates on *Pothos scandens* L. Pharmacogn. Commn. 2018; 8(4): 138-145.

[26] Apichai PAS, Sooklert K, Satirapipatkul C and Sukrong S. Anticancerous activity of selected *Colocatia gigantia*. J Med Assoc Thai. 2015; 98(1): 98-106.

[27] Gireesha J and Raju NS. Ethno botanical study of medicinal plants in BR Hills region of Western Ghats, Karnataka. Asian J Plant Sci Res. 2013; 3(5):36-40.

[28] Sathiyaraj R, Sarvalingam AA, Balachandran A and Reddy RK. Diversity of Ethnomedicinal Plants in Bodamalai Hills Eastern Ghats, Namakkal District, Tamil Nadu. J Plant Sci. 2015; 3(2): 77-84.

[29] Jones CE and Jones C. Indigenous knowledge and bioprospecting; International conference, 2002 April 21-24; Maquarie University, Sydney, Australia. 2002.

[30] WHO. Inter-regional Workshop on Intellectual Property Rights in the Context of Traditional Medicine, Bangkok, December 2000.

[31] Akshay KR, Sudharani N, Anjali KB and Deepak TM. Biodiversity and strategies for conservation of rare, endangered and threatened medicinal plants researchand reviews. J pharmacogn phytochem. 2014; 2 (3): 12-20.

[32] Michael RW, Rands WM, AdamsLB, Stuart HM, *et al*. Biodiversity Conservation: Challenges Beyond. Science.2010; 329: 1298-1303.

[33] Ashok KP and Tripathi YC. Ethnobotany and its relevance in contemporary research. J Med Plants Studies. 2017; 5 (3): 123-129.

[34] Sharma S and Thokchou R. A review on endangered medicinal plants of India and their conservation. J Crop and Weed. 2014; 10(2):205-218.

[35] Giday MZ, Asfaw and Woldu Z. Ethno medicinal study of plants used by Shekoethnic group of Ethiopia. J Ethnopharmacol.2010; 132:75-85.

[36] Durairaj P and Kamaraj M. Ethnobotanical Studies on Plant Resources of Trichirapalli District Tamilnadu, India. Int J Human Art Med Sci. 2013; 1(3); 17-30.

[37] Shah A, Bharati KA, Ahmad J and Sharma MP. New ethnomedicinal claims from Gujjar and Bakerwals tribes of Rajouri-Poonch districts of Jammu and Kashmir, India. J Ethnopharmacol.2015; 166:119-128.

Chapter 13

Traditional Usage of Plants of Costus Species in Assam, India

Biman Bhuyan, Dipak Chetia and Prakash Rajak

Abstract

Customary use of plants in the treatment of ailments in Assam, India is a typical situation. Ethno medicinal study was led in a few topographically unique zones of the state and utilization of plants from Costus species were reported. The extent of study chose for the investigation range across seven organizational regions spread across Assam, India. The regions include Dibrugarh, Golaghat, Tinsukia, Dhemaji, Karbi Anglong, Goalpara and Kokrajhar. Different plants were reported and plants fitting with the said species were chosen for determining the relevance concerning its use in customary medication. The survey divulged that plants associated to three species of the genus Costus namely *Costus speciosus*, *Costus pictus* and *Costus scaber* were espied to be primarily ubiquitous in traditional medicine in the discrete contemplated regions. The species were predominantly utilized as prime ingrediants in hepatoprotactive and anti-diabetic formulations. *Costus speciosus* was perceived to be chiefly used in the treatment of hepatic disorders and ailments. *Costus pictus* was observed to be used customarily in the upper Assam region bordering Nagaland for treating diabetes and *Costus scaber* was being used in the area bordering Arunachal Pradesh for tending people with jaundice, snake bite etc. The research climaxed with the profiling of the costus species as annotated from the ethnomedicinal survey.

Keywords: Costus, Ethnomedicine, Assam, *Costus speciosus*, *Costus pictus*, *Costus scaber*

1. Introduction

Customary medical understanding is undergoing augmented consideration globally in health sector. The importance of traditional medicine in catering the health needs cannot be undermined. The herbal medicine sector commercially is already booming with the annual turnover crossing billions of dollars. With the passage of time newer knowledge is being incorporated substantially thereby highlighting the significance of documentation aspects pertaining to these medicinal plants and practices associated with herbal medicine.

Documentation based upon ethnomedicinal survey along with interaction with local healers practicing traditional system of medicine can be said to be the basis for establishing a systematic protocol for validating traditional medical knowledge.

2. Ethnomedicinal survey area

Assam was selected as the targeted study area due to the rich diversity in flora, fauna and above all due to the presence of diverse ethnic groups with a

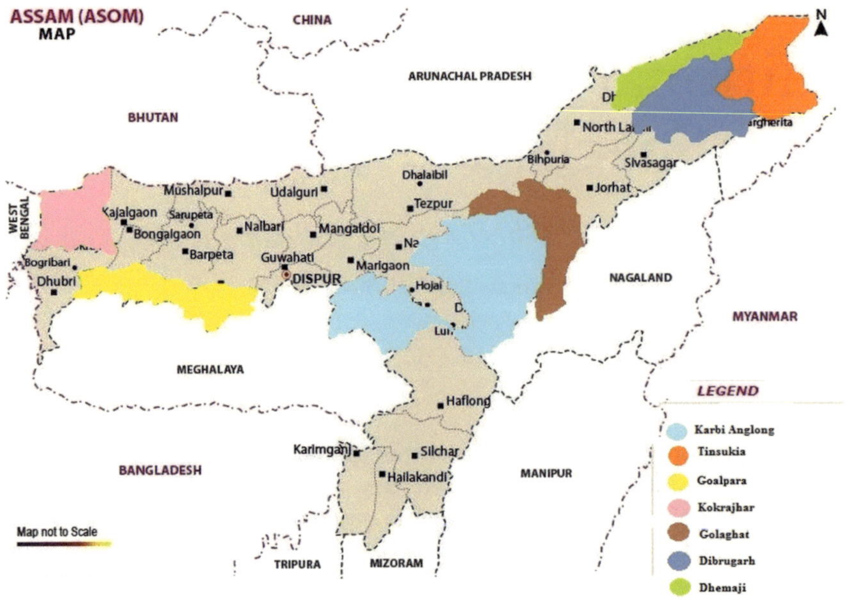

Figure 1.
Map of Assam showing different districts where ethnomedicinal survey was conducted.

wide array of traditional practices. Several geographically distinct zones, encompassing seven administrative districts spread across Assam were considered for the study (**Figure 1**).

The selected areas in which the ethnomedicinal survey was done are as follows:

1. Nagakhelia village and Jokai area, Dibrugarh

2. Naojan and Borghoria area, Golaghat

3. Laipuli, Tinsukia

4. Majarbari village and Sissiborgaon, Dhemaji

5. Kathkatia village, Karbi Anglong

6. Dhupdhora, Goalpara

7. Dotma, Kokrajhar

2.1 Survey area: Dibrugarh

Dibrugarh is known as the Tea City of North-East.To the north and east lays Dhemaji and Tinsukia district respectively. South-east and south-west parts of Dibrugarh are bounded by Tirap and Sivsagar district [1–4].

Two places in Dibrugarh district were selected for ethnomedicinal survey viz. Nagakhelia and Jokai. Nagakhelia is a small village, consisting of around hundred households under Barbaruah block of Dibrugarh district lays about 6 km from Dibrugarh University [5]. The village is located on the banks of river Brahmaputra and the area boasts of thick vegetation which serves as a prime source of medicinal plant materials for the local healers of the area practicing traditional medicine.

Jokai comes under Barbaruah block in Dibrugarh district. It is located about 10 km south from Dibrugarh University. It is also home to the over twelve hectare Jokai reserve forest within which Jokai Botanical Garden cum Germplasm Centre is located. The reserve forest is endowed with different flora species of medicinal, oil bearing and aromatic plants. It also has diverse fauna species like flying squirrel, black panther and leopard including various species of butterflies and fishes. The villages surrounding the forest areas in Jokai has a rich heritage of prescribing traditional medicine, mostly from plants for many types of ailments like jaundice, diabetes, malaria, fever, skin infection etc.

2.2 Survey area: Golaghat

Golaghat is an important district of upper Assam having its own historical and cultural heritage. Golaghat is bordered by the Brahmaputra River in north, towards south lays Nagaland, whereas in the east it is bounded by Jorhat district and the western side lays Karbi Anglong and Nagaon district [4]. The major rivers of the district are Brahmaputra, Dhansiri, Kakodonga, Doyand, Gelabil and Diplolu [6]. The vast geography of Golaghat district also includes tropical evergreen and semi evergreen forest; tropical grassland in Kaziranga National Park and swampy vegetation. The topography of Golaghat is dominated by a diverse array of flora and fauna [7, 8].

Borghoria and Naojan were the areas selected for ethnomedicinal survey in Golaghat district. Borghoria village and Naojan are located about 30 km and 60 km from Golaghat town and about 2.5 km and nearly 70 km from Numaligarh Refinary township, respectively. Naojan, due to its close proximity to Barpathar, an archeological site where the remains of an 8th century temple made of square bricks and a stone inscription of Brahmi characters belonging to the 5th century were excavated along with the hot water springs and Garampani Wildlife Sanctuary of Garampani, has a very rich abundance of diverse flora and fauna. Borghoria situated in the vicinity of Dhansiri river has an exposure to vast and varied natural resources. Traditional healers around the area are mainly engaged in agricultural activities and prescriptions of traditional medicine by these healers are done on philanthropic basis [3].

2.3 Survey area: Tinsukia

Tinsukia is situated in the northernmost portion of Assam [2]. The district is surrounded on three sides by Arunachal Pradesh. The south part is ecompassed by Dibrugarh. As the district falls in the far east of North-East region of Assam (India), it is a part of global bio-diversity hot spot and has great biodiversity significance [9, 10]. The high biological diversity found in the district is often related to its forest cover, which is categorized into tropical wet evergreen forests. The important sanctuary located in the district is Dibru-Chaikhowa Sanctuary. It has an area of 640 sq. km and is famous for rare, endangered animals and birds such as white-winged wood duck, elephant, tiger, sambar, buffalo, aquatic avifauna and wild white horse. The other protected areas and important forests are Dum Duma-Dangori-Kumsong Reserve Forests, Tirap-Burihidihing, Sadiya plains, Upper Dihing (East) and Upper Dihing (West).

Ethnomedicinal survey in Tinsukia districted was conducted in Laipuli area. Laipuli is located at a distance of around 6 km from Tinsukia town [3].

2.4 Survey area: Dhemaji

Situated in the northern bank of the mighty river Brahmaputra, Dhemaji can be suitably described to be located in one of the remote area of north eastern region of India. In its northern and eastern end the state of Arunachal Pradesh lies. The western part is bounded by Lakhimpur district followed by river Brahmaputra in the South. Dhemaji has a total geographical area of 3237 sq. km [1–4].

Two places selected for the ethnomedicinal survey in Dhemaji district were Majarbari and Sissiborgaon.

2.5 Survey area: Karbi Anglong

The district of Karbi Anglong is located in the central Assam region. The eastern part is surrounded by Golaghat district, in its west lies the state of Meghalaya and Morigaon district, the north is bounded by Nagaon and Golaghat district whereas North Cachar Hills and the state of Nagaland is located towards south. Karbi Anglong district is home to thick forest cover having numerous species of flora and fauna. It is to be noted that a new district, West Karbi Anglong was curved out from erstwhile Karbi Anglong district on 15th of August, 2015 [1, 3].

The district can be broadly divided into two physiographic units' viz. hills and plains. About 85 percent of the district is covered by hills [4]. Environmental and topology studies of Karbi Anglong specify a great degree of diversity among the existing plant and animal species. The forest areas serves as the natural gene bank of important types and sub types pertaining to various species.

Kathkatia village located in Silonijan of Karbi Anglong district was selected for the ethnomedicinal survey [11].

2.6 Survey area: Goalpara

Goalpara is sited towards the southern bank of Brahmaputra River. The district is surrounded by the state of Meghalaya in the South, towards east lays Kamrup district, the western end is bounded by Dhubri district and, the northern part is covered by the mighty Brahmaputra. In 1983, Goalpara Civil sub-division was separated from original Goalpara district to form the present Goalpara district [1, 2].

Dhupdhara selected for the ethnomedicinal survey, is a village in Rongjuli circle in Goalpara district of Assam. It is located about 58 km east of district headquarter Goalpara and 13 km from Rangjuli [3, 4].

2.7 Survey area: Kokrajhar

Kokrajhar district is the entry point to the NER of India. It is boardered by Bhutan in the north, followed by the district of Dhubri in its south, whereas Bongaigaon and West Bengal is situated in the east and west directions.

On the 1st of July, 1983 the Kokrajhar Sub-division was upgraded into Kokrajhar district with headquarter at Kokrajhar town [3]. The district is situated in a humid sub-tropical climate, which is the characteristic of the lower Brahmaputra Valley of Assam. The district also has one of the largest concentrations of forest in the state. About 55% of the total geographical area of the district is under reserved forest. The Bhutan hills are the source of a number of rivers that flow through the district and act as tributaries of the mighty Brahmaputra that flows from east to west far from the southern boundary of Kokrajhar district [4].

Dotma village in Kokrajhar district of Assam was selected for the survey for ethnomedicinal documentation. It is located about 17 km towards North from District head quarters Kokrajhar, 188 km from State capital Dispur towards East. Dotma is bounded by Kokrajhar town towards East, Kachugaon towards west, Rupshi towards west, Chapor-Salkocha towards west. Kokrajhar, Bilasipara, Bongaigaon, Gauripur are the nearby towns to Dotma [12].

3. Documentation of medicinal plants in the surveyed areas

Plants surveyed in Dibrugarh region were documented on the basis of interview and questionnaire with the traditional healers with emphasis on the part of the plants and their applications in treating different diseases and disorders (**Table 1**).

Plants in the surveyed areas of Golaghat district were subjected to documentation on the basis of interview and questionnaire with the traditional healers with

District	Plant name		Part used	Use/applications
	Botanical name	**Local name**		
Dibrugarh	*Asparagus racemosus*	Sotmul	Root	Kidney stone
	Averrhoa carambola	Kordoi	Leaves, Fruit	Jaundice
	Bonnaya brachiata	Horu Kasidoria	Leaves	Wound healing
	Cassia fistula	Sonaru	Bark	Fever, Deworming
	Caesalpinia bonducella	Letaguti	Seed	Wound healing
	Cassia tora	Bilokhoni	Leaves	Skin infection, Snake bite, Joint pain
	Centella asiatica	Barmanimuni	Whole plant	Wound healing, Well being
	Cleodendrum viscosum	Dhapat tita	Leaves, Root	Malaria, Diabetes, Jaundice, Skin infection
	Costus speciosus	Jomlakhuti	Rhizome	Jaundice
	Coscorus olitorius	Meetha Pat	Leaves	Body pain, dysentery, piles, fever
	Cucumis sativus	Tiyanh	Leaves, Fruit	Bleeding nose, Diabetes
	Dillenia indica	Ow tenga	Fruit	Constipation, Stomach trouble
	Drymaria cordata	Laijabori	Aerial part	Fever, stomach ache
	Eupatorium cannabinum	Tongloti	Root	Tooth ache
	Euphorbia nerifolia	Hiju	Latex	Asthma
	Hiptage benghalensis	Madhoi maloti	Root	Asthma
	Houttuynia cordata	Mosonduri	Leaves	Constipation
	Leucas apsera	Durum bon	Aerial parts	Cough, Fever
	Momordica dioica	Bhat kerela	Root	Urinary problems
	Murrya koenigii	Narashinha	Leaves, Tender aerial parts	Stomachic
	Naravelia zylenica	Gorob choi	Aerial parts	Tooth ache, Skin infection
	Paederia foetida	Bhedai lota	Aerial parts	Stomach problem, Constipation, Joint pain
	Physalis peruviana	Kopalphoota	Aerial parts	Jaundice
	Polygonum chinense	Modhuhuleng	Aerial parts	Stomach trouble, Dysentery
	Rosa centifolia	Tezi gulap	Flower	Eye infection
	Sapindus mukorossi	Monisal	Fruit	Tonsillitis
	Sarcochlamys pulcherrima	Mesaki	Leaves	Infection, Diarrhea, Dysentery
	Spondias pinnata	Omora	Fruit	Acidity, Stomach trouble
	Stereospermum chelonoides	Paroli	Leaves	Skin infection
	Stephania hernandifolia	Tubuki lota	Leaves	Wound healing
	Syzygium jambolanum	Kola jamuk	Seed	Diabetes, Stomach trouble
	Sida rhombifolia	Hunbarial	Leaves	Body pain, Joint pain
	Vitex negundo	Pochotia	Leaves	Fever, Cough

Table 1.
Some of the medicinal plants used in Dibrugarh district and their allied applications.

emphasis on the part of the plants and their applications in treating different diseases and disorders. Some of the plants are listed in **Table 2**.

Plants in Tinsukia district, surveyed areas were documented on the basis of interview and questionnaire with the traditional healers with emphasis on the part of the plants and their applications in treating different diseases and disorders (**Table 3**).

District	Plant name		Part used	Use/applications
	Botanical name	Local name		
Golaghat	*Achasma loroglossum*	Kor Phool	Rhizome	Tooth ache
	Aegle mermelos	Bel	Leaves, Fruit	Kidney problem, Dysentery
	Adiantum capillus	Chuli dhekia	Aerial part	Wounds, Infection, Tooth ache
	Averrhoa carambola	Kordoi	Fruit	Jaundice, Diarrhea, Dysentery
	Ageratum conyzoides	Gandhalibon	Leaves	Cuts and wound
	Alpinia allughos	Tora	Rhizome	Stomach trouble, Joint pain
	Alternanthera sessilis	Mati Kanduri	Aerial part	Constipation
	Baccuarea sapida	Leteku	Fruit	Stomach problem
	Borreria hispda	Dolicha Bon	Leaves	Tooth ache, Gum swelling
	Bryophyllum calycinum	Dupor tenga	Leaves	Leaves Kidney stone
	Cissus repens	Bogi tenga	Leaves	Menstrual discomfort
	Clenogyne dichotoma	Patidoi	Stem	Support in fracture
	Costus speciosus	Jomlakhuti	Rhizome	Jaundice, Diabetes
	Costus pictus	Leteki	Aerial parts	Diabetes
	Cinnamomum bejalghota	Patihunda	Leaves	Asthma, Cough
	Clitoria ternatea	Aparijita	Root, Flower	Fever, Snake bite, Infection of skin
	Croton bonplandianum	Bonoria jaifal	Seed	Laxative
	Cissampelos pareira	Tubuki lota	Leaves	Diabetes
	Eclipta alba	Kehraj sesu	Leaves	Blood clotting
	Heydichium coronarium	Pakhila phool	Rhizome	Joint pain
	Hydrocotyl sibthropioides	Horu manimuni	Whole plant	Fever, Stomach problem
	Leucas aspera Durun	Durun Bon	Leaves	Snake bite, Sinusitis
	Litsea salicifolia	Dighloti	Leaves	Insect repellent
	Phyllanthus niririi	Bon Amlokhi	Shoot	Stomach trouble, Urinary problem
	Polygonum chinense	Madhu huleng	Aerial parts	Diarrhea
	Sarochlamys pulcherrima	Mesaki	Aerial parts	Tapeworm infection
	Sida rhombifolia	Hunbariol	Root	Helps in child birth for pregnant women
	Smilax perfoliata	Tikoni barua	Leaves, Root	Wound healing
	Styrex serulatum	Lota madhuri	Shoot	Anti infective
	Triumfetta rhomboidea	Bon Agora	Aerial parts	Insect repellent
	Xanthozylum nitidum	Tejmuri	Stem	Fractured bone

Table 2.
Some of the medicinal plants used in Golaghat district and their allied applications.

District	Plant name		Part used	Use/applications
	Botanical name	Local name		
Tinsukia	*Abroma augusta*	Gorokhia korai	Root	Urinary disorders
	Abrus precatorius	Latumoni	Root	Urinary disorders
	Achyranthes aspera	Bionihakuta	Leaves, Root	Wound, Sore throat, Cough and Cold
	Acorus calamus	Bosh	Rhizome	Acidity
	Amaranthus spinosus	Hatikhutura	Root, Aerial parts	Diarrhea, Increases milk output in lactating mother
	Amaranthus tricolor	Bishalya karani	Leaves	Wound healing
	Alternanthera sessilis	Mati kanduri	Aerial parts	Dysentry, Stomach trouble
	Caesalpinia bonduc	Letaguti	Seed	Fever, Body pain
	Caryota urens	Sewa	Root	Increases milk output in lactating mother
	Cascabela thevetia	Karabi	Seed, Bark, Latex	Anti-infective, Diabetes, Fever
	Celtis tetrandra	Hukuta	Tender Aerial parts	Relieves pain after child birth
	Centalla asiatica	Bormanimuni	Whole plant	Health tonic, Memory enhancer
	Cinnamomum bejolghata	Patihonda	Leaves	Diabetes
	Ipomoea aquatic	Kolmou	Leaves	Diabetes
	Cissus quadriangularis	Harjura lota	Stem, Tendrils	Wound, Fracture
	Citrus grandis	Robab tenga	Fruit	Jaundice, Deworming
	Clerodendron colebrookianum	Nephafu	Leaves	Hypertension
	Costus pictus	Leteki	Leaves	Diabetes, Blood purification
	Costus speciosus	Jomlakhuti	Rhizome, Leaves	Jaundice, snake bite
	Croton joufra	Gochmahudi	Leaves	Menstrual discomfort
	Curanga amada	Bhui tita	Leaves	Fever, Malaria
	Curcuma amada	Aam ada	Rhizome	Diarrhea, Dysentery
	Cuscuta reflexa	Akashi lota	Stem	Jaundice, Wound healing
	Garcinia cowa	Kuji thekera	Fruit	Diarrhea, Dysentery
	Garcinia lancifolia	Rupahi thekera	Fruit	Gastric discomfort, Diarrhea
	Hibiscus sabdarifolia	Tengamora	Aerial parts	Diarrhea, Dysentery
	Houttuynia cordata	Mosondori	Leaves, Tender shoot	Flatulence, Diarrhea, Dysentery
	Lasia spinosa	Sengmora	Rhizome, Aerial parts	Menstrual discomfort
	Lindernia pursilla	Gakhiroti bon	Whole plant	Increases milk output in lactating mother
	Lygodium flexuosum	Kopou dhekia	Leaves	Fungal infection
	Malastoma malabathricum	Phutuki	Leaves	Wound healing
	Mussandra roxburghii	Hukloti	Aerial parts	Stomach problems
	Vetivera zizanoides	Birina	Root	Rheumatic pain

Table 3.
Some of the medicinal plants used in Tinsukia district and their allied applications.

Plants in Dhemaji district selected areas were documented on the basis of interview and questionnaire with the traditional healers with emphasis on the part of the plants and applications in treating different diseases and disorders (**Table 4**).

Documentation of plants in Karbi Anglong district, surveyed areas was then done on the basis of interview and questionnaire with the traditional healers with emphasis on the part of the plants and their applications in treating different diseases and disorders (**Table 5**).

Documentation of plants in the surveyed region of Goalpara district was initiated on the basis of interview and questionnaire with the traditional healers with

District	Plant name		Part used	Use/applications
	Botanical name	**Local name**		
Dhemaji	*Abroma augusta*	Ui-sipak	Leaves	Cuts and wound healing
	Ageratum conyzoides	Namnyin/ Gunduabon	Aerial parts	Aids blood clotting, Wound healing
	Alternanthera sessilis	Patang oying	Aerial parts	Jaundice, Body ache
	Bombax ceiba	Singgi	Leaves	Wound healing
	Catharanthus roseus	Sada Bahar	Leaves	Diabetes
	Calotropis gigantean	Akon	Leaves, Latex	Wound healing, Body ache
	Caesalpinia cucullatum	Tezmuri	Leaves	Tooth ache, Fever
	Chromolaena odorata	Jarmanibon	Leaves, Root	Snake bite, Anti infective
	Cissus quadrangularis	Gomset sori	Aerial parts, Tendrils	Tendrils Joining of fractured bone
	Costus scaber	Keuri	Leaves	Snake bite, wounds
	Costus speciocus	Peki jigjig	Rhizome	Jaundice, UTI
	Cyclosorus extensus	Rukji	Leaves	Increases milk output in lactating mother
	Desmodium laxiforum	Bhuter chira	Aerial parts	Infection, Menstrual discomfort
	Eryngium foetidum	Bormang ori	Leaves	Appetizer, stomach problems
	Ficus hispida	Takpi	Fruit	Jaundice
	Garcinia lanceifolia	Rupohi tehekera	Fruit	Jaundice, Diarrhea
	Houttuynia cordata	Musondri	Leaves	Optimizes stomach function
	Ipomoea aquatic	Mou	Leaves	Jaundice, Diabetes
	Mentha arvensis	Takemare	Leaves	Stomach trouble
	Mimosa pudica	Yuptap	Root	Deworming
	Musa velutina	Doge kopak	Flower	Diarrhea, Dysentry
	Litseacitr ata	Mezangkori	Bark	Asthma, Cough
	Solanum nigrum	Loshkosi	Leaves	Jaundice
	Tylophora asthamatica	Jangli pikran	Leaves, Roots	Purify blood, Stops white vaginal discharge
	Oxalis corniculata	Tengsi	Leaves	Hypertension, Diabetes, Stomach upset
	Zanthoxylum nitidum	Rikom	Aerial parts	Anti infective

Table 4.
Some of the medicinal plants used in Dhemaji district and their allied applications.

District	Plant name		Part used	Use/applications
	Botanical name	Local name		
Karbi Anglong	*Acmella paniculata*	Bapchuki	Leaves, Flower	Stomach ache, Acidity
	Abelmoschus moschatus	Arnam hanserong	Leaves, Fruit	Snake bite
	Abrus precatorius	Chuselok	Leaves	Fever, Asthma, Joint pain
	Abutilon indicum	Mir-at	Leaves, Flower	Snake bite, Insect bite
	Acacia pennata	Themra/Khemra	Leaves, Bark	Snake bite
	Alpinia galangal	Phrikan gnek	Leaves, Rhizome	Stomach ache, Improves digestion
	Alternanthera sessilis	Raeaba	Aerial parts	Fever, Infection
	Amorphophalus bulbifer	Hen salku	Leaves, Flower	Piles, Irregular bowel movement
	Arisaema tortuosum	Chamua	Leaves, Tuber	Piles, Irregular bowel movement
	Calamus rotang	Pri	Aerial parts	Snake bite
	Cassia tora	Bapduli	Leaves, Flower	Joint pain, Improves bowel movement
	Costus pictus	Tui	Leaves	Diabetes, Jaundice
	Costus speciosus	Ai-upo	Leaves, Rhizome	Jaundice, Snake bite
	Cycas pectinata	Or-oh	Aerial parts	Acidity, Heart burn
	Lasia spinosa	Chusot	Aerial parts	Piles, Irregular bowel movement
	Laportea cremulata	Bap kangsam	Fruit, Flower	Scorpion bite
	Murraya koenigii	Thengsakso	Leaves	Acidity, Fever
	Olax acuminate	Hanboka	Leaves	Wound healing
	Oroxylum indicum	Nopak ban	Leaves, Flower	Intestinal worm, Stomach ache
	Paederia foetida	Rekang nemthu	Leaves	Acidity
	Physalis peruviana	Thebongkang	Leaves, Fruit	Stomach ache, Deworming
	Phlogocanthus thyriflorus	Titaful	Flower	Fever, Jaundice
	Solanum torvum	Bhekuri tita	Leaves, Fruit	Anti infective
	Spondias pinnata	Siming	Leaves, Flower	Acidity, Diarrhea
	Tagetes erecta	Mir kadomphui	Leaves, Flower	Anti infective, Wound healing, Improves digestion
	Vitex negundo	Vorke abap	Leaves, Flower	Fever, Ache, Malaria

Table 5.
Some of the medicinal plants used in Karbi Anglong district and their allied applications.

District	Plant name		Part used	Use/applications
	Botanical name	Local name		
Goalpara	*Abroma augusta*	Dadhubedang	Leaves	Stomach ache, Ringworm infestation
	Acalypha indica	Muktaborcha	Leaves	Asthma, Bronchitis
	Calamus rotang	Batbelai	Leaves	Eye infection
	Clerodendrum bracteatum	Vate gakha	Leaves	Memory tonic
	Calotropis gigantia	Aakon	Leaves, Bark	Snake bite, Asthma
	Deeringia amaranthoides	Matak tuka	Leaves	Wound, Sore
	Euphorbia hirta	Dudh bon	Shoot, Latex	Infection
	Ficus hispida	Domuru	Leaves	Jaundice
	Murraya koenigii	Narasinghabelai	Leaves, Tender aerial parts	Fever, Stomach upset
	Nelumbo nucifera	Podum	Rhizome	Menstrual discomfort
	Ocimum sanctum	Dhulungshi	Leaves	Cough, Fever
	Paederia foetida	Bhadalilewa	Leaves	Diarrhea, Dysentry
	Polyalthia longifolia	Debdaru	Bark	Menstrual discomfort
	Solanum integrifolium	Tita Bhekri	Fruit	Malaria, Fever, Jaundice, Diabetes
	Terminalia tomentosa	Amra	Fruit	Diabetes, Stomach upset
	Vitex negundo	Pasatia	Leaves	Body pain, Wound, Fever

Table 6.
Some of the medicinal plants used in Goalpara district and their allied applications.

District	Plant name		Part used	Use/applications
	Botanical name	Local name		
Kokrajhar	*Benincasa hispida*	Kumbra	Fruit, Leaves	Diabetes, Acidity
	Canarium bengalensis	Dhuna	Leaves,	Bark Joint pain
	Chromolaena odorata	Bangrilewa	Leaves	Stomache ache, dysentery
	Chrystella parasitica	Daokhumwi	Young aerial part	Wound healing
	Clerodendum infortunatum	Lwkwna	Leaves	Jaundice, Wound healing
	Clitonia ternatea	Nilkantha	Leaves	Fever, antiseptic
	Costus speciosus	Buritokon	Rhizomes, Leaves	Jaundice, Snake bite
	Corchorus capsularis	Patw	Leaves, Root	Fever, Diarrhea
	Datura stramonium	Datura	Leaves, Fruits	Tooth ache, Heartburn, Asthma
	Emblica officinalis	Amla	Fruit	Tonic, Stomachic
	Laportea crenulata	Koma	Leaves, Root	Heartburn, Fever, Cuts and Wound
	Leucas plukenetii	Khangsinsa	Leaves	Sinusitis, Pain
	Nyctanthes arbortristis	Sephali	Leaves, Flower	Antihelmintic
	Ocimum sanctum	Tulsi	Leaves	Cough relief, Asthma
	Paederia foetida	Bhedalilewa	Leaves	Diarrhea, Constipation
	Scoparia dulcis	Bongpang rakeb	Whole plant	Kidney stone, Diarrhea, Fever
	Xanthium strumarium	Agara	Root, Leaves	Fever, Joint pain

Table 7.
Some of the medicinal plants used in Kokrajhar district and their allied applications.

emphasis on the part of the plants and their applications in treating different diseases and disorders (**Table 6**).

Plants in surveyed areas of Kokrajhar district were documented on the basis of interview and questionnaire with the traditional healers with emphasis on the part of the plants and their applications in treating different diseases and disorders (**Table 7**).

4. Profiles of Costus species used predominantly in traditional medicine in the surveyed areas

The ethnomedicinal survey conducted in the different areas revealed the prominent use of the species belonging to the genus costus. The species were *Costus speciosus*, *Costus scaber* and *Costus pictus*. Therefore botanical and pharmacognostic profiling of the said species were done accordingly.

4.1 *Costus speciosus* (J. Konig) Smith

Costus speciosus (**Figure 2**) is an erect plant, up to 2.7 meters high; root stock is tuberous; stem is sub-woody at the base. Leaves have an average dimensions of (15–30) cm × (5.7–7.5) cm and are sub sessile, oblong, spirally arranged with silky-pubescent base [13, 14]. The flowers are present in very dense spikes having ovate bracts that are mucronate and bright red in color. The corolla have short tube with lobes which are ovate-oblong subequal. Flower lips are white with yellow center with crisped, concave, disk with a tuft of hair at the base. Fruits are capsule, globosely trigonus and are red in color. The seeds are black with white aril. Flowering time in Indian condition is August to October [13, 15].

It is a herb occurring in the moist and wet evergreen areas of the Indo-Malayan region and Sri Lanka along with Brazil, Bolivia, Colombia, Peru, Mexico etc. Within India it occurs from Central and Eastern Himalayas to Southern India [15, 16].

4.2 *Costus scaber*

Costus scaber (**Figure 3**) is an erect plant, up to 4 meters high; root stock is tuberous; stem is sub-woody at the base. Leaf shape is elliptical with entire margin and

Figure 2.
C. speciosus (J. Konig) Smith collected from Nagakhelia village, Dibrugarh.

Figure 3.
Costus scaber collected from Dhemaji (insert: flower specimen).

are spirally arranged around the stem. The primary bracts are borne on the inflorescence in spiral phyllotaxy. One flowered cincinni occur in the axils of these bracts. Each cincinnus consists of an axis bearing a terminal flower [17]. The floral organs are formed sequentially starting with calyx. Flowering time in Indian condition is October to December.

It is mainly distributed in the neo tropical regions. Within India its geographical distribution is in the sub-Himalayan tract from Kangra district of Himachal Pradesh eastwards to Arunachal Pradesh; and in the Western ghats in Maharastra, Goa, Karnataka, Kerala and Tamil Nadu.

4.3 *Costus pictus* D. Don

Costus pictus (**Figure 4**) is a plant that goes upto 3 meters in height; it has tuberous root with a nearly woody base. The leaf arrangement is spiral with an elliptical shape. Leaf bears rigid and rubbery morphology. Spiral phyllotaxy is observed in

Figure 4.
C. pictus D. Don collected from Naojan, Golaghat.

the primary bracts. The external appearance of the flowers as depicted in **Figure 4** is primarily are creamy colored along with pink stripes initiating from the base. The plant generally bears flower between the months of August and October.

This plant is mainly distributed in the neo tropical regions [18, 19]. In India it found in the sub-Himalayan tract from Himachal Pradesh to Arunachal Pradesh; and in the Western ghats in Goa, Kerala and Tamil Nadu.

5. Conclusion

The state of Assam, popularly known as the land of the red river and blue hills is home to a diverse array of flora and fauna. Assam falls in one of the great migration routes of mankind of different groups who over the centuries have come and settled down. Every community has its own traditional rituals, customs and herbal remedies which have been molded by the geographical location and the environmental factors where they reside. The abundant natural resources in encompassing location form the basis for the characteristic food habits and related medicinal practices of each community. By their experience, the knowledge of herbal remedies was transferred to generation after generation as folk medicine.

A study was conceived based on the aforesaid facts with intent to scientifically analyze different folkloric healing practices encompassing various medicinal plants. Subsequently an ethno medicinal survey was conducted across the state of Assam for compiling information with respect to traditional medicine. Thereafter, plants belonging to Costaceae family were selected for scientific validation studies owing to their predominant use among the traditional healers in the surveyed regions particularly in upper Assam for treating ailments like jaundice, diabetes etc.

Three plants belonging to the costus genus were identified *viz. Costus scaber, Costus speciosus* and *Costus pictus* for the study. *Costus speciosus* locally known as 'Jomlakhuti' in Dibrugarh, Golaghat and Tinsukia district; 'Peki jigjig' in Dhemaji; 'Ai-upo'in Karbi Anglong district and 'Buritokon' in Kokrajhar district, the rhizomes, leaves are primarily used for treating liver aliments, diabetes, UTI, snake bite respectively. *Costus scaber* locally known as 'Keuri' in Dhemaji district, the leaves are used in the treatment of snake bite and wound healing. *Costus pictus* locally known as 'Leteki' in Golaghat and Tinsukia district and 'Tui' in Karbi Anglong district, the aerial parts and leaves are used traditionally in the treatment of diabetes, for blood purification and jaundice respectively.

Therefore, it can be safely concluded that species belonging to this genus are traditionally used in the mitigation of various ailments particularly diabetes. Furthermore, *in vivo* and *in vitro* studies are warrented against these species so as to elucidate viable phyto components as a future prespective.

Acknowledgements

The authors thankfully acknowledges the traditional healers of Dotma, Kokrajhar district; Dhupdhora, Goalpara district; Laipuli, Tinsukia district; Kathkatia village of Silonijan of Karbi Anglong district; Naojan and Baragharia village of Golaghat district; Nagakhelia village and Jokai area of Dibrugarh district; Sissiborgaon, Barmukuli and Majarbari village of Dhemaji district of Assam who helped by sharing their valuable information regarding the methodology of usage of different plant species used in the treatment of ailments. The authors also acknowledge Mrs. Monika Kuli of Barmukuli village of Dhemaji district, Mrs. Sarala

Rabha of Dhupdhora of Goalpara district, Mrs. Minu Borah and Mr. Dhruba Borah of Baragharia village and Mrs. Purnima Borah of Jyotinagar, Golaghat district, Mrs. Savitri Sonowal of Jokai and Mr. Anil Bhuyan and Mr. Ripul Bhuyan of Nagakhelia village of Dibrugarh district, Mrs. Kareng Rongpi of Silonijan of Karbi Anglong district, Dr. Pranjit Narzaree, Ms. P. Narzaree of Kokrajhar district for their immense help regarding the collection of information in the conducted ethnomedicinal survey. The conducted study was not funded by any organization whether government, semi government or private funding bodies whatsoever.

Conflict of interest

"The authors declare no conflict of interest."

Author details

Biman Bhuyan*, Dipak Chetia and Prakash Rajak
Department of Pharmaceutical Sciences, Dibrugarh University,
Dibrugarh, Assam, India

*Address all correspondence to: bimanbhuyan01@dibru.ac.in

References

[1] Barua GC. Ahom-Buranji (with parallel English translation), From the Earliest Time to the End of Ahom Rule. Calcutta: Baptist Mission Press; 1930. p. 43-67.

[2] Gait E. A History of Assam. 2nd ed. Assam, Gauhati: Lawyer's book stall; 1962. p. 5-86.

[3] Rustomji N. Imperilled Frontiers: India's North-Eastern Borderlands. New York:OUP; 1983. p. 39-56.

[4] Sengupta S. Deori N, Sing SK. People of India: Assam (Anthropological Survey of India). Calcutta: Seagull Books; 2003. p. 189-195.

[5] District Census Handbook, Village and Town wise Primary Census Abstract, Series 19, Part XII B. Assam: Directorate of Census Operations, Govt. of India; 2011. p. 60.

[6] District at a Glance: Golaghat. Assam, Golaghat: Dy. Director, Economics & Statistics; 2011. p. 1-19.

[7] Gogoi P. A detail study of the flora of Golaghat sub-division and its neighbouring areas, vol. I and II, Ph. D. Thesis. Guwahati: University of Gauhati; 1981. p. 2-18.

[8] Mahanta PK, Gogoi P. Ethnobotanical studies on Assam, Survey of useful vegetables. Adv Plant Sc, 1988; 1(2):329-334.

[9] Myers N. Threatened biotas: "hot spots" in tropical forests. Environmentalist 1988; 8:187-208.

[10] Myers N. The biodiversity challenge: expanded hot-spots analysis. Environmentalist 1990; 10:243-256.

[11] Development Scenario of Karbi Anglong District. Guwahati: Directorate of Economics and Statistics; 2004. p. 2-34.

[12] Statistical Hand Book. Guwahati: Directorate of Economics and Statistics, Government of Assam; 2008. p. 1-15.

[13] Dutta AC, Dutta TC. Botany. 6th ed. Oxford: Oxford University Press; 1998. p. 599.

[14] Sudhir K. The Medicinal Plants of North East India. Jodhpur: Scientific Publishers; 2002. p. 70.

[15] Basu BD, Kirtikar KK. Indian Medicinal Plants, vol 4. New Delhi: Oscar Publication; 1975. p. 24-40.

[16] Wagner WL, Herbst DR, Sohmer SH. Manual of the flowering plants of Hawaii. Revised ed. Honolulu: University of Hawaii Press; 1999. p. 1381, 898.

[17] Kirchoff BK. Inflorescence and Flower Development in Costus scaber (Costaceae). Can J Bot 1988; 66(2):339-345.

[18] Jiang BQ. Banksea speciosa J. Flora of China, vol. 24. China: König in Retzius; 2001. p. 321.

[19] Sukhdev SH, Dev D, Rakesh KV. Compendium of Medicinal and Aromatic Plants, ASIA, Vol 2. New Delhi: United Nations Industrial Development Organization and the International Centre for Science and High Technology; 2006. p. 58-192.

Chapter 14

Benefaction of Medicinal Plant *Uraria picta*

Harsha Kashyap

Abstract

Medicinal plants are very significant as they not only maintain the health and vitality but most importantly also cure the various ailments in humans and animals without causing any toxic side effects. These are readily available and cost effective therapeutic agents. Uraria picta was first proposed by Desvaux, (1813), is highly medicinal and critically endangered plant species found throughout India and other parts of the world like Africa, Australia, Philippines, Malaysia, Japan, Nigeria etc. This herb is full of antiseptic, anti-inflammatory, antimicrobial, anti-emetic, aphrodisiac, analgesic, cardiovascular and expectorant properties. Due to its high therapeutic use and growing need, the plant is becoming rare and endangered, therefore it is necessary to create awareness of this plant to support its propagation in large numbers. This herb also shows properties of anti-cancer and anti-cholinergic properties hence can manage depressions, anxiety, sleeping problems. Analgesic property helps in reducing body pain.

Keywords: Medicinal plants, health, cost effective, therapeutic, endangered, awareness

1. Introduction

Medicinal plants are very significant as they not only maintain the health and vitality but most importantly also cure the various ailments in humans and animals without causing any toxic side effects. These are readily available and cost effective therapeutic agents. India is known as the "Botanical Garden of the World" as it produces largest medicinal plant wealth. Over the past few decades, herbal medicines from the medicinal plants have been accepted universally. Therefore the efforts are being taken all over the globe in exploring and documenting the ethnomedicinal and pharmacological scientific research data on plants [1].

Uraria picta was first proposed by Desvaux, (1813), is highly medicinal and critically endangered plant species found throughout India and other parts of the world like Africa, Australia, Philippines, Malaysia, Japan, Nigeria etc. [2, 3]. The plant has been extensively used in Ayurvedic system to treat various ailments. All the plant parts are packed with therapeutic properties from flowers, leaves to the fruits and seeds. This herb is full of antiseptic, anti-inflammatory, antimicrobial, anti-emetic, aphrodisiac, analgesic, cardiovascular and expectorant properties [3]. This herb is a major ingredient of well-established Ayurvedic medicine called Dashamoola which is formulation of 10 herbs. The description of the plant can also be seen in classical ancient books of Ayurveda Medicines like Charak Samhita, Sushrut Samhita and Vaghabhatta. There are other formulations in Ayurveda where

IntechOpen

Figure 1.
U. picta whole plant.

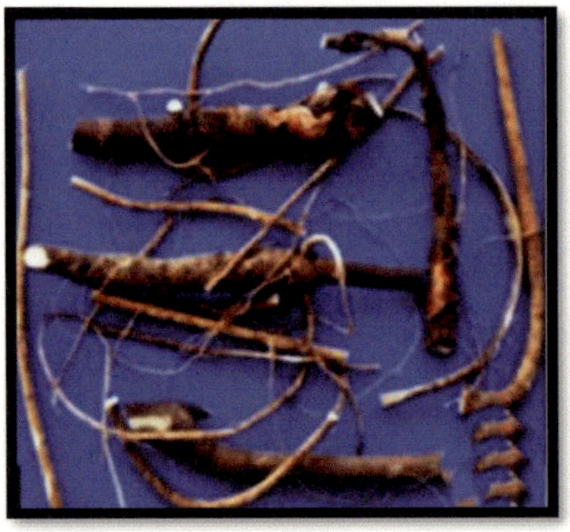

Figure 2.
U. picta plant roots.

Prishnaparni is used like Abana, Amrutharishtha, Angamarda prashamana kashaya churna, Vyaghri taila, Madhyama Narayana taila, Dasamularishtha.

U. picta is found to be an erect [4, 5], branched, perennial herb [6, 7] (**Figures 1** and **2**).

Classification of *U. picta* [8]:

Kingdom	Plantae
Phylum	Tracheophyta
Class	Magnoliopsida
Order	Fabales
Family	Fabaceae
Genus	*Uraria*
Species	*Picta*

2. Morphology

Stem: Suffruticose undershrubs which is 1–1.9 m tall stout stem, hooked hairs.

Leaves: Alternate, compound, unipinnate, imperipinnate, 15–30 cm long approximately.

Leaflate: Subcoriaeous, 6–20 x1–4 cm rounded at base acute green blotched with white and glabrous above minutely pubescent beneath, entire margin.

Stipels: Subulate, 4.5 mm long lower leaves often 1–3 foliolate sometime 5–9 foliolate leaflet, sub orbibicular or oblong.

Flowers: Flowers in close fascicles along the rachis of spicate, cyllendric racemes 10–30 cm long curving upwards with ages pubescent. Standard 5–6 mm long pink or purple.

Pods: Pods glabrous 4–6 jointed joints smooth, polished.

Seeds: Seeds ovate, 2 mm long light brown in color.

Roots: Stout, nodulated, branched.

Synonyms.

There are various synonyms of *U. picta* that are *D. picta* Roxb., *Hedysarum pictum* Jacq., *U. aphrodisiaca* Welw., *U. leucantha* Span., *U. linearis* Hassk.

3. Geographical distribution

U. picta is widely distributed throughout India [2]. It is commonly found in dry grasslands, growing densely and producing poorly viable seeds and it also extend upto Tarai Region of the Himalayas [9]. Apart from India it can also be found in various parts of Asia including China, Japan, Bangladesh, Pakistan, Bhutan and Nepal. The plant is also found in regions of Africa like Nigeria, Egypt, Ethiopia, Congo, South Africa, Queensland Australia [10], and in Philippines, Malaysia [3].

4. Plantation and agroclimatic factors

The plant can grow well in tropical and subtropical regions. It can germinate in humus and sand in comparison to red earth [5]. Loam to clay-loam soil is suitable for its cultivation and the species can grow in the soil pH upto 8.5 [9]. Okusanya et al. (1991) found that the species showed a significantly better growth in wet and moist soil conditions than dry and waterlogged conditions. They also reported that the species responded identically to pH range 3.5, 5.5 and 7.5. The plant also has salinity tolerance [5].

5. Ethnomedicinal importance

According to Charak Samhita this is one of the major ingredient for Dashamoola, Angamardhaprashmana- group of herbs that help to relieve bodyaches, Shothahara-group of herbs having anti-inflammatory properties, Sandhaneeya-group of herbs that are used in fractures and dislocations. Charak suggested liquid formulation cooked with Prishnaparni for diarrhea, cooked with Prishnaparni, parched paddy and processed with *S. cordifolia* for bleeding piles and hemorrhage. According to Sushruta, for gout, milk cooked with Prishnaparni added with honey was given, and promoting adhesion of fractured bones, powdered roots of the Prishnaparni were given [11, 12]. The root decoction of this plant is being used to treat Fever, cough and cold [2, 4, 11, 13]. Yusuf et al. (2007) suggested leaf paste can be applied on boil to burst [14]. Igboechi et al. (1989) studied the ethnomedicinal properties of the plant in Nigeria, can be used in the control of ectoparasites in men and domestic animals [15]. Plant was also assessed for acaricidal activity on *Ixodes ricinus*.

According to Billore et al. (2004) the plant can be used for gynecological disorders [16]. The plant is also effective for the treatment of gonorrhea [17]. Leaves of the plant are used as a diuretic, aphrodisiac, general antiseptic and to cure oral sores [5]. Whole plant shows the antivenom activity against *Echis carinata* [2, 18]. Extract of dried arboreal parts of plant species reported to show antimicrobial properties [7, 19]. This herb also used to cure malarial fever. Also shows cardiovascular properties. This also used to maintain the good HDL level in body and lowers level of LDL and maintains the healthy blood pressure in body. This herb also shows properties of anti-cancer and anti-cholinergic properties hence can manage depressions, anxiety, sleeping problems. Analgesic property helps in reducing body pain.

6. Phytochemical constituents

Phytochemical constituents show specific physiological action individually or in combination with other constituents on human body [20]. Phytochemical constituents include both Primary and secondary metabolites and hence form the backbone of the modern medicine [21]. Important Phytochemical constituents of *U. picta* include alkaloids, flavonoids, steroids, terpenoids, pterocarpans, saponins, phenols, tannins, carbohydrates, proteins, cardiac glycosides etc. [22, 23]. Various mineral components or inorganic nutrients are also required for the human health. This herb is packed with essential minerals that are Sodium, Magnesium, Potassium, Calcium and Phosphorus [24]. Among these phytochemicals, flavanoids, alkaloids and pterocarpans are the bioactive compounds (**Figure 3**).

Quantitative phytochemical study by Madhikatti, 2011 shows that the *U. picta* is packed with primary and secondary metabolites [23] (**Figure 4**).

Saxena et al. 2014 studied the mineral contents (**Figure 5.**) of the *U. picta* in different plant parts which showed the ample amount of mineral elements. This becomes useful when requirements of these minerals in the body are considered [20].

7. Pharmacological properties

The medicinal value of the plant is characterized by the phytochemical constituents and their pharmacological properties. Alkaloids possess various pharmacological properties such as antiarrhythmic, anticholinergic, analgesic, antitumor, antihypertensive, antipyretics, antimalarial, stimulant, anti-HIV, antileukemic etc., and are often utilized as recreational drugs [20, 25]. Flavonoids are the commonly found polyphenolic compounds in the human diet and these are present all over the plant. The pharmacological property of flavonoids include CNS activity, cardiotonic, lipid lowering, antiulcer, hepatoprotective, anti-inflammatory, antineoplastic, antimicrobial, antioxidant and hypoglycemic activity. Intake of food containing Flavonoids lowers risk of certain free radical related pathophysiology [26]. Therapeutic uses of terpenoids include antimicrobial, antifungal, antiviral, antihyperglycemic, anti-inflammatory, antioxidants, antiparasitic, immunomodulatory, and as skin permeation enhancer. Luo et al. (1999) have reported terpenoids can decrease the blood sugar level in animals. Steroids can possess the analgesic properties [27]. They are also used to treat a variety of inflammatory diseases and conditions. Cardiac Glycosides also show therapeutic properties and are used in the treatment of congestive heart failure and cardiac arrhythmia [28]. Phenols or phenolic compounds show antimicrobial, antioxidant, anticarcinogenic, or antimutagenic and anti-inflammatory effects. Also contribute to their inducing apoptosis by arresting cell cycle, regulating carcinogen metabolism and ontogenesis expression,

Alkaloids Isolated From Aerial Parts Of Plant	
5-Methoxy N, N-Dimethyl Tryptamine	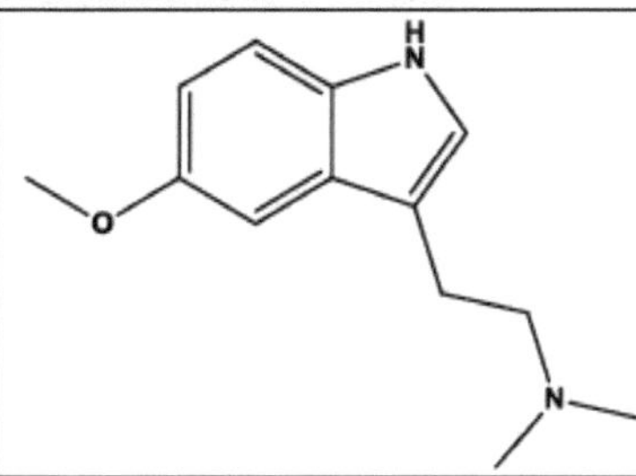
B-Carbolinium Cation	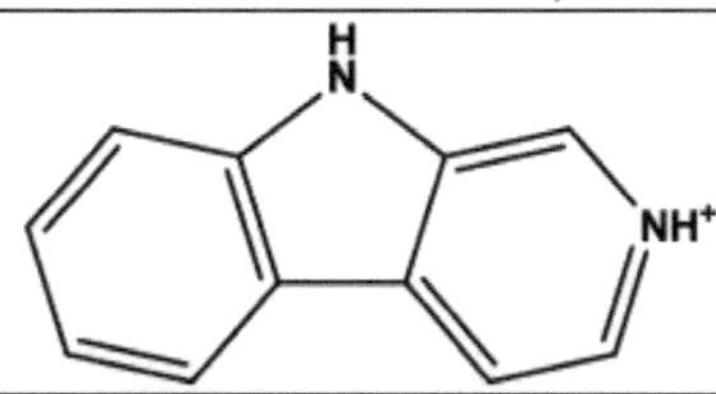

Flavones Isolated From Aerial Parts Of Plant	
4,5,7-Trihydroxy-8-Prenylflavone 4-O-A-L-Rhamnopyranosyl-(1→6)-B-D-Glucopyranoside	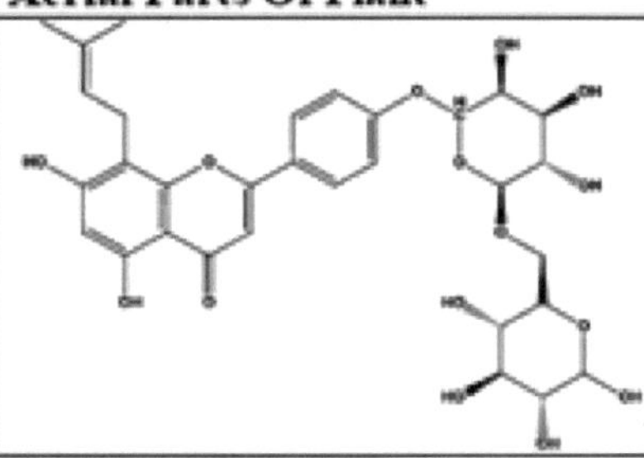
8-C-Prenyl-5, 7, 5-Trimethoxy-3, 4-Methylenedioxyflavone	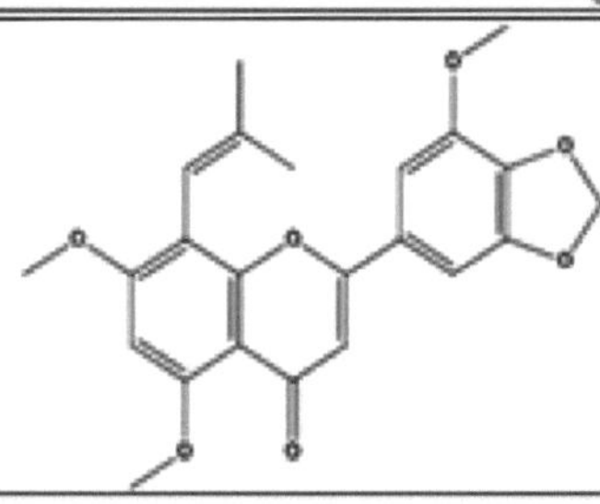
Rutin	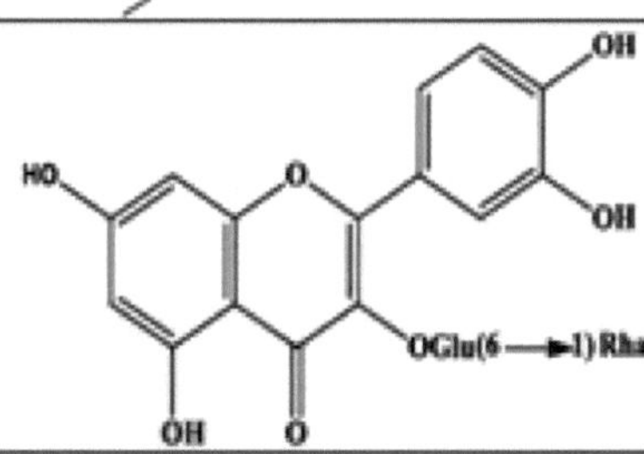
Quercetin-7-*O*-B-D-Glucopyranoside	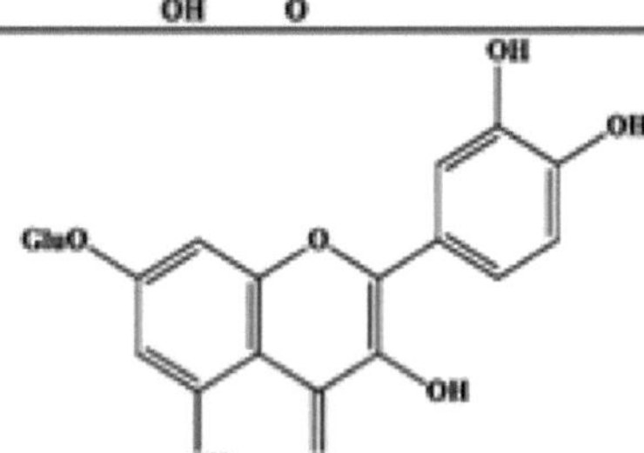

Figure 3.
Chemical structures of active phytochemical constituents representatives isolated from U. picta.

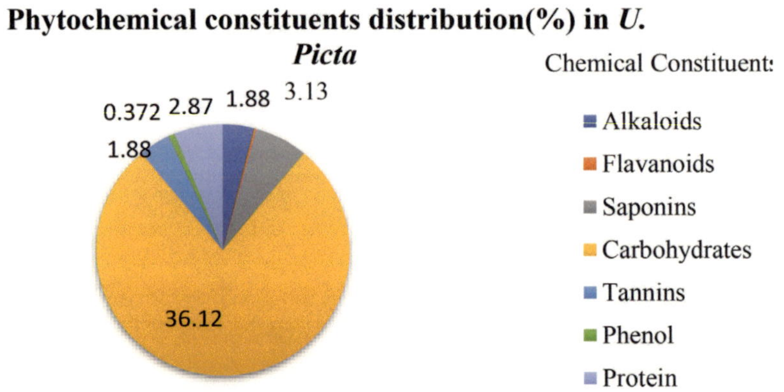

Figure 4.
Graphical representation of the quantitative data of phytochemical constituents.

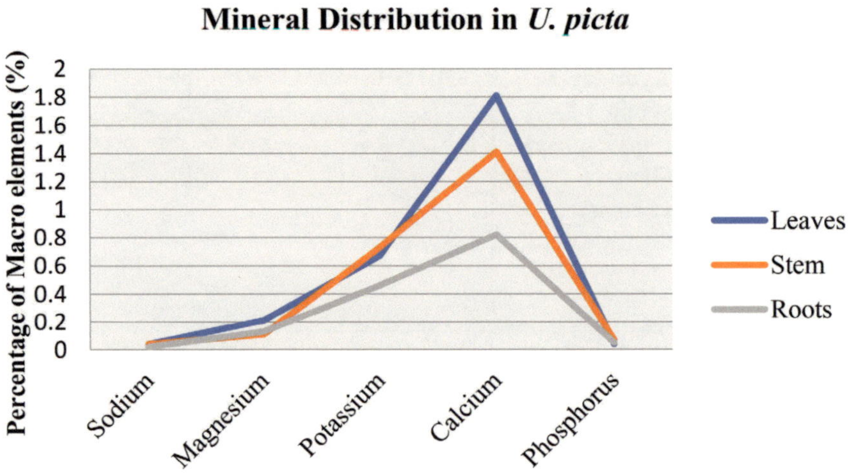

Figure 5.
Macro elements (mineral content) in different plant parts of U. picta.

inhibiting DNA binding and cell adhesion, migration, proliferation or differentiation, and blocking signaling pathways [29]. Saponins are being used commercially as dietary supplements and nutraceuticals in traditional medicine preparations. They also show hypocholesterolemic and antidiabetic properties [20]. As all these phytochemical constituents are present in the *U. picta,* all the pharmacological properties associated with these chemical compounds are found to be present in the plant as well. The earlier study conducted by Odubanjo et al. (2013) shows the plant possess the "Anticholinesterase property". They analyzed this property against two enzymes namely AchE (Acetylcholinesterase) and BchE (Butyrylcholinesterase). Here the aqueous extract of the plant and tested for the amount of IC50 Value. The result of the experiment showed the ample amount of dose dependent AchE and BchE inhibitory activity at the highest tested concentration [30]. Numerous studies have been done on "Antioxidant effect" of *U. picta*. Patel and Kamariya et al. (2011) studied the antioxidant behavior of aqueous extract of the plant. They analyzed the result statistically using regression method [31]. The significant antioxidant effect was calculated based on IC50 Value in presence of phenolic, flavonoid, sterol and terpene derivatives [8]. Methanol extract of *U. picta* showed the "Myocardial protection" [8] upon Rat Ischemic Reperfusion Injury Model by invigorating Muscarinic Receptors. To explore the hepatoprotective effect of *U. picta*, Hem, et al.

(2017) used PCM-induced liver injury model. The experiment showed the ability of the extract to reduce the serum liver enzymes level ALT (Alanine Transaminase) and AST (Aspartate AminoTransferase) in the blood. They also studied the "Anti-Inflammatory properties" of the plant in dose dependent manner where the aqueous decoction of roots and aerial parts showed the significant activity against egg albumin-induced and formalin induced rats paw edema. Both pre-clinical and clinical experiments suggested that the plant possesses the quick fracture healing effect due to deposition of phosphorous and calcium [22, 32].

8. Conclusion and future aspects

This report is based on the comprehensive study conducted on *U. picta,* which summarizes the botany, geographical distribution, propagation, phytochemical constituents and pharmacological properties of plant. This shows that the plant treasures a great medicinal wealth as each part of the plant reportedly has various phytochemical constituents having their respective pharmacological properties. These pharmacological properties are providing the evidence to various ethnomedicinal uses of the plant which have been in practice in many continents for centuries. Hence the whole plant plays an important area of research and developmental properties for pharmacologists and researchers. Due to its high therapeutic use and growing need, the plant is becoming rare and endangered; therefore it is necessary to create awareness of this plant to support its propagation in large numbers.

The development of modern drugs from less toxic plant products with proven medicinal properties is now being supported globally. There is no doubt that the products of this plant consecrate bright prospects as a reliable cure for various ailments.

Acknowledgements

The author is grateful to Professor Aditya Shastri, Vice-Chancellor, Banasthali Vidyapith, Rajasthan and Bioscience and Biotechnology Department, Banasthali Vidyapith for providing research facilities and constant motivation.

Conflict of interest

The authors declare no conflict of interest.

Author details

Harsha Kashyap
Department of Bioscience and Biotechnology, Banasthali Vidyapith, Rajasthan, India

*Address all correspondence to: dr.harshakashyap@gmail.com

References

[1] Ganie SA and Yadav SS. *Holoptelea integrifolia* (Roxb.) Planch: a review of its ethnobotany, pharmacology, and phytochemistry. BioMed research International 2014; 1-12. doi. org/10.1155/2014/401213

[2] Kirtikar KR, Basu BD. Indian Medicinal Plants (Vol. 1). International Book Distribution, Dehradun. 1995;72:308-562. ISBN 10: 8121103541

[3] Bhattacharya A, Dutta AK. Medicinal and Aromatic Plant Science and Biotechnology. Global Science Journal. 2010;4(1):1-4.

[4] Burkill HM. The Useful Plants of West Tropical Africa (Vol. 3) In (Ed. Burkill HM.). Royal Botanic Gardens, London. 1985;895. http://plants.jstor. org/upwta/2_580

[5] Okusanya OT, Lakanmi OO. and Oyesiku OO. Germination ecology of the woody herb *Uraria picta,* from Southern Nigeria. Journal of Tropical Ecology. 1991;7:139-146. DOI: https:// doi.org/10.1017/S0266467400005204

[6] Anand A, Lognay G, Wathelet B, Malaisse F. Micropropagation of *Uraria picta*, a medicinal plant, through axillary bud culture and callus regeneration. In vitro Cellular and Developmental Biology- Plant. 1998;34:136-140. DOI: 10.1007/ BF02822778

[7] Rahman MM, Gibbons S, Gray AI. Isoflavones from *Uraria picta* and their antimicrobial activity. Phytochemistry. 2007;68:1692-1697. DOI: 10.1016/j. phytochem.2007.04.015

[8] Azmi L, Rastogi C, Shukla I, Verma P, Kant P, Rao Ch V. Review on *Uraria picta* - A Traditionally Medicinal Plant of India: A Herbal Benefaction. Elixir Application Botany. 2017;106:46806-46812.

[9] AYUSH. Agro-techniques of Selected Medicinal Plants (Vol. 1). National Medicinal Plants Board, Government of India. 2008;211-213. ISBN 978-81-7993-154-7

[10] Batianoff GN, Neldner VJ, Singh S. Vascular Plant census and floristic analysis of serpentine landscapes in Central Queensland. The Proceedings of the Royal Society of Queensland. 2000. 109:1-30.

[11] Khare CP. Indian Herbal Remedies: Rational Western Therapy, Ayurvedic and Other Traditional Usage, Botany. (Ed. Khare CP.) Germany: Springer, Verlag, Berlin and Heidelberg Gmbh & Co. Kg. 2003. ISBN 978-3-642-18659-2.

[12] Pisharath DM. Comparative Pharmacognostic and Phytochemical Analysis of Four Species commonly used as *Prsniparni* in *Dasamula,* an Ayurvedic Formulation. International Journal for Green Pharmacy. 2008;1-42.

[13] Singh AK, Raghubanshi AS, Singh JS. Medicinal ethnobotany of the tribals of Songhati of Sonbhadra district, UP, India. Journal of Ethnopharmacology. 2002;81(1):31-41. DOI: 10.1016/s0378-8741(02)00028-4

[14] Yusuf M, Wahab MA, Yousuf MD, Chowdhury JU, Begum J. Some tribal medicinal plants of Chittagong hill tracts, Bangladesh. Bangladesh Journal of Plant Taxonomy. 2007; 14(2):117-128. **DOI:** https://doi.org/10.3329/bjpt. v14i2.531

[15] Igboechi AC, Osazuwa EO, Igwe UE. Laboratory evaluation of the acaricidal properties of extracts from *Uraria picta* (Leguminosae). J. Ethnopharmacol. 1989; 26(3): 293-298. DOI: 10.1016/0378-8741(89)90102-5

[16] Billore KV, Yelne MB, Dennis TJ, Chaudhari BG. Database on Medicinal

Plants Used in Ayurveda (Vol. 6), Central Council for Research in Ayurveda and Siddha, New Delhi. 2004;1: 314-320.

[17] Jain SK, Defillips RA. Medicinal Plants of India (Vol.1), Alganao MI; Reference Publication. 1991; 372.

[18] Allen ON, Allen KE. The Leguminosae, University of Wisconsin Press. Madison, W. 1981;672-673.

[19] Osazuwa EO, Igboechi AC. Anti-microbial activity of a chemical isolate from the leaves of *Uraria picta*. Phytotherapy Research. 2006;2:204-206. doi.org/10.1002/ptr.2650020413

[20] Saxena HO, Soni A, Mohammad N, Choubey SK. Phytochemical Screening and Elemental Analysis in Different Plant Parts of *Uraria picta* desv: A Dashmul Species. Journal of Chemical and Pharmaceutical Research. 2014;6(5): 756-760. ISSN : 0975-7384

[21] Goh SH, Chuah CH, Mok JSL, Soepadmo E. Malaysian Medicinal Plants for the Treatment of Cardiovascular Diseases. Selangor Darul Ehsan: Pelanduk Publication. Kaula Lumpur, Malaysia. 1995. ISBN 10: 9679785157

[22] Sagwan S, Rao DV, Sharma RA. Phytochemical Evaluation and Quantification of Primary Metabolites of *Maytenus emarginata* (Willd.) Ding Hou. Journal of Chemical and Pharmaceutical Research. 2010;2(6): 46-50. ISSN No: 0975-7384

[23] Madhikatti UB. Phytochemical and pharmacological investigations on Uraria picta and its Substitutes. Rajiv Gandhi University of Health Sciences, Karnataka. 2011. http://hdl.handle.net/123456789/5289

[24] Njoku PC, Akumefula MI. Phytochemical and Nutrient Evaluation of *Spondias mombin* Leaves. Pakistan Journal of Nutrition. 2007;6(6):613-615. **DOI:** 10.3923/pjn.2007.613.615

[25] Tadzabia K, Maina HM, Maitera ON, Ezekiel JS. Evaluation of phytochemical and elemental contents of *Haematostaphis barteri* leaves and stem bark in Hong local government area of Adamawa state, Nigeria. Journal of Chemical and Pharmaceutical Research. 2013;5(9):150-1567. ISSN: 0975-7384

[26] Duthie GG, Duthie SJ, Kyle AM. Plant Polyphenols in Cancer and Heart Disease: Implications as Nutritional Antioxidants. Nutrition Research Reviews. 2000;13:79-106. DOI: 10.1079/095442200108729016

[27] Luo J, Cheung J, Yevich E. Novel Terpenoid-Type Quinines Isolated from *Pycnanthus angolensis* of Potential Utility in the Treatment of Type-2 Diabetes. Journal of Pharmacology and experimental therapeutics. 1999;288: 529-534.

[28] Malairajan P, Gopalakrishnan G, Narasimhan S, Veni KJK. Analgesic Activity of Some Indian Medicinal Plants. Journal of Ethnopharmacology. 2006;19: 425-428. DOI: 10.1016/j.jep.2006.03.015

[29] Sharma HL, Sharma KK. Drug Therapy of Heart Failure. Principle of Pharmacology, (1st Ed Sharma KK). Paras Publishers. Hyderabad. 2007;13(3):977. doi.org/10.1177/1074248408320286

[30] Odubanjo VO, Oboh G, Ibukun EO. Antioxidant and anticholinesterase activities of aqueous extract of *Uraria picta* (Jacq.) DC; Department of Biochemistry, Federal University of Technology, Akure, Nigeria. African Journal of Pharmacy and Pharmacology. 2013;7(41):2768-2773. doi.org/10.5897/AJPP2013.3899

[31] Patel BD, Kamariya YH, Patel MB. Antioxidant Potential Of Aqueous

Extract Of Entire Plant Of *Uraria picta* Desv. International Journal for Pharmaceutics. 2011;3(4):92-96.

[32] Jain V, Prasad V, Pandey R. Wound Healing Profile of *Desmodium gangeticumin* Different Wound Models. Journal of Plant Sciences. 2006;1(3): 247-253.

Chapter 15

Spices-Reservoir of Health Benefits

Cheryl Sachdeva and Naveen Kumar Kaushik

Abstract

Spices contribute to the quality, nutritive value, and flavor of food. Since ancient times, they hold a great medicinal value. Their antimicrobial, antiviral, antibacterial, anti-inflammatory, and other numerous properties have made them a potent source of therapeutic agents. Phytochemical analysis revealed presence of active constituents such as eugenol, curcumin, carotenoids in clove, turmeric, saffron respectively that explains the efficacious nature of these spices. Owing to their easy availability and consumption, it is advised to make spices daily part of our diet though in balanced amount as sometimes excess usage bear few consequences. Evaluating multiple benefits offered by these as immunity boosters especially in times of pandemic and incorporating them in our routine diet would improve disease management strategies. This chapter discusses the reservoir of activities exhibited by few spices along with the components responsible for these activities. Here, we also discussed their negative effects if at all.

Keywords: curcumin, spices, antiviral, clove, immune booster

1. Introduction

Spices are plant-derived substances and play a crucial role in cooking. They are responsible for bringing out complex and rich flavors of food with their color, taste, aroma and making a cuisine distinct and popular. Without them kitchen looks empty and food gives a feeling of Stone Age. But do these spices just provide aroma or piquancy to food? Or is there any underlying advantage behind their regular consumption with food. Moreover, their usage depends upon habitat, time, and weather conditions. It is evident from both traditional as well as modern literature that vast medicinal properties offered by the spices makes them legendary and a vital part of daily diet. Pharmacological and molecular studies revealed that oils and alkaloids produced by most of the spices possess antimicrobial, anti-parasitic, immune booster, antioxidant and other important biological properties [1–4]. Their antioxidant and bactericidal activities prevents rancidity thereby increasing storage life of food [5]. Turmeric, with curcumin as principle ingredient, has been quite extensively used in the treatment of disorders such as amenorrhea, inflammation, hepatitis, arthritis [6, 7]. Clove, cinnamon, are rich source of antioxidants exhibiting wide range of pharmacological effects [8].

As spices are rich in immunity boosting ingredients, their daily consumption would lead to development of long-lasting immune protection and might maintain

IntechOpen

safe drug bullets at a certain basal level in blood stream to tackle the infection. Since the outbreak of COVID-19, the first line of defense is body's immune system. Therefore, regular usage of spices not only enhances the body's immunity to fight against the infection but also provides a prophylactic therapy to prevent and minimize the chances of infection. Out of wide variety of spices, pharmaco-potential of some of them has been discussed here.

2. Turmeric

Turmeric (rhizome of *Curcuma longa* L.), native to Tropical South Asia, is used as a condiment, food preservative and a traditional remedy for various diseases. This spice of Zingiberaceae family is widely cultivated in tropics and is known by different names such as Haldi, bhadra, pitika, mehagni, terre merite etc. To date various compounds of turmeric have been identified such as monoterpenes, sesquiterpenes, curcuminoids, alkaloids, sterols. Among these, most abundant is curcumin (77%) which is responsible for characteristic yellow color of turmeric and exhibits a wide spectrum of biological effects viz. antidiabetic, antimicrobial, anti-inflammatory, etc. (**Figure 1**) [9].

Anti-inflammatory and antioxidant effects of curcumin have proved to be beneficial against neurological diseases. Curcumin has the ability to bind amyloid β (Aβ) inhibiting fibrils formation [10] and, also enhance its cellular uptake [11], circumvent plaque deposition [12] thereby preventing Alzheimer's disease. Furthermore, curcumin is capable of decreasing Aβ serum levels and attenuating inflammation in Alzheimer's disease mouse models [13] along with rescuing altered neuritic morphology around Aβ plaques [14]. Additionally, studies have shown that curcumin decreased Huntington protein aggregation [15], suppressed cell death relieving disease symptoms [16]; modulated accumulation of α-synuclein which is the prime reason for Parkinson's disease [17]. It has been reported that intravenous and oral administration of curcumin modulates dopamine related damage, induces microglial activation, and improves locomotion [18]. Curcumin has, further, shown to increase docosahexaenoic acid (DHA) levels [19]; improve learning and mental ability in

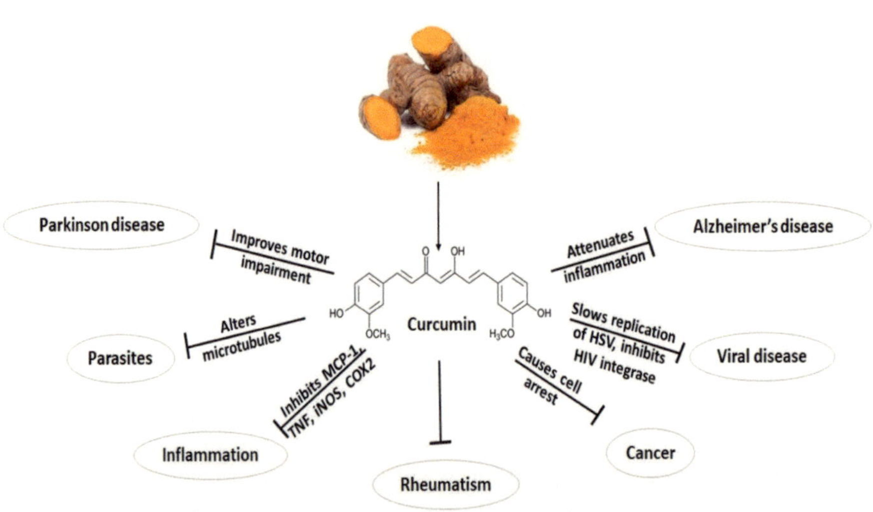

Figure 1.
Biological properties and chemical constituents of Turmeric.

scopolamine induced amnesia in mice [20]. Moreover, numerous other compounds with potent antioxidant activity have also been isolated from turmeric [21].

Curcumin has been known to downregulate production of inflammatory cytokines such as MCP- 1, TNF α, IL6, IL-1β both in vitro and in vivo [22–24]. However, recent study have suggested that efficacy to reduce levels of pro-inflammatory cytokines is enhanced on administering liposomal curcumin [25]. Studies have reported upregulation of heme oxygenase-1 reducing oxidative stress and providing protection against acute vascular inflammation in vitro as well as in vivo [26, 27].

Anti-rheumatic effects of curcumin have also been reported as curcumin decreased β3 and β7 integrins expression (adhesion molecules) ultimately decreasing joint inflammation; downregulated expression of chemokines, pro-inflammatory cytokines and growth-related oncogene/keratinocyte chemoattractant [28]. Further, randomized trials have also shown the efficacy of oral administration of curcumin in treatment of Rheumatoid Arthritis [29, 30].

Curcumin is effective in the management of virus infections as it inhibited Zika and Chikungunya virus at a concentration of 5 μM (IC50 = 1.9 and 3.89 μM respectively) [31], herpes simplex virus [32] and dengue virus [33]. Further, it inhibited human immunodeficiency virus (HIV) integrase and protease suggesting its protective effects against AIDS [34, 35]. The pandemic, COVID-19 related mortality is mainly due to acute respiratory distress syndrome with extensive cytokine storm. It has been reported that curcumin upregulates peroxisome proliferator-activated receptor-γ induction leading to inhibition of nuclear factor-κB (NFkb) signaling eventually decreasing cytokine storm [36] which suggests that curcumin might ameliorate COVID associated symptoms. Bioinformatic analysis have further shown ability of curcumin to interact with ACE2 receptor [37] and main protease [38] thereby fighting against COVID-19.

Curcumin displays antiparasitic and anti-cancer effects too. At a dose of 5 μM, it altered *P. falciparum* microtubules leading to reduction of 70–90% of parasitemia (IC_{50} = 50 μM) [39] and further at 100 mg/kg showed 80–90% decrease in *P. berghei* parasitemia [40]. Curcumin, in synergistic effect with Mitomycin C (5 μmol/L) arrested growth of MCF 7 breast cancer cell lines at G_0/G_1 phase of the cell cycle at a concentration of 40 μmol/L [41], decreased sensitivity of NFκB in human pancreatic cells lines BxPC-3, Capan-1, Capan-2, ASPC-1, and HS766-T (73–95% inhibition) [42], induced apoptosis [43]; inhibited cyclo-oxygenase 2 (COX 2- its overexpression leads to carcinogenesis) production in HT-29 colon cancer cell lines [44]. This prompts the requirement of detailed investigation to understand the potential of curcumin in cancer biology. Apart from curcumin, non-curcuminoids have also been reported to exhibit potential anticancer activities too [45].

Regardless of its demonstrated efficacy, purified curcumin has also been reported as pan assay interference compounds (PAINS) that show activity by interfering with assay readouts [46]. Curcumin exhibits PAIN properties such as fluorescence interference [47], aggregation [48], metal chelation [49], redox reactivity [50]. It is a highly unstable compound as it degrades rapidly in alkaline solutions [51]. Another drawback of curcumin is its poor bioavailability, however, number of formulations of curcumin with enhanced bioavailability and absorption are now available such as BioPerine-20x [52], BCM-95CG [52], Longvida-67x [53], Meriva-29x [54]. Furthermore, it is advised, traditionally, to consume turmeric powder with warm milk and ghee (Milk fat) as it is believed that this combination boosts immunity, purifies blood, beats everyday fatigue and anxiety, relieves cold and cough, which all are requirement to fight against COVID-19. Moreover, this combination might also enhance bioavailability of curcumin due to the constituents of milk and ghee such as casein, fats, iodine, phosphorus, calcium, vitamins etc.

3. Cinnamon

Cinnamon has a sweet, warm taste and is derived from dried central part of bark of *Cinnamomum zeylanicum* Blume (family Lauraceae). It is native to the Caribbean, South America, and Southeast Asia and is used throughout the world for its astounding properties such as anti-inflammatory, antidiabetic, antimicrobial, anticarcinogenic effects (**Figure 2**) [55]. Biochemical investigation of cinnamon revealed presence of camphor, linalool, cinnamaldehyde (major constituent), terpinen-4-ol, 1,8-cineole, α-cadiene, safrole, α-cadinol, germacrene D, γ-muurolene, α- terpineol, eugenol, 1,6-octadien-3-ol, 3,7-dimethyl,1-phenyl-propanr-2,2-diol diethanoate, etc. [56].

Cinnamon has been reported to exhibit antioxidant effects. In vitro studies showed free radical scavenging activity of methanolic extracts of cinnamon against 2,20 -azinobis- 3-ethylbenzothiazoline-6-sulphonic acid (ABTS) radical and Diphenylpicrylhydrazyl (DPPH) radical cations [57]. Intake of 100 mg/30 ml of cinnamon tea for 2 weeks showed reduction in the levels of 2-thiobarbituric acid reactive substances (TBARS) in plasma by 38% and increase in total antioxidant power 21% in a clinical study [58].

Antidiabetic and cholesterol lowering effects of aqueous extracts of cinnamon (AEC) have also been reported where it reduced fasting glucose levels (at 250 mg) from 1.14 mg/ml to 1.02 mg/ml in patients with impaired fasting glucose [59]; at a dose of 500 mg/kg for two months decreased glucose levels ($p < 0.005$) along with increasing in insulin sensitivity [60] and at 200 mg/kg reduced levels of LDL cholesterol, triglycerides and total cholesterol and increased levels of HDL-cholesterol in diabetic rats and hyper-lipidemic albino rabbits [61]. Further, decline in gastric acid secretion by 60% and reduction in gastric hemorrhagic lesions in rats was observed on pre-treatment with 250 mg/kg and 500 mg/kg of AEC [62].

Peterson et al., reported cinnamon as a potent anti-alzheimer agent as AEC inhibited tau aggregation (aggregation destabilizes microtubules causing Alzheimer's disease) [63].

In addition, cinnamon possesses antimicrobial and anticancer activities. Ethanolic extracts of cinnamon exhibited anti-microbial properties against

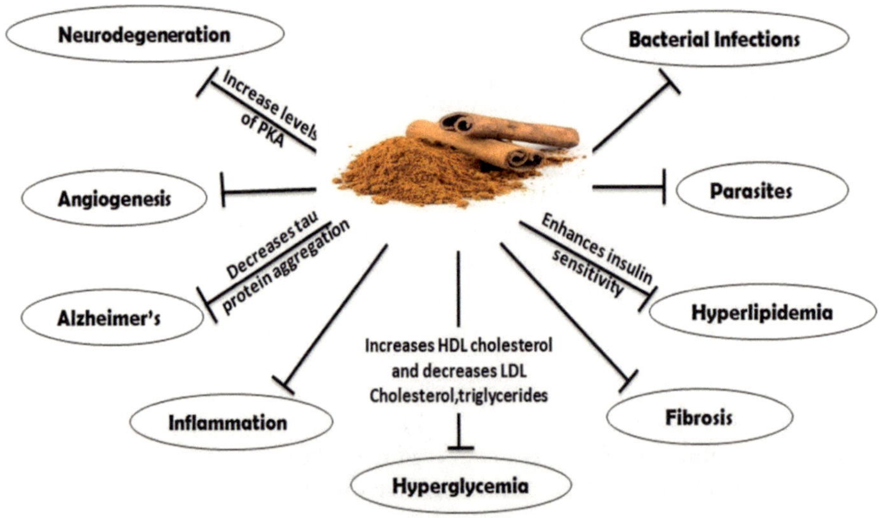

Figure 2.
Biological properties of Cinnamon.

Listeria monocytogenes (MIC 0.4 mg/ml other MICs) [64]. Further, essential oil of cinnamon also showed inhibitory effects against *Candida albicans* (MIC 7.81 μl/ml) [65]; nosocomial *P. aeruginosa* isolate (MIC- 1.9 μl/ml) [66]; *Anopheles tessellatus* (LD50: 0.33 μg/mL) and *Culex quinquefasciatus* (LD50: 0.66 μg/mL) [67]. Furthermore, AEC inhibited vascular endothelial growth factor (VEGF) and induced growth of vessels in aortic ring of rat ex vivo at a dose of 25 and 50 μg/ml suppressing angiogenesis [68]. Constituents of cinnamon essential oil, trans-cinnamaldehyde and its analogue - 4-hydroxy-3-methoxy-trans-cinnamaldehyde are effective inhibitors of bacterial acetyl-CoA carboxylase [69].

Cinnamon is considered as a strong immunity booster. At a dosage of 10 mg/kg, it significantly increased serum immunoglobulin levels whereas at 100 mg/kg dosage, along with boosting humoral immunity, cinnamon increased antibody titer and phagocytic index too thereby increasing cell-mediated immunity [70]. Cinnamon reportedly exhibits immunomodulatory properties as well. Studies have suggested that cinnamon decreased fibrotic symptoms and pro-inflammatory cytokines on treatment with 4.5 ml/kg dose [71] and 0.8 g/kg dose for 12 weeks [72] respectively in colitis infected mice models. Suppression of pro-inflammatory cytokines such as TNF-a, IL-1b, and IL-6 along with inhibition of nitric oxide secretion were also observed in BV 2 microglial cells on treatment with 50 μg/ml of ethanolic cinnamon extract [73]. Further, cinnamon bark is capable of decreasing in IFN-γ levels, enhancing of IL2 secretion thereby inhibiting cell death [74]. Studies have also suggested that cinnamaldehyde inhibits PI3K, NF-B activation and PDK1 thereby regulating monocyte\macrophage-mediated immune responses [74].

Recent bioinformatic analysis showed the possible effectiveness of molecules isolated from cinnamon against COVID-19 [75]. The ability to reduce pro-inflammatory cytokines and strong in silico investigation suggests the probable potential of cinnamon in fight against COVID-19.

Though cinnamon exhibits a wide range of health benefits, it is important to keep a check on the quantity of cinnamon consumed. A large amount can cause a dramatic drop in the blood sugar levels. In addition, high levels of cinnamon can cause rapid increase in heart rate, liver toxicity [76, 77].

4. Fennel

Fennel (seeds of *Foeniculum vulgare* Mill), originally cultivated in Mediterranean region, is used throughout the world for its licorice-like flavor. Belonging to class Magnoliopsida and family Apiaceae, fennel is known by different names such as suanf, sweet fennel, florence fennel, finocchio and is a concentrated source of minerals. This perennial herb is long loved for its culinary use. The essential oil of fennel constitutes anethole, estragole as major components and limonene, fenchone and others as minor components (**Figure 3**) [78, 79].

Fennel has been used for a long time for medicinal purposes (**Figure 3**) [80, 81]. It is a potent antioxidant agent. Essential oil of fennel seeds (FS) showed 45.05% DPPH radical scavenging activity along with 48.80–70.35% inhibition of peroxidation [81]. Ethanol and water extracts of FS (100 μg/ml) inhibited peroxidation in linoleic acid system by 77.5 and 99.1% respectively [82]. Parejo et al., isolated phenolic compounds from fennel viz. rosmarinic acid, kaempferol-3- O-glucoside, eriodictyol-7-Orutinoside, caffeoylquinic acid, quercetin-3-O-galactoside and observed decrease in absorbance of DPPH by 50% (IC50 in μg/mL = 1.17, 8.21, 24.78, 3.82, 7.52 respectively) [83].

Further, essential oil from fennel fruit showed inhibitory effects on growth of *Paenibacillus* larvae (MIC = 250 μg/ml) [84] and mycelial growth of *S. sclerotiorum*

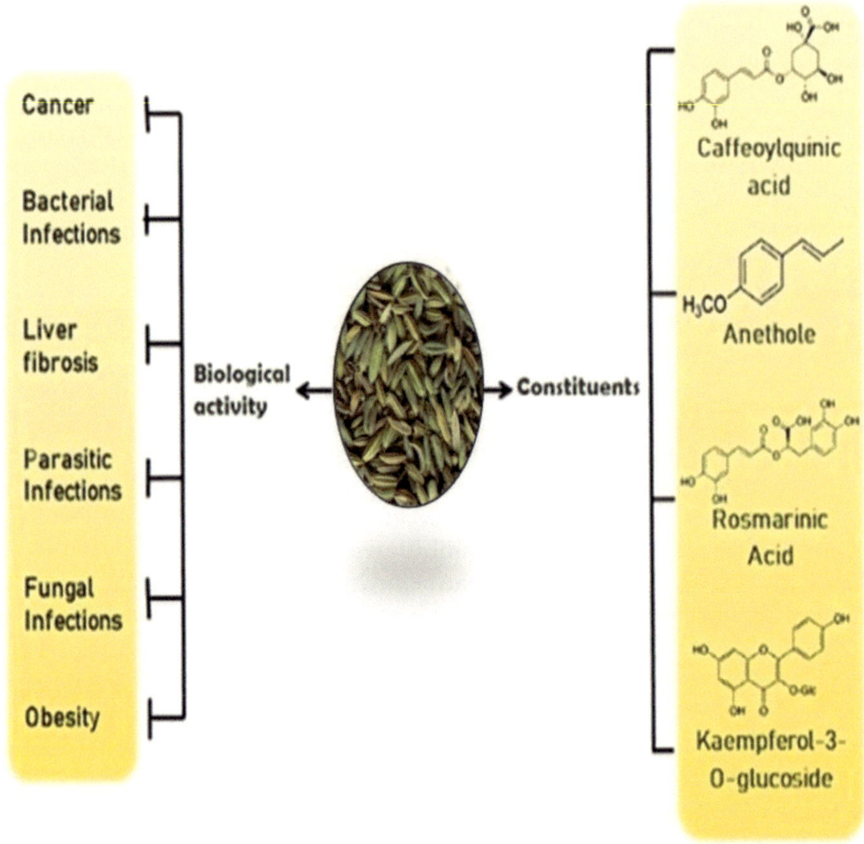

Figure 3.
Biological properties and chemical constituent of Fennel.

(MIC = 0.2 μg/ml) [85]. Similar inhibitory effects were observed against *C. quinquefasciatus* (LC_{50} = 70.85); *A. gambiae* (LC_{50} = 44.74) [86] and larvae of *C. pipiens* (90% inhibition at 60 mg/L) [87]. Hexane extracts of FS showed potent inhibitory effects against *E. coli* (MIC = 12.5 μg/mL), *S. typhi* (MIC =15 μg/mL), *S. aureus* (MIC = 10 μg/mL) [88].

Recent study reported that 300 μg/μl FS ethanolic extract inhibited Influenza virus H5N1 by 82.8% [89]. Moreover, Alazadeh et al., showed that oral administration of FS extract in capsular form ameliorated knee ostroarthritis [90]. In addition, methanol extract of FS showed significant anticancer potential against liver cancer cell line Hepg-2 (IC50 = 27.96 μg/mL) and breast cancer cell line MCF-7 (IC50 = 15.78 μg/mL) [80]. Özbek et al., 2003 studied hepatoprotective effects of fennel and observed that a dose of 0.4 ml/kg of fennel oil showed significant protective role against liver fibrosis (CCl_4 induced) in rats [91].

It is safe to consume fennel and since it provides protection against flu, cough, it is advised to take fennel with warm milk.

5. Clove

Clove [*Syzygium aromaticum* (L.) Merr. & L.M. Perry] is one of the most valuable spices of family Myrtaceae and is a native of Indonesia, albeit found all around the globe. Major bioactive component of clove is eugenol. Other components

DOI: http://dx.doi.org/10.5772/intechopen.96471

include phenolic acids such as gallic acid, gallic acid derivatives, caffeic acids, salicylic acids; flavonoids such as quercetin, kaempferol (**Figure 4**) [92].

Traditionally used to prevent nausea, enhance blood circulation and liver function, clove is commonly applied for toothache relief and has been long known as a medicine for numerous ailments (**Figure 4**). Miyazawa & Hisama, isolated dehydrodieugenol and trans-coniferyl aldehyde from ethyl acetate extract of clove bud and observed significant activity of both, at a concentration of 0.6 μmol/mL and 1.2 μmol/ml respectively, against mutagens 4-nitroquinolin 1-oxide and N-methyl-N′-nitro-N-nitrosoguanidine [93]. Furthermore, eugenol and eugenol acetate extracted from aroma extract of clove buds inhibited 99% of hexanal oxidation [94].

Eugenol also exhibits remarkable antimicrobial, antiparasitic, antiviral activities. Essential oil of clove has reportedly inhibited growth of *S. aureus, H. influenzae* (MIC = .0125 ml/ml each), *K. pneumoniae* (MIC = .050 ml/ml) [95], *C. albicans* (MIC = 2.5 μg/ml), *L. monocytogenes* (MIC = 5 μg/ml), *Y. enterocolitica* (MIC = 2.5 μg/ml) [96]. It has been reported that clove decreases ergosterol (cell membrane component) inhibiting growth of *C. albicans* (MIC = 0.64 μg/ml) [97]. Methanol extract, ether soluble fraction (ES), ethyl acetate soluble fraction (EAS) and acetone soluble fraction (AS) of clove buds showed inhibitory effects on *Bacillus cereus* (MIC =250 μg/disc each). Further, EAS and AS also inhibited *Micrococcus luteus* and *Shigella dysenteriae* respectively (MIC = 62.5 μg/disc each) [98]. In addition, antiparasitic effects were reported by Bagavan et al., who observed growth inhibition of chloroquine resistant *P. falciparum* on treatment with ethyl acetate and methanol extracts of clove (IC50 = 13 μg/ml and 6.25 μg/ml

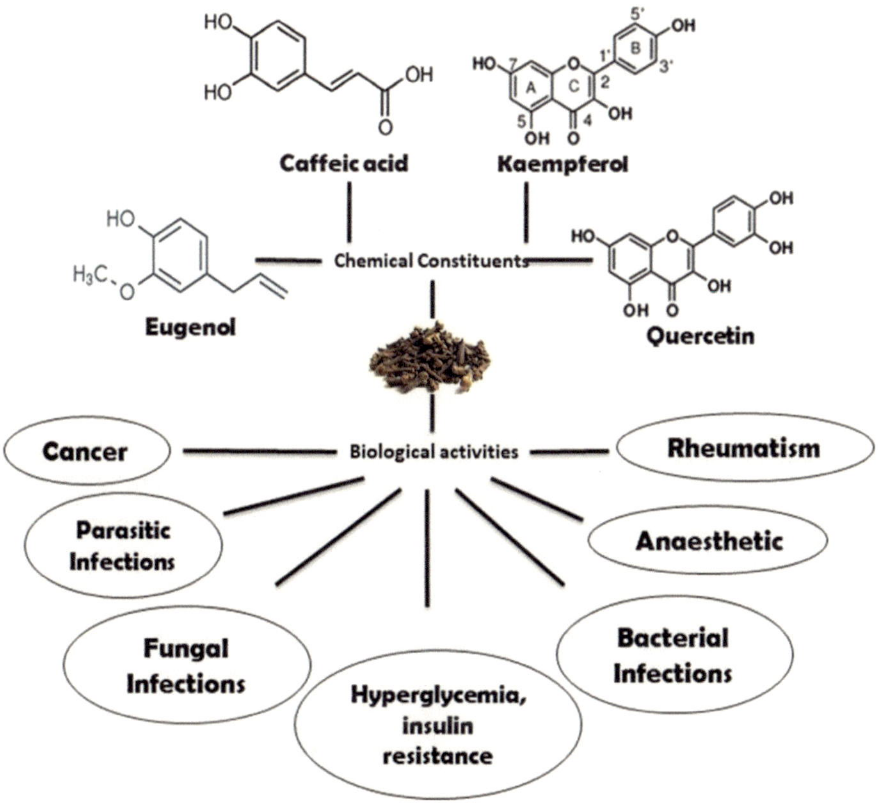

Figure 4.
Chemical constituents and biological properties of Clove.

respectively) [99]. Hypoglycemic effects of clove have also been identified. Methanol extract of clove activated AMP-activated protein kinase (AMPK) pathway, a regulator of cellular energy homeostasis thereby regulating glucose metabolism [100]. A recent report has pointed out that it relieves insulin resistance [101].

Consumption of clove is safe, and it offers numerous benefits, however, studies have shown that clove oil increases clotting time [102] which might increases the risk of bleeding in case of bleeding disorders.

6. Cardamom

Cardamom [*Elettaria cardamomum* (L.) Maton, also known as "Queen of Spices"] is an aromatic spice commonly used as a flavoring agent. This perennial herb belongs to Zingiberaceae family. It is native to Western Ghats of Southern India and widely cultivated in countries such as Sri Lanka, North America, Guatemala, New Guinea, and Thailand. Essential oil of cardamom constitutes α- terpinyl acetate, 1,8-cineole, linalool, limonene, eugenol, safrole (**Figure 5a**) [103, 104].

Cardamom is a valuable in relieving against ischemic heart disease. It showed protective effects against cardiac dysfunction associated with oxidative stress. A dose of 100 and 200 mg/kg of cardamom extract showed cardioprotective effects in albino rats who were induced with myocardial infarction due to isoproterenol [105]. Further, cardamom-oil maintained cholesterol homeostasis by potentially reducing cholesterol levels in hypercholesterolemic conditions and restoring atherogenicity index [106]. In addition, a dose of 1.5 g of cardamom powder for 12 weeks (two times a day) decreased the blood pressure in hypertensive individuals by 19 mmHg in systolic and 12 mmHg in diastolic BP [107].

Cardamom is an effective immunomodulatory agent due to its anti-inflammatory effects. It is capable of downregulating pro-inflammatory cytokines (**Figure 5b**) [108]; suppressing T helper (TH)1 cytokine release and enhancing

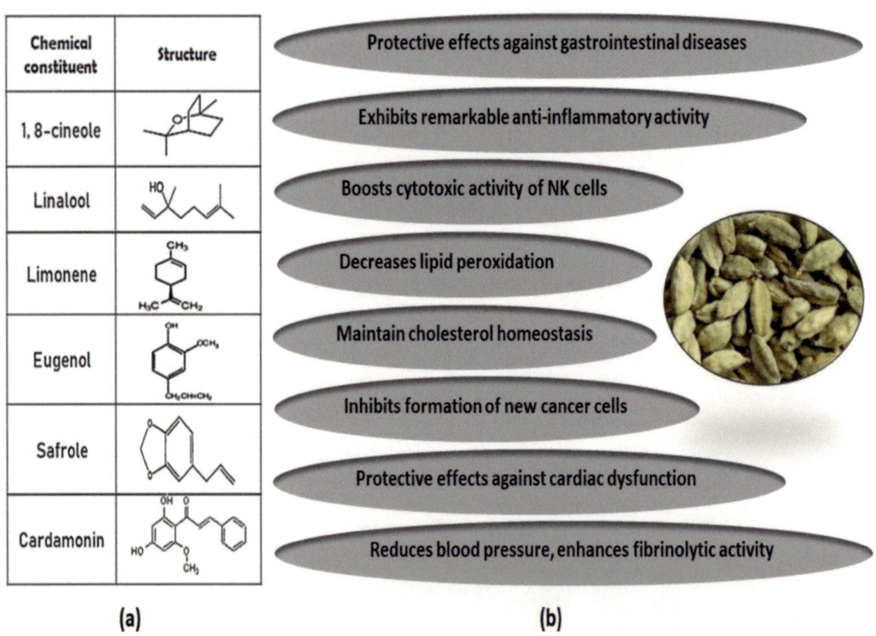

Figure 5.
(a) Chemical constituents of Cardamom; (b) Biological properties of Cardamom.

T helper (TH)2 cytokine release [109]. Methanol and petroleum ether extract of cardamom reduced 70% and 50% lesions in ethanol-induced ulcer mice model at 500 mg/kg and 100 mg/kg respectively exhibiting gastroprotective effects [110]. A constituent of cardamom, cardamonin inhibited the formation of new cancer stem cells in case of breast cancer by effectively suppressing the up-regulation of IL-8 and, MCP-1 cytokines and activation of NF-κB pathway [111]. Antibacterial effects of cardamom has also been demonstrated [108, 112].

Recently, its efficacy against toxic effects of uranium has been reported. Uranium increases sodium and calcium ion levels, decreases and phosphate ion levels, inhibits $Na+/K+$ ATPase increasing the influx of into the neurons, and thereby damaging central nervous system. Aqueous extracts of cardamom (250 mg/kg) significantly increased the levels of phosphate and potassium ions and decreased the levels of calcium and sodium ions in albino rats administered with uranyl acetate dehydrate (40 mg/kg) [113]. Furthermore, protective effects against neurological disorders have also been reported as oral administration with 100 and 200 mg/kg of cardamom oil improved behavioral patterns, inhibited amyloid-β expression, declined in oxidative stress and inhibited of acetylcholinesterase in Wister rats [114, 115].

7. Red chili

Out of the wide variety of species under genus *Capsicum* like frutescens, pubscens, baccatum etc., dried fruits of *Capsicum annum* L. are the most commonly used spice of Solanaceae family. Commonly known as red chili, lal mirch, this plant is native to Central and South America and is cultivated in many parts of the world such as India, China. Biochemical analysis of red chili led to the identification of capsaicinoids (capsaicin, dihydrocapsaincin) (**Figure 6**), antioxidant vitamins (ascorbic acid, vitamin E), carotenoids (β - carotene, β cryptoxanthine) and several organic acids and minerals [116, 117]. Red chili has a phenolic substance, capsaicin responsible for its pungent smell and irritant properties. Capsaicin excites nociceptors that induce pain along with rise in temperature giving a sensation of heat (pungency). As spicy is not one of the five basic tastes viz. sweet, sour, bitter, umami and salty, heat produced by capsaicin is considered as taste of red chili.

Capsaicin also exhibits anti-inflammatory properties [118]. It showed protective effects against gastric mucosal injury (ethanol-induced) in rats [119].

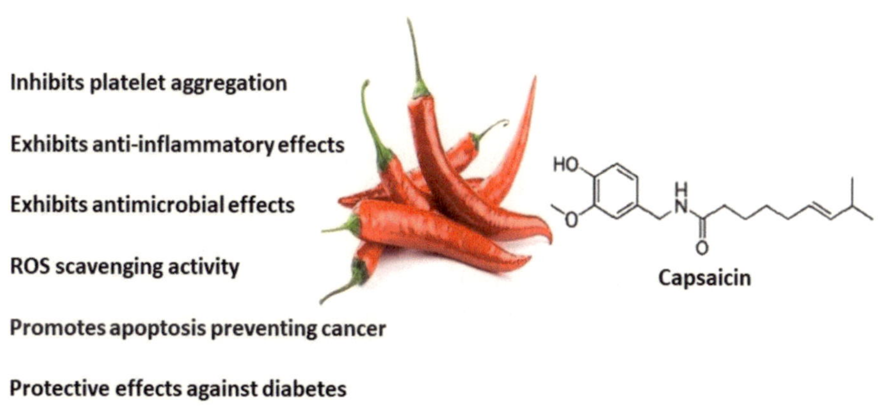

Figure 6.
Biological properties and chemical constituent of Red Chili.

Capsaicin treatment ameliorated lipid peroxidation, inhibited myeloperoxidase activity in gastric lesions (ethanol induced) in rats [120].

Dried red chili powder (1% w/v) showed inhibitory effects against *Listeria monocytogenes* [121]. Tropical application of capsaicin (0.075%) reduced fat accumulation in mesenteric and epididymal adipose tissue by increasing expression of adipokines [122]. It was also found that a dose of 200 mg/kg of capsaicin decreases plasma triglyceride levels and fasting glucose in obese mouse model kept on high fat diet [123].

Earlier studies suggested the risk of oral cavity [124] and stomach [125] cancer on consumption of red chili, however, recent reports suggested that consumption of red chili is safe and does not increase risks of cancer [126]. Numerous reports have suggested protective effects of capsaicin against cancer. Capsaicin showed inhibitory effects on the growth of human KB cancer cells by promoting apoptosis at a concentration of 200–250 μM [127]. Further, capsaicin led to the formation of reactive oxidative species through mitochondria (at a dose of 150 μM) causing loss in mitochondrial membrane potential in BxPC-3 and AsPC-1, human pancreatic cancer cell lines [128]; inhibited activation of NF-kB and AP-1, transcription factors responsible for cellular proliferation and malignant formation, in mice model [129]. In addition, capsaicin inhibited growth of MCF breast cancer cell lines by causing cell cycle arrest at S phase and induced poly(ADP-ribose) polymerase-1 (PARP-1) cleavage (apoptosis is marked by the cleavage of PARP-1) by activating caspase-7 which is involved in apoptosis (**Figure 6**) [130]. Capsaicin and cisplatin, in a synergistic manner, arrested the growth of SNU-668, human gastric cell line at G1/S phase [131].

8. Black cumin

Seeds of *Nigella sativa* L., an annual flowering plant of Ranunculaceae family, have been of extensive use as a spice. Commonly known as black cumin, it is native to South and Southwest Asia and has diverse medicinal applications. Phytochemical analysis of black cumin seeds has showed presence of thymoquinone (TQ), para-cymene, and carvone, linoleic acid, oleic acid, palmitic acid, and stearic acid (**Figure 7**) [132].

Black cumin has been reported to exhibit anti-inflammatory properties. Intra-peritoneal injection black cumin essential oil reduced inhibited carrageenan-induced paw oedema in rats thereby relieving inflammation [133]. Aqueous extracts of black cumin up-regulated the secretion of T-helper 2 cells and suppressed the secretion of pro-inflammatory cytokines viz. IL-6, TNFα, and NO [109]. Furthermore, black El-Mahmoudy et al., isolated TQ from essential oil of black cumin seeds and showed that TQ reduced nitrite accumulation and decreased inducible nitric oxide synthase levels (responsible for NO production) in rat peritoneal macrophages suggesting anti-inflammatory and cytoprotective effects of black cumin [134]. Methanol extract of seeds of black cumin (1 mg/ml) protected erythrocytes against protein degradation and loss of deformability induced due to peroxide [135] and has proved to stimulate innate humoral immune responses [136]. Proteins purified from black cumin exhibit potent antioxidant activities [137]. Furthermore, neuroprotective effects of black cumin have also been reported [138].

In addition, black cumin exhibits anti-bacterial, anti-viral, anti-helminthic effects. Black cumin seed oil (BSO) showed protective effects against murine cytomegalovirus (MCMV) that targets liver and spleen. Treatment with BSO showed approximately 38% and 20% decrease in viral load in liver and spleen

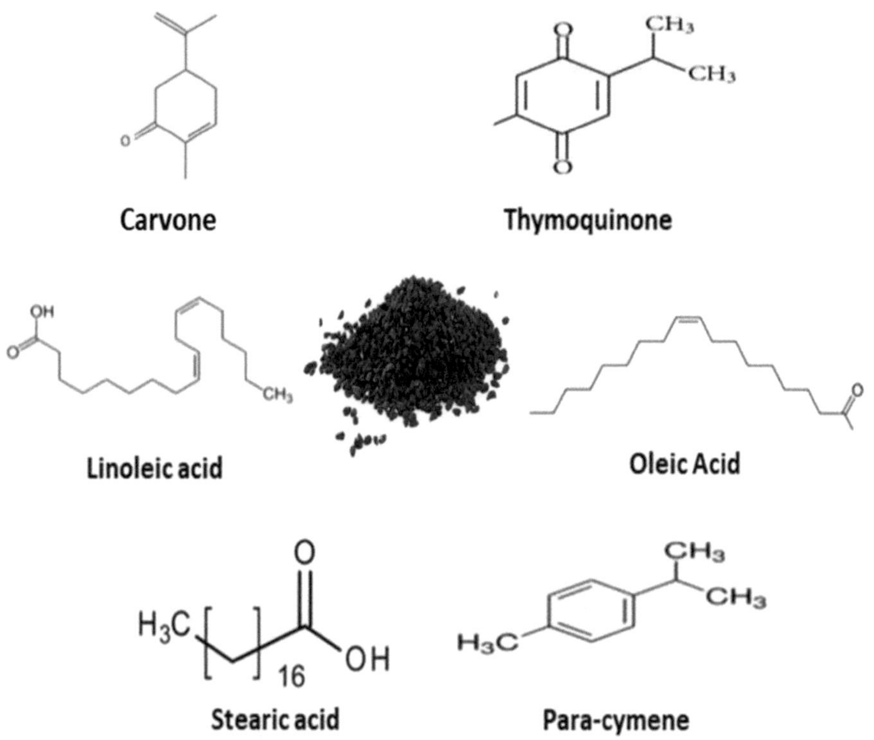

Figure 7.
Chemical constituents of Black Cumin.

respectively in MCMV infected mice [139]. Decrease in the number of *S. mansoni* worms in liver and eggs in both liver and intestine of mice has been observed after treatment with black cumin oil at a dose of 2.5 ml/kg and 5 ml/kg [140]. TQ has proved to be an effective bactericidal agent as it inhibited growth of *Staphylococcus aureus* (BIC_{50} = 22 μg/ml) [141], *Salmonella typhii* [142] *Streptococcus mutans* [143]. Further, black cumin inhibited the *P. yoelii* parasitemia by 94% [144].

Black cumin has protective effects against diabetes. Black cumin seeds (BCS) led to the decrease in elevated levels of glucose with increase in GSH levels and further inhibited liver damage induced by lipid peroxidation in diabetic rabbits [145]. In addition, improvement of glucose homeostasis in patients with type 2 diabetes on administration of black cumin (2 g/day for 3 months) with hypoglycemic drugs have also been reported [146].

Furthermore, aqueous extracts of black cumin significantly up-regulated cytotoxic activity of natural killer cells against YAC-1 tumor cells [109]. Oral administration of TQ at a dose of 0.01% suppressed benzo(a)pyrene induced forestomach tumor in mice by 70% [147] exhibiting anti-tumor effects.

Minor toxicological effects have also been reported, however, numerous studies demonstrated diverse therapeutic effects of black cumin and TQ and have supported its safe consumption [132].

9. Conclusion

Spices are rich source of bioactive components with innumerable beneficial attributes that have been verified and accepted by modern world in the past few decades. These are nowadays considered as a crucial & natural component of our daily diet. Consumption of spices aids in combating diseases when they are at their peak, for instance, best remedy for stomach infections are fennel seeds; turmeric is the tonic for fever-related diseases. Antimicrobial activities of spices make them valuable in hot climates as they prevent food spoilage. Although they lower the risk of various diseases such as diabetes, cancer, etc., there are some contradictions about their use. Despite all the pleiotropic effects offered by the spices, further evaluation about their mechanism of action is mandated to validate their clinical effects and their amount of consumption.

Conflict of interest

The authors declare no conflict of interest.

Author details

Cheryl Sachdeva and Naveen Kumar Kaushik*
Amity Institute of Virology and Immunology, Amity University Uttar Pradesh, Noida, Uttar Pradesh, India

*Address all correspondence to: inaveenkaushik@gmail.com; nkkaushik@amity.edu

References

[1] Baratta MT, Dorman HJD, Deans SG, Figueiredo AC, Barroso JG, Ruberto G. Antimicrobial and antioxidant properties of some commercial essential oils. Flavour Fragr J. 1998;13. DOI:3.0.CO;2-T

[2] Mohamad RH, El-Bastawesy AM, Abdel-Monem MG, Noor AM, Al-Mehdar HAR, Sharawy SM, et al. Antioxidant and anticarcinogenic effects of methanolic extract and volatile oil of fennel seeds (*Foeniculum vulgare*). J Med Food. 2011;14:986-1001.

[3] Dolati M, Rezaei K, Vanak ZP, Movahed S. Study of the Effects of Essential Oils of Cumin, Savory and Cardamom as Natural Antioxidants on the Flavor and Oxidative Stability of Soybean Oil During the Storage. J Essent Oil Bear Plants. 2016;19:176-84. DOI:10.1080/0972060X.2014.935030

[4] Sharma V, Singh P, Rani A. Antimicrobial Activity of *Trigonella foenum-graecum* L . (Fenugreek) Keywords : Eur J Exp Biol. 2017;7:1-4.

[5] Teneva D, Denkova Z, Goranov B, Denkova R, Kostov G, Atanasova T, et al. Chemical composition and antimicrobial activity of essential oils from black pepper , cumin , coriander and cardamom against Against Some Pathogenic Microorganisms [2016]. Acta Univ Cibiniensis Ser E FOOD Technol. 2016;XX.

[6] Aggarwal BB, Surh Y-J, Shishodia S. The Molecular Targets and Therapeutic Uses of Curcumin in Health and Disease. 2014.

[7] Aggarwal B. Curcumin : The Indian solid gold. 2015.

[8] Özcan MM, Arslan D. Antioxidant effect of essential oils of rosemary, clove and cinnamon on hazelnut and poppy oils. Food Chem. 2011;129:171-4. DOI:http://dx.doi.org/10.1016/j.foodchem.2011.01.055

[9] Wickenberg J, Ingemansson SL, Hlebowicz J. Effects of *Curcuma longa* (turmeric) on postprandial plasma glucose and insulin in healthy subjects. Nutr J. 2010;9:43. DOI:10.1186/1475-2891-9-43

[10] Ono K, Hasegawa K, Naiki H, Yamada M. Curcumin has potent anti-amyloidogenic effects for Alzheimer's ?-amyloid fibrils in vitro. J Neurosci Res. 2004;75:742-50. DOI:10.1002/jnr.20025

[11] Zhang L, Fiala M, Cashman J, Sayre J, Espinosa A, Mahanian M, et al. Curcuminoids enhance amyloid-β uptake by macrophages of Alzheimer's disease patients. J Alzheimer's Dis. 2006;10:1-7. DOI:10.3233/JAD-2006-10101

[12] Kim H, Park B-S, Lee K-G, Choi CY, Jang SS, Kim Y-H, et al. Effects of Naturally Occurring Compounds on Fibril Formation and Oxidative Stress of β-Amyloid. J Agric Food Chem. 2005;53:8537-41. DOI:10.1021/jf051985c

[13] Wang Y-J, Thomas P, Zhong J-H, Bi F-F, Kosaraju S, Pollard A, et al. Consumption of Grape Seed Extract Prevents Amyloid-β Deposition and Attenuates Inflammation in Brain of an Alzheimer's Disease Mouse. Neurotox Res. 2009;15:3-14. DOI:10.1007/s12640-009-9000-x

[14] Garcia-Alloza M, Borrelli LA, Rozkalne A, Hyman BT, Bacskai BJ. Curcumin labels amyloid pathology in vivo , disrupts existing plaques, and partially restores distorted neurites in an Alzheimer mouse model. J Neurochem. 2007;102:1095-104. DOI:10.1111/j.1471-4159.2007.04613.x

[15] Hickey MA, Zhu C, Medvedeva V, Lerner RP, Patassini S, Franich NR,

et al. Improvement of neuropathology and transcriptional deficits in CAG 140 knock-in mice supports a beneficial effect of dietary curcumin in Huntington's disease. Mol Neurodegener. 2012;7:12. DOI:10.1186/1750-1326-7-12

[16] Chongtham A, Agrawal N. Curcumin modulates cell death and is protective in Huntington's disease model. Sci Rep. 2016;6:18736. DOI:10.1038/srep18736

[17] Sharma N, Nehru B. Curcumin affords neuroprotection and inhibits α-synuclein aggregation in lipopolysaccharide-induced Parkinson's disease model. Inflammopharmacology. 2018;26:349-60. DOI:10.1007/s10787-017-0402-8

[18] Tripanichkul W, Jaroensuppaperch E. Curcumin Protects Nigrostriatal Dopaminergic Neurons and Reduces Glial Activation in 6-Hydroxydopamine Hemiparkinsonian Mice Model. Int J Neurosci. 2012;122:263-70. DOI:10.3109/00207454.2011.648760

[19] Sugasini D, Lokesh BR. Curcumin and linseed oil co-delivered in phospholipid nanoemulsions enhances the levels of docosahexaenoic acid in serum and tissue lipids of rats. Prostaglandins, Leukot Essent Fat Acids. 2017;119:45-52. DOI:10.1016/j.plefa.2017.03.007

[20] Khalid A, Shakeel R, Justin S, Iqbal G, Shah SAA, Zahid S, et al. Pharmacological Effects of Turmeric on Learning, Memory and Expression of Muscarinic Receptor Genes (M1, M3 and M5) in Stress-induced Mouse Model. Curr Drug Targets. 2017;18. DOI:10.2174/1389450118666170315120627

[21] Akter J, Hossain MA, Takara K, Islam MZ, Hou D-X. Antioxidant activity of different species and varieties of turmeric (Curcuma spp): Isolation of active compounds. Comp Biochem Physiol Part C Toxicol Pharmacol. 2019;215:9-17. DOI:10.1016/j.cbpc.2018.09.002

[22] Jain SK, Rains J, Croad J, Larson B, Jones K. Curcumin Supplementation Lowers TNF-α, IL-6, IL-8, and MCP-1 Secretion in High Glucose-Treated Cultured Monocytes and Blood Levels of TNF-α, IL-6, MCP-1, Glucose, and Glycosylated Hemoglobin in Diabetic Rats. Antioxid Redox Signal. 2009;11:241-9. DOI:10.1089/ars.2008.2140

[23] Aggarwal BB, Gupta SC, Sung B. Curcumin: an orally bioavailable blocker of TNF and other pro-inflammatory biomarkers. Br J Pharmacol. 2013;169:1672-92. DOI:10.1111/bph.12131

[24] Park S, Lee LR, Seo JH, Kang S. Curcumin and tetrahydrocurcumin both prevent osteoarthritis symptoms and decrease the expressions of pro-inflammatory cytokines in estrogen-deficient rats. Genes Nutr. 2016;11:2. DOI:10.1186/s12263-016-0520-4

[25] Bulboacă AE, Boarescu PM, Bolboacă SD, Blidaru M, Feştilă D, Dogaru G, et al. Comparative Effect Of Curcumin Versus Liposomal Curcumin On Systemic Pro-Inflammatory Cytokines Profile, MCP-1 And RANTES In Experimental Diabetes Mellitus. Int J Nanomedicine. 2019;14:8961-72.

[26] Xiao Y, Xia J, Wu S, Lv Z, Huang S, Huang H, et al. Curcumin Inhibits Acute Vascular Inflammation through the Activation of Heme Oxygenase-1. Oxid Med Cell Longev. 2018;2018:1-12. DOI:10.1155/2018/3295807

[27] Motterlini R, Foresti R, Bassi R, Green CJ. Curcumin, an antioxidant and anti-inflammatory agent, induces heme oxygenase-1 and protects endothelial cells against oxidative stress. Free Radic Biol Med. 2000;28:1303-12. DOI:10.1016/S0891-5849(00)00294-X

[28] Funk JL, Oyarzo JN, Frye JB, Chen G, Lantz RC, Jolad SD, et al. Turmeric Extracts Containing Curcuminoids Prevent Experimental Rheumatoid Arthritis. J Nat Prod. 2006;69:351-5. DOI:10.1021/np050327j

[29] Chandran B, Goel A. A Randomized, Pilot Study to Assess the Efficacy and Safety of Curcumin in Patients with Active Rheumatoid Arthritis. Phyther Res. 2012;26:1719-25. DOI:10.1002/ptr.4639

[30] Amalraj A, Varma K, Jacob J, Divya C, Kunnumakkara AB, Stohs SJ, et al. A Novel Highly Bioavailable Curcumin Formulation Improves Symptoms and Diagnostic Indicators in Rheumatoid Arthritis Patients: A Randomized, Double-Blind, Placebo-Controlled, Two-Dose, Three-Arm, and Parallel-Group Study. J Med Food. 2017;20:1022-30. DOI:10.1089/jmf.2017.3930

[31] Mounce BC, Cesaro T, Carrau L, Vallet T, Vignuzzi M. Curcumin inhibits Zika and chikungunya virus infection by inhibiting cell binding. Antiviral Res. 2017;142:148-57. DOI:10.1016/j.antiviral.2017.03.014

[32] Zhu L, Ding X, Zhang D, Yuan C, Wang J, Ndegwa E, et al. Curcumin inhibits bovine herpesvirus type 1 entry into MDBK cells. Acta Virol. 2015;59:221-7. DOI:10.4149/av_2015_03_221

[33] Padilla-S L, Rodríguez A, Gonzales MM, Gallego-G JC, Castaño-O JC. Inhibitory effects of curcumin on dengue virus type 2-infected cells in vitro. Arch Virol. 2014;159:573-9. DOI:10.1007/s00705-013-1849-6

[34] Vajragupta O, Boonchoong P, Morris GM, Olson AJ. Active site binding modes of curcumin in HIV-1 protease and integrase. Bioorg Med Chem Lett. 2005;15:3364-8. DOI:10.1016/j.bmcl.2005.05.032

[35] Prasad S, Tyagi AK. Curcumin and its analogues: A potential natural compound against HIV infection and AIDS. Food Funct. 2015;6:3412-3419.

[36] Jacob A, Wu R, Zhou M, Wang P. Mechanism of the Anti-inflammatory Effect of Curcumin: PPAR- γ Activation. PPAR Res. 2007;2007:1-5. DOI:10.1155/2007/89369

[37] Shanmugarajan D, P. P, Kumar BRP, Suresh B. Curcumin to inhibit binding of spike glycoprotein to ACE2 receptors: computational modelling, simulations, and ADMET studies to explore curcuminoids against novel SARS-CoV-2 targets. RSC Adv. 2020;10:31385-99. DOI:10.1039/D0RA03167D

[38] Rajagopal K, Varakumar P, Baliwada A, Byran G. Activity of phytochemical constituents of *Curcuma longa* (turmeric) and *Andrographis paniculata* against coronavirus (COVID-19): an in silico approach. Futur J Pharm Sci. 2020;6:104. DOI:10.1186/s43094-020-00126-x

[39] Chakrabarti R, Rawat PS, Cooke BM, Coppel RL, Patankar S. Cellular Effects of Curcumin on *Plasmodium falciparum* Include Disruption of Microtubules. Bejon P, editor. PLoS One. 2013;8:e57302. DOI:10.1371/journal.pone.0057302

[40] Reddy RC, Vatsala PG, Keshamouni VG, Padmanaban G, Rangarajan PN. Curcumin for malaria therapy. Biochem Biophys Res Commun. 2005;326:472-4. DOI:10.1016/j.bbrc.2004.11.051

[41] Zhou Q, Wang X, Liu X, Zhang H, Lu Y, Su S. Curcumin enhanced antiproliferative effect of mitomycin C in human breast cancer MCF-7 cells in vitro and in vivo. Acta Pharmacol Sin. 2011;32:1402-10. DOI:10.1038/aps.2011.97

[42] Li L, Aggarwal BB, Shishodia S, Abbruzzese J, Kurzrock R. Nuclear

factor-kappaB and IkappaB kinase are constitutively active in human pancreatic cells, and their down-regulation by curcumin (diferuloylmethane) is associated with the suppression of proliferation and the induction of apoptosis. Cancer. 2004;101:2351-62. DOI:10.1002/cncr.20605

[43] Yuliani S, Widyarini S, Mustofa, Partadiredja G. Turmeric extract inhibits apoptosis of hippocampal neurons of trimethyltin-exposed rats. Bratisl Lek Listy. 2017;118:142-8.

[44] Goel A, Boland CR, Chauhan DP. Specific inhibition of cyclooxygenase-2 (COX-2) expression by dietary curcumin in HT-29 human colon cancer cells. Cancer Lett. 2001;172:111-8. DOI:10.1016/S0304-3835(01)00655-3

[45] Nair A, Amalraj A, Jacob J, Kunnumakkara AB, Gopi S. Non-Curcuminoids from Turmeric and Their Potential in Cancer Therapy and Anticancer Drug Delivery Formulations. Biomolecules. 2019;9:13. DOI:10.3390/biom9010013

[46] Nelson KM, Dahlin JL, Bisson J, Graham J, Pauli GF, Walters MA. The Essential Medicinal Chemistry of Curcumin. J Med Chem. 2017;60:1620-37. DOI:10.1021/acs.jmedchem.6b00975

[47] Priyadarsini KI. Photophysics, photochemistry and photobiology of curcumin: Studies from organic solutions, bio-mimetics and living cells. J Photochem Photobiol C Photochem Rev. 2009;10:81-95. DOI:10.1016/j.jphotochemrev.2009.05.001

[48] Duan D, Doak AK, Nedyalkova L, Shoichet BK. Colloidal Aggregation and the in Vitro Activity of Traditional Chinese Medicines. ACS Chem Biol. 2015;10:978-88. DOI:10.1021/cb5009487

[49] Chin D, Huebbe P, Frank J, Rimbach G, Pallauf K. Curcumin may impair iron status when fed to mice for six months. Redox Biol. 2014;2:563-9. DOI:10.1016/j.redox.2014.01.018

[50] Schneider C, Gordon ON, Edwards RL, Luis PB. Degradation of Curcumin: From Mechanism to Biological Implications. J Agric Food Chem. 2015;63:7606-14. DOI:10.1021/acs.jafc.5b00244

[51] Kharat M, Du Z, Zhang G, McClements DJ. Physical and Chemical Stability of Curcumin in Aqueous Solutions and Emulsions: Impact of pH, Temperature, and Molecular Environment. J Agric Food Chem. 2017;65:1525-32. DOI:10.1021/acs.jafc.6b04815

[52] Antony B, Merina B, Iyer V, Judy N, Lennertz K, Joyal S. A pilot cross-over study to evaluate human oral bioavailability of BCM-95 ® CG (Biocurcumax TM), a novel bioenhanced preparation of curcumin. Indian J Pharm Sci. 2008;70:445. DOI:10.4103/0250-474X.44591

[53] Gota VS, Maru GB, Soni TG, Gandhi TR, Kochar N, Agarwal MG. Safety and Pharmacokinetics of a Solid Lipid Curcumin Particle Formulation in Osteosarcoma Patients and Healthy Volunteers. J Agric Food Chem. 2010;58:2095-9. DOI:10.1021/jf9024807

[54] Pia A. Management of chronic anterior uveitis relapses: efficacy of oral phospholipidic curcumin treatment. Long-term follow-up. Clin Ophthalmol. 2010;1201. DOI:10.2147/OPTH.S13271

[55] Rao PV, Gan SH. Cinnamon: A Multifaceted Medicinal Plant. Evidence-Based Complement Altern Med. 2014;2014:1-12. DOI:10.1155/2014/642942

[56] Jantan I, Ling YE, Romli S, Ayop N, Ahmad AS. A Comparative Study of the Constituents of the Essential Oils of Three Cinnamomum Species from

Malaysia. J Essent Oil Res. 2003;15:387-91. DOI:10.1080/10412905.2003.9698618

[57] Mathew S, Abraham TE. Studies on the antioxidant activities of cinnamon (*Cinnamomum verum*) bark extracts, through various in vitro models. Food Chem. 2006;94:520-8. DOI:10.1016/j.foodchem.2004.11.043

[58] Ranjbar A, Ghasmeinezhad S, Zamani H, Malekirad AA, Baiaty A, Mohammadirad A, et al. Antioxidative stress potential of *Cinnamomum zeylanicum* in humans: a comparative cross-sectional clinical study. Therapy. 2006;3:113-7. DOI:10.2217/14750708.3.1.113

[59] Roussel A-M, Hininger I, Benaraba R, Ziegenfuss TN, Anderson RA. Antioxidant Effects of a Cinnamon Extract in People with Impaired Fasting Glucose That Are Overweight or Obese. J Am Coll Nutr. 2009;28:16-21. DOI:10.1080/07315724.2009.10719756

[60] Anderson RA, Zhan Z, Luo R, Guo X, Guo Q, Zhou J, et al. Cinnamon extract lowers glucose, insulin and cholesterol in people with elevated serum glucose. J Tradit Complement Med. 2016;6:332-6. DOI:10.1016/j.jtcme.2015.03.005

[61] Hassan S, Barthwal R, Nair M, Haque S. Aqueous Bark Extract of *Cinnamomum zeylanicum*: A Potential Therapeutic Agent for Streptozotocin-Induced Type 1 Diabetes Mellitus (T1DM) Rats. Trop J Pharm Res. 2012;11. DOI:10.4314/tjpr.v11i3.12

[62] Alqasoumi S. Anti-secretagogue and antiulcer effects of Cinnamon *Cinnamomum zeylanicum* in rats. J Pharmacogn Phyther. 2012;4:53-61. DOI:10.5897/JPP12.023

[63] Peterson DW, George RC, Scaramozzino F, LaPointe NE, Anderson RA, Graves DJ, et al. Cinnamon Extract Inhibits Tau Aggregation Associated with Alzheimer's Disease In Vitro. J Alzheimer's Dis. 2009;17:585-97. DOI:10.3233/JAD-2009-1083

[64] Bayoub K, Baibai T, Mountassif D, Retmane A, Abdelaziz S. Antibacterial activities of the crude ethanol extracts of medicinal plants against *Listeria monocytogenes* and some other pathogenic strains. African J Biotechnol. 2010;9:4251-8.

[65] Sharma M, Bhatia A. Inactivation of Candidia albicans in culture media by eight spices native to Indian subcontinent. Int J Pharm Sci Rev Res. 2012;16:125-9.

[66] Kaskatepe B, Kiymaci ME, Suzuk S, Erdem SA, Cesur S, Yildiz S. Antibacterial effects of cinnamon oil against carbapenem resistant nosocomial Acinetobacter baumannii and *Pseudomonas aeruginosa* isolates. Ind Crops Prod. 2016;81:191-4. DOI:10.1016/j.indcrop.2015.11.058

[67] Samarasekera R, Kalhari KS, Weerasinghe IS. Mosquitocidal Activity of Leaf and Bark Essential Oils of Ceylon *Cinnamomum zeylanicum*. J Essent Oil Res. 2005;17:301-3. DOI:10.1080/10412905.2005.9698909

[68] Kim E-C, Kim HJ, Kim T-J. Water extract of *Cinnamomum cassia* suppresses angiogenesis through inhibition of VEGF receptor 2 phosphorylation. Biosci Biotechnol Biochem. 2015;79:617-24. DOI:10.1080/09168451.2014.993917

[69] Meades G, Henken R, Waldrop G, Rahman M, Gilman S, Kamatou G, et al. Constituents of Cinnamon Inhibit Bacterial Acetyl CoA Carboxylase. Planta Med. 2010;76:1570-5. DOI:10.1055/s-0030-1249778

[70] Niphade SR, Asad M, Chandrakala GK, Toppo E,

Deshmukh P. Immunomodulatory activity of *Cinnamomum zeylanicum* bark. Pharm Biol. 2009;47:1168-73. DOI:10.3109/13880200903019234

[71] Hagenlocher Y, Satzinger S, Civelek M, Feilhauer K, Köninger J, Bischoff SC, et al. Cinnamon reduces inflammatory response in intestinal fibroblasts in vitro and in colitis in vivo leading to decreased fibrosis. Mol Nutr food Res. 2017;

[72] Hagenlocher Y, Hösel A, Bischoff S, Lorentz A. Cinnamon extract reduces symptoms, inflammatory mediators and mast cell markers in murine IL-10−/− colitis. J Nutr Biochem. 2016;30:85-92.

[73] Ho S-C, Chang K-S, Chang P-W. Inhibition of neuroinflammation by cinnamon and its main components. Food Chem. 2013;138:2275-82. DOI:http://dx.doi.org/10.1016/j.foodchem.2012.12.020

[74] Kim BH, Lee YG, Lee J, Lee JY, Cho JY. Regulatory effect of cinnamaldehyde on monocyte/macrophage-mediated inflammatory responses. Mediators Inflamm. 2010;2010.

[75] Prasanth DSNBK, Murahari M, Chandramohan V, Panda SP, Atmakuri LR, Guntupalli C. In silico identification of potential inhibitors from Cinnamon against main protease and spike glycoprotein of SARS CoV-2. J Biomol Struct Dyn. 2020;1-15. DOI:10.1080/07391102.2020.1779129

[76] Kawatra P, Rajagopalan R. Cinnamon: Mystic powers of a minute ingredient. Pharmacognosy Res. 2015;7:1. DOI:10.4103/0974-8490.157990

[77] Shinde P, Patil P, Bairagi V. Herbs In Pregnancy And Lactation: A Review Appraisal. Int J Pharm Sci Res. 2012;3:3001-6.

[78] García-Jiménez N, Péerez-Alonso MJ, Velasco-Negueruela A. Chemical Composition of Fennel Oil, *Foeniculum vulgare* Miller, from Spain. J Essent Oil Res. 2000;12:159-62. DOI:10.1080/10412905.2000.9699487

[79] Diao W-R, Hu Q-P, Zhang H, Xu J-G. Chemical composition, antibacterial activity and mechanism of action of essential oil from seeds of fennel (*Foeniculum vulgare* Mill.). Food Control. 2014;35:109-16. DOI:10.1016/j.foodcont.2013.06.056

[80] Mohamad RH, El-Bastawesy AM, Abdel-Monem MG, Noor AM, Al-Mehdar HAR, Sharawy SM, et al. Antioxidant and Anticarcinogenic Effects of Methanolic Extract and Volatile Oil of Fennel Seeds (*Foeniculum vulgare*). J Med Food. 2011;14:986-1001. DOI:10.1089/jmf.2008.0255

[81] Anwar F, Ali M, Hussain AI, Shahid M. Antioxidant and antimicrobial activities of essential oil and extracts of fennel (*Foeniculum vulgare* Mill.) seeds from Pakistan. Flavour Fragr J. 2009;24:170-6. DOI:10.1002/ffj.1929

[82] Oktay M, Gülçin İ, Küfrevioğlu Öİ. Determination of in vitro antioxidant activity of fennel (*Foeniculum vulgare*) seed extracts. LWT - Food Sci Technol. 2003;36:263-71. DOI:10.1016/S0023-6438(02)00226-8

[83] Parejo I, Viladomat F, Bastida J, Schmeda-Hirschmann G, Burillo J, Codina C. Bioguided Isolation and Identification of the Nonvolatile Antioxidant Compounds from Fennel (*Foeniculum vulgare* Mill.) Waste. J Agric Food Chem. 2004;52:1890-7. DOI:10.1021/jf030717g

[84] Gende LB, Maggi MD, Fritz R, Eguaras MJ, Bailac PN, Ponzi MI. Antimicrobial Activity of *Pimpinella anisum* and *Foeniculum vulgare* Essential

Oils Against Paenibacillus larvae. J Essent Oil Res. 2009;21:91-3. DOI:10.1080/10412905.2009.9700120

[85] Soylu S, Yigitbas H, Soylu EM, Kurt Ş. Antifungal effects of essential oils from oregano and fennel on Sclerotinia sclerotiorum. J Appl Microbiol. 2007;103:1021-30. DOI:10.1111/j.1365-2672.2007.03310.x

[86] Runyoro D, Ngassapa O, Innocent E, Sangeda R, Lukuba T. Larvicidal activity of essential oils from spices sold at Kariakoo market in Dar es Salaam, Tanzania, against *Anopheles gambiae* Giles ss and *Culex quinquefasciatus* Say. J Complement Med Drug Discov. 2016;6:1. DOI:10.5455/spatula.20160613034220

[87] Zoubiri S, Baaliouamer A, Seba N, Chamouni N. Chemical composition and larvicidal activity of Algerian *Foeniculum vulgare* seed essential oil. Arab J Chem. 2014;7:480-5. DOI:10.1016/j.arabjc.2010.11.006

[88] Roby MHH, Sarhan MA, Selim KA-H, Khalel KI. Antioxidant and antimicrobial activities of essential oil and extracts of fennel (*Foeniculum vulgare* L.) and chamomile (*Matricaria chamomilla* L.). Ind Crops Prod. 2013;44:437-45. DOI:10.1016/j.indcrop.2012.10.012

[89] Dorra N, El-Berrawy M, Sallam S, Mahmoud R. Evaluation of Antiviral and Antioxidant Activity of Selected Herbal Extracts. J High Inst Public Heal. 2019;49:36-40. DOI: 10.21608/jhiph.2019.29464

[90] Alazadeh M, Azadbakht M, Niksolat F, Asgarirad H, Moosazadeh M, Ahmadi A, et al. Effect of sweet fennel seed extract capsule on knee pain in women with knee osteoarthritis. Complement Ther Clin Pract. 2020;40:101219. DOI:https://doi.org/10.1016/j.ctcp.2020.101219

[91] Özbek H, Uğraş S, Dülger H, Bayram İ, Tuncer İ, Öztürk G, et al. Hepatoprotective effect of *Foeniculum vulgare* essential oil. Fitoterapia. 2003;74:317-9. DOI:10.1016/S0367-326X(03)00028-5

[92] Shan B, Cai YZ, Sun M, Corke H. Antioxidant Capacity of 26 Spice Extracts and Characterization of Their Phenolic Constituents. J Agric Food Chem. 2005;53:7749-59. DOI:10.1021/jf051513y

[93] Miyazawa M, Hisama M. Antimutagenic Activity of Phenylpropanoids from Clove (*Syzygium aromaticum*). J Agric Food Chem. 2003;51:6413-22. DOI:10.1021/jf030247q

[94] Lee K-G, Shibamoto T. Antioxidant property of aroma extract isolated from clove buds [*Syzygium aromaticum* (L.) Merr. et Perry]. Food Chem. 2001;74:443-8. DOI:10.1016/S0308-8146(01)00161-3

[95] Fabio A, Cermelli C, Fabio G, Nicoletti P, Quaglio P. Screening of the antibacterial effects of a variety of essential oils on microorganisms responsible for respiratory infections. Phyther Res. 2007;21:374-7. DOI:10.1002/ptr.1968

[96] Trajano VN, Lima E de O, Souza EL de, Travassos AER. Inhibitory effect of the essential oil from *Eugenia caryophyllata* Thumb leaves on coalho cheese contaminating microorganisms. Ciência e Tecnol Aliment. 2010;30:1001-6. DOI:10.1590/S0101-20612010000400025

[97] Pinto E, Vale-Silva L, Cavaleiro C, Salgueiro L. Antifungal activity of the clove essential oil from *Syzygium aromaticum* on Candida, Aspergillus and dermatophyte species. J Med Microbiol. 2009;58:1454-62. DOI:10.1099/jmm.0.010538-0

[98] Never Zekeya, Francis Shahada MC. In vitro Antibacterial and Antifungal Activity of Tanzanian Bersama abyssinica. Int J Sci Res. 2014;3:1150-4.

[99] Bagavan A, Rahuman AA, Kaushik NK, Sahal D. In vitro antimalarial activity of medicinal plant extracts against *Plasmodium falciparum*. Parasitol Res. 2011;108:15-22. DOI:10.1007/s00436-010-2034-4

[100] Tu Z, Moss-Pierce T, Ford P, Jiang TA. Syzygium aromaticum L. (Clove) Extract Regulates Energy Metabolism in Myocytes. J Med Food. 2014;17:1003-10. DOI:10.1089/jmf.2013.0175

[101] Ghaffar S, Afridi SK, Aftab MF, Murtaza M, Hafizur RM, Sara S, et al. Clove and Its Active Compound Attenuate Free Fatty Acid-Mediated Insulin Resistance in Skeletal Muscle Cells and in Mice. J Med Food. 2017;20:335-44. DOI:10.1089/jmf.2016.3835

[102] Chegu K, Mounika K, Rajeswari M, Vanibala N, Sujatha P, Sridurga P, et al. In Vitro Study Of The Anticoagulant Activity Of Some Plant Extracts. WORLD J Pharm Pharm Sci. 2018;7:904-13.

[103] Singh G, Kiran S, Marimuthu P, Isidorov V, Vinogorova V. Antioxidant and antimicrobial activities of essential oil and various oleoresins Of*elettaria cardamomum* (seeds and pods). J Sci Food Agric. 2008;88:280-9. DOI:10.1002/jsfa.3087

[104] Ashokkumar K, Murugan M, Dhanya MK, Warkentin TD. Botany, traditional uses, phytochemistry and biological activities of cardamom [*Elettaria cardamomum* (L.) Maton] – A critical review. J Ethnopharmacol. 2020;246:112244. DOI:10.1016/j.jep.2019.112244

[105] Goyal S, Sharma C, Mahajan U, Patil C, Agrawal Y, Kumari S, et al. Protective Effects of Cardamom in Isoproterenol-Induced Myocardial Infarction in Rats. Int J Mol Sci. 2015;16:27457-69. DOI:10.3390/ijms161126040

[106] Nagashree S, Archana KK, Srinivas P, Srinivasan K, Sowbhagya HB. Anti-hypercholesterolemic influence of the spice cardamom (*Elettaria cardamomum*) in experimental rats. J Sci Food Agric. 2017;97:3204-10. DOI:10.1002/jsfa.8165

[107] Verma SK, Jain V, Katewa SS. Blood pressure lowering, fibrinolysis enhancing and antioxidant activities of Cardamom (*Elettaria cardamomum*). Indian J Biochem Biophys. 2009;46:503-6.

[108] Souissi M, Azelmat J, Chaieb K, Grenier D. Antibacterial and anti-inflammatory activities of cardamom (*Elettaria cardamomum*) extracts: Potential therapeutic benefits for periodontal infections. Anaerobe. 2020;61:102089. DOI:10.1016/j.anaerobe.2019.102089

[109] Majdalawieh AF, Carr RI. In Vitro Investigation of the Potential Immunomodulatory and Anti-Cancer Activities of Black Pepper (*Piper nigrum*) and Cardamom (*Elettaria cardamomum*). J Med Food. 2010;13:371-81. DOI:10.1089/jmf.2009.1131

[110] Jamal A, Javed K, Aslam M, Jafri MA. Gastroprotective effect of cardamom, *Elettaria cardamomum* Maton. fruits in rats. J Ethnopharmacol. 2006;103:149-53. DOI:10.1016/j.jep.2005.07.016

[111] Jia D, Tan Y, Liu H, Ooi S, Li L, Wright K, et al. Cardamonin reduces chemotherapy-enriched breast cancer stem-like cells in vitro and in vivo. Oncotarget. 2016;7:771-85. DOI:10.18632/oncotarget.5819

[112] Kaushik P, Goyal P, Chauhan A, Chauhan G. In Vitro Evaluation of Antibacterial Potential of Dry FruitExtracts of *Elettaria cardamomum* Maton (Chhoti Elaichi). Iran J Pharm Res IJPR. 2010;9:287-92.

[113] Abdel-Rahman M, Rezk MM, Kader SA. The role of cardamom on the hazardous effects of depleted uranium in cerebellum and midbrain of albino rats. Toxicol Environ Health Sci. 2017;9:64-73. DOI:10.1007/s13530-017-0305-5

[114] Auti ST, Kulkarni YA. Neuroprotective Effect of Cardamom Oil Against Aluminum Induced Neurotoxicity in Rats. Front Neurol. 2019;10:399. DOI:10.3389/fneur.2019.00399

[115] Abu-Taweel GM. Effects of Perinatal Cardamom Exposure on Social Behavior, Anxiety, Locomotor Activity, Blood Biochemical Parameters and Brain Acetylcholinesterase of Mice Offspring. Curr Pharm Biotechnol. 2020;21:1316-24. DOI:10.2174/1389201021666191216160546

[116] Marín A, Ferreres F, Tomás-Barberán FA, Gil MI. Characterization and Quantitation of Antioxidant Constituents of Sweet Pepper (*Capsicum annuum* L.). J Agric Food Chem. 2004;52:3861-9. DOI:10.1021/jf0497915

[117] Materska M, Perucka I. Antioxidant Activity of the Main Phenolic Compounds Isolated from Hot Pepper Fruit (*Capsicum annuum* L.). J Agric Food Chem. 2005;53:1750-6. DOI:10.1021/jf035331k

[118] Kim C-S, Kawada T, Kim B-S, Han I-S, Choe S-Y, Kurata T, et al. Capsaicin exhibits anti-inflammatory property by inhibiting IkB-a degradation in LPS-stimulated peritoneal macrophages. Cell Signal. 2003;15:299-306. DOI:10.1016/S0898-6568(02)00086-4

[119] Kang JY, Teng CH, Wee A, Chen FC. Effect of capsaicin and chilli on ethanol induced gastric mucosal injury in the rat. Gut. 1995;36:664-9. DOI:10.1136/gut.36.5.664

[120] Park J-S, Choi M-A, Kim B-S, Han I-S, Kurata T, Yu R. Capsaicin protects against ethanol-induced oxidative injury in the gastric mucosa of rats. Life Sci. 2000;67:3087-93. DOI:10.1016/S0024-3205(00)00890-0

[121] Leuschner RGK, Ielsch V. Antimicrobial effects of garlic, clove and red hot chilli on *Listeria monocytogenes* in broth model systems and soft cheese. Int J Food Sci Nutr. 2003;54:127-33. DOI:10.1080/0963748031000084070

[122] Lee G-R, Shin MK, Yoon D-J, Kim A-R, Yu R, Park N-H, et al. Topical Application of Capsaicin Reduces Visceral Adipose Fat by Affecting Adipokine Levels in High-Fat Diet-Induced Obese Mice. Obesity. 2013;21:115-22. DOI:10.1038/oby.2012.166

[123] Kang J-H, Tsuyoshi G, Le Ngoc H, Kim H-M, Tu TH, Noh H-J, et al. Dietary Capsaicin Attenuates Metabolic Dysregulation in Genetically Obese Diabetic Mice. J Med Food. 2011;14:310-5. DOI:10.1089/jmf.2010.1367

[124] Notani PN, Jayant K. Role of diet in upper aerodigestive tract cancers. Nutr Cancer. 1987;10:103-13. DOI:10.1080/01635588709513945

[125] Archer VE, Jones DW. Capsaicin pepper, cancer and ethnicity. Med Hypotheses. 2002;59:450-7. DOI:10.1016/S0306-9877(02)00152-4

[126] Yang Y, Zhang J, Weiss NS, Guo L, Zhang L, Jiang Y, et al. The consumption of chili peppers and the risk of colorectal cancer: a matched case-control study. World J Surg Oncol. 2019;17:71. DOI:10.1186/s12957-019-1615-7

[127] Lin C-H, Lu W-C, Wang C-W, Chan Y-C, Chen M-K. Capsaicin induces cell cycle arrest and apoptosis in human KB cancer cells. BMC Complement Altern Med. 2013;13:46. DOI:10.1186/1472-6882-13-46

[128] Pramanik KC, Boreddy SR, Srivastava SK. Role of Mitochondrial Electron Transport Chain Complexes in Capsaicin Mediated Oxidative Stress Leading to Apoptosis in Pancreatic Cancer Cells. Polymenis M, editor. PLoS One. 2011;6:e20151. DOI:10.1371/journal.pone.0020151

[129] Han SS, Keum Y-S, Seo H-J, Chun K-S, Lee SS, Surh Y-J. Capsaicin suppresses phorbol ester-induced activation of NF-κB/Rel and AP-1 transcription factors in mouse epidermis. Cancer Lett. 2001;164:119-26. DOI:10.1016/S0304-3835(01)00378-0

[130] Chang H, Chen S, Chien S, Kuo S, Tsai H, Chen D. Capsaicin may induce breast cancer cell death through apoptosis-inducing factor involving mitochondrial dysfunction. Hum Exp Toxicol. 2011;30:1657-65. DOI:10.1177/0960327110396530

[131] Huh H-C, Lee S-Y, Lee S-K, Park NH, Han I-S. Capsaicin Induces Apoptosis of Cisplatin-Resistant Stomach Cancer Cells by Causing Degradation of Cisplatin-Inducible Aurora-A Protein. Nutr Cancer. 2011;63:1095-103. DOI:10.1080/01635581.2011.607548

[132] Yimer EM, Tuem KB, Karim A, Ur-Rehman N, Anwar F. *nigella* sativa L. (Black Cumin): A Promising Natural Remedy for Wide Range of Illnesses. Evidence-Based Complement Altern Med. 2019;2019:1-16. DOI:10.1155/2019/1528635

[133] Hajhashemi V, Ghannadi A, Jafarabadi H. Black cumin seed essential oil, as a potent analgesic and antiinflammatory drug. Phyther Res. 2004;18:195-9. DOI:10.1002/ptr.1390

[134] El-Mahmoudy A, Matsuyama H, Borgan M., Shimizu Y, El-Sayed M., Minamoto N, et al. Thymoquinone suppresses expression of inducible nitric oxide synthase in rat macrophages. Int Immunopharmacol. 2002;2:1603-11. DOI:10.1016/S1567-5769(02)00139-X

[135] Suboh SM, Bilto YY, Aburjai TA. Protective effects of selected medicinal plants against protein degradation, lipid peroxidation and deformability loss of oxidatively stressed human erythrocytes. Phyther Res. 2004;18:280-4. DOI:10.1002/ptr.1380

[136] Celik Altunoglu Y, Bilen S, Ulu F, Biswas G. Immune responses to methanolic extract of black cumin (*Nigella sativa*) in rainbow trout (*Oncorhynchus mykiss*). Fish Shellfish Immunol. 2017;67:103-9. DOI:10.1016/j.fsi.2017.06.002

[137] Trigui I, Zarai Z, Chevance S, Cheikh-Rouhou S, Attia H, Ayadi MA. Physicochemical properties, antioxidant activity and in vitro gastrointestinal digestion of purified proteins from black cumin seeds. Int J Biol Macromol. 2019;126:454-65. DOI:10.1016/j.ijbiomac.2018.12.198

[138] Sahak MKA, Kabir N, Abbas G, Draman S, Hashim NH, Hasan Adli DS. The Role of *Nigella sativa* and Its Active Constituents in Learning and Memory. Evidence-Based Complement Altern Med. 2016;2016:1-6. DOI:10.1155/2016/6075679

[139] Salem ML, Hossain MS. Protective effect of black seed oil from *Nigella sativa* against murine cytomegalovirus infection. Int J Immunopharmacol. 2000;22:729-40. DOI:10.1016/S0192-0561(00)00036-9

[140] Mahmoud M., El-Abhar H., Saleh S. The effect of *Nigella sativa*

oil against the liver damage induced by *Schistosoma mansoni* infection in mice. J Ethnopharmacol. 2002;79:1-11. DOI:10.1016/S0378-8741(01)00310-5

[141] Chaieb K, Kouidhi B, Jrah H, Mahdouani K, Bakhrouf A. Antibacterial activity of Thymoquinone, an active principle of *Nigella sativa* and its potency to prevent bacterial biofilm formation. BMC Complement Altern Med. 2011;11:29. DOI:10.1186/1472-6882-11-29

[142] Utami AT, Pratomo B, Noorhamdani. Study of Antimicrobial Activity of Black Cumin Seeds (*Nigella sativa* L.) Against Salmonella typhi In Vitro. J Med Surg Pathol. 2016;01. DOI:10.4172/2472-4971.1000127

[143] Rostinawati T, Karipaya S, Iskandar Y. Antibacterial Activity of Ethanol Extract of *Nigella sativa* L. Seed Against Streptococcus mutans. IOP Conf Ser Earth Environ Sci. 2019;334:012050. DOI:10.1088/1755-1315/334/1/012050

[144] Okeola VO, Adaramoye OA, Nneji CM, Falade CO, Farombi EO, Ademowo OG. Antimalarial and antioxidant activities of methanolic extract of *Nigella sativa* seeds (black cumin) in mice infected with Plasmodium yoelli nigeriensis. Parasitol Res. 2011;108:1507-12. DOI:10.1007/s00436-010-2204-4

[145] Meral I, Yener Z, Kahraman T, Mert N. Effect of *Nigella sativa* on Glucose Concentration, Lipid Peroxidation, Anti-Oxidant Defence System and Liver Damage in Experimentally-Induced Diabetic Rabbits. J Vet Med Ser A. 2001;48:593-9. DOI:10.1046/j.1439-0442.2001.00393.x

[146] Kaatabi H, Bamosa AO, Badar A, Al-Elq A, Abou-Hozaifa B, Lebda F, et al. *Nigella sativa* Improves Glycemic Control and Ameliorates Oxidative Stress in Patients with Type 2 Diabetes Mellitus: Placebo Controlled Participant Blinded Clinical Trial. Ye J, editor. PLoS One. 2015;10:e0113486. DOI:10.1371/journal.pone.0113486

[147] Badary OA, AI-Shabanah OA, Nagi MN, AI-Rikabi AC, Elmazar MMA. Inhibition of benzo(a)pyrene-induced forestomach carcinogenesis in mice by thymoquinone. Eur J Cancer Prev. 1999;8:435-40. DOI:10.1097/00008469-199910000-00009

Chapter 16

Historical Evidence and Documentation of Remedial Flora of Azad Jammu and Kashmir (AJK)

Fozia Abasi, Muhammad Shoaib Amjad and Huma Qureshi

Abstract

Determining the pharmacognostic specifications of medicinal plants used in several drugs is very necessary and actually crucial. Ethnobotany has significant role in understanding the active relations between the biological diversity and cultural systems. Azad Jammu and Kashmir (AJK) is gifted with variety of medicinal plants. The theme of this chapter is to present information about wild medicinal plants in different areas of Azad Jammu and Kashmir. Common woody species are *Diospyros lotus, Taxus wallichiana, Viburnum cylindricum*, and perennial herbs comprise *Geranium nepalense, Oxalis acetosella* and *Androsace umbellata. Betula utilis, Berberis lycium, Cedrus deodara, Abies pindrow, Pinus wallichiana, Juglans regia* and *Salix* species with large number of herbal diversity at elevations are common. Most of people use wild plants as traditional food and medicine. This ethnic flora not only plays important role in human health care but it is also an important source for present and future drug development. There is need for correct documentation, conservation of plants samples in herbarium of research institutes, and growing plants in gardens.

Keywords: medicinal plants, Himalaya Kashmir, ethno-veterinary resources, ethno mycological data, plant parts, sustainable use

1. Introduction

Azad Jammu and Kashmir valley extends between 34°22′25 North latitude and 73°28′14 East longitude. Muzaffarabad is capital city of Kashmir and total area covered by Kashmir valley is 13,297 square kilometers. Estimated population of Azad Jammu and Kashmir is about 4-million. Mean maximum temperature was documented during summer (16 °C–24 °C) while −4 °C was recorded mean minimum temperature during winter. AJK is rich in diversity of plants because of its expanded habitations, such as streams, springs, nullahs, lakes, rivers, steep mountain slopes and roads, waste lands and cultivated fields, etc.

2. Geographical conditions and topography

The area of valley can be divided into two geographical zones; East and North are mostly hilly and mountainous categorized by undulating terrain, deep ravines,

Figure 1.
(a) Sudhen Gali district Bagh (b) Tolipeer district Poonch (c) Neelum valley (d) Areng khel Neelum valley (e) Leepa valley (f) Las Dana.

and rugged (Neelum, Muzaffarabad, Hattian, Bagh, Haveli, Poonch, and Sudhnoti) while South and West are valleys and plains (Kotli, Mirpur, Bhimber) (**Figure 1**).

3. Flora and plant diversity

In AJK, vegetation can be divided into four groups:

i. Subtropical vegetation is further divided into Dry scrub forest vegetation and Pine forest vegetation

ii. Temperate forest vegetation further divided into Moist Temperate and Dry Temperate Forest vegetation

iii. Sub alpine vegetation

iv. Alpine vegetation

The Himalaya Kashmir is documented as worldwide epicenter of endemism and plant diversity. Accordance to the report of Pei [1], in Himalayan range, total number of plant species is about 25,000 and total number of angiosperms in Kashmir Himalaya is about 3,054 [2]. About 80% endemic angiosperms is in Pakistan are confined to Northern and Western mountains [3, 4]. 70–80% of population in this region depends on traditional medicines for health care and in Himalayan ranges; at least 70% of the medicinal plants and animals in the region consists of wild species [5]. A total of 104 medicinal plant species including tree, shrubs and herb species used ethnobotanically by the local people of Muzaffarabad were reported from Machyara National Park Muzaffarabad [6, 7]. Most of People living in mountains regions use plants in different ways such as medicines, fire wood, timber wood, food, fodder etc. [8].

4. Historical evidence of wild plants usage

Medicinal plants are considered as safe medication and it is also naturally valuable remedy for many human sufferings in rural and remote hilly regions of Kashmir [9]. Due to the lack of advanced medicinal services, usage of flora as ethno medicine is renowned. Traditional curative usage of herbal plants by indigenous populations of AJK has been stated ([10] a&b; [11]). Saghir et al. [12] found 53 plant species useful mostly as medicinal, fuel, fodder, fruit, timber and vegetables reported from Chikar and allied areas of District Muzaffarabad. Gorsi and Shahzad [13] documented medicinal flora and suggested regeneration work to save the traditional knowledge about plants of Dirkot. Ishtiaq et al. [14] stated that plants are indirectly related to the culture and they stated 36 plant species used for the treatment of various diseases in Samahni valley. Khan et al. [15] indicated that the inhabitants of Poonch Valley utilized 169 plant species for more than 30 domestic needs. Ajaib et al. [16] provided ethnobotanical data on medicinal flora of district Kotli by reporting 38 species of shrubs. Saqib et al. [17] studied the medicinal flora of mountainous areas of AJK. Some of medicinally important plant species include *Saussurea lappa, Aconitum heterophyllum, Jurinea dolomea, Bistorta amplexicaule, Plectranthus rugosus,, Geranium wallichianum, Ajuga bracteosa, Taraxacum offinci-nale, Quercus incana, Berberis lyceum* and *Viola canescens* [18]. 70% of the therapeutic flora in the area comprise of wild species; 70–80% inhabitants dependent on traditional medications [19]. People of Azad Jammu & Kashmir are still dependent mainly on medicinal plants for folk remedies, hence creating immense pressure on native vegetation by overexploiting them, particularly in the mountainous region of Kashmir [20].

5. Documentation on remedial flora of Azad Jammu and Kashmir (AJK)

The original printed data of plants as medication initiating from the Himalayas date back to ancient scripts of the Rigveda, monitored by Auryveda (600–100 BC) and Atharveveda (2000–1000 BC). Northern mountains of Pakistan located at intersection of three Mountain ranges i.e., Himalaya, Karakorum and Hindu Kush are well recognized for their biodiversity [21]. Azad Jammu and Kashmir is endowed with productive variety of medicinal plants. It has been stated on many curative practices of plants by the indigenous populations [10, 14, 22, 23]. For above 10,000 classes of curative and scented plants, 600 million folks exist in in Himalayan section. In Himalayan ranges, 70% of therapeutic flora comprise of wild species [19]. Northern regions including Kashmir are in pressure from indigenous people and tourists. Primary reasons include unselective displacing and storing systems of remedial plants. Therefore, therapeutic tradition needs to be recognized and protected. Hundreds of species are currently endangered for the reason of excessive harvesting. Northern mountainous areas have several climatic and vegetation regions. These diverse natural regions have distinctive ethnobotanical vital plants that are significant for the economy of a nation. For traditional medications People of AJK are generating massive stress on flora by damaging those [20]. In north-western zones of Pakistan, several ethnobotanical trainings have been convoyed and which have assembled evidence on the usage of therapeutic flora [4]. The valuable ethnobotanical data is declining owing to the deficiency of awareness and information.

6. Folklore of wild plants in medicine

Azad Jammu and Kashmir is gifted with dynamic variety of medicinal plants. Below, we discuss some wild fruits and vegetables commonly used by indigenous people of AJK. Main wild fruits of the valley are *Ficus palmata Malus pumila, Prunus persica, Prunus cerasus, Morus alba, Diospyros lotus, Rubus fruticosus, Vitis vinifera, Viburnum foetens* and *Punica granatum*. Fruit of *Juglans regia* L. (Juglandaceae) is used as dry fruit. Fruit also remove gall bladder stones and is aphrodisiac. Fruits of *Morus nigra* L. (Moraceae) are dried and sold in market as a dry fruit. Fresh fruit is ground and used as tonic and for cough and throat irritation. Fruits of *Rubus ellipticus* Smith (Rosaceae) are edible. *Withinia somnifera* (L.) Dunal (Solanaceae) is used in Ayurvedic medicinal purposes and fruits are edible. Fruits of *Zanthoxylum alatum* DC. (Rutaceae) are aromatic, condiment and carminative and are used in sauce. They are also used for the treatment of piles. *Ziziphus nummularia* (Burm.f.) Wight & Arn. (Rhamnaceae) fruit is edible and laxative and leaves are used as fodder for goat. *Punica granatum* L. (Punicaceae) is used as treatment for Cancer, Osteoarthritis and other diseases. It has been used in natural and holistic medicine to treat sore throats, coughs, urinary infections, digestive disorders, skin disorders, arthritis. *Pyrus pashia* L. (Rosaceae) fruit is superlative to eat when it is slightly decaying. It is set apart from the cultivated pears by having a grittier quality. The fully ripe fruit has a reasonable taste and, when bletted, is sweet and very pleasant to eat. *Viburnum grandiflorum* Wall. (Caprifoliaceae) fruit is edible used against malaria [24] (**Table 1**). Miscellaneous uses of plants in the area comprise spices and condiments, ornamental plant species, vegetables and pot herbs, s agricultural tools, basket making, cosmetics, dish cleaner, house decoration, feed, field fencing, furniture, narcotics, packing material, curing snake and scorpion bite, soil binder, sticks and handles, shade tree, herbal tea and for making utensils. Maswak made from the roots of *J. regia* and branches of *A. modesta, O. ferruginea* and *Z. alatum* is used for cleaning their teeth. Plants are used as a major source of veterinary medicine. Interest of such use in the veterinary sector has resulted primarily from the increasing cost of livestock maintenance and the introduction of new technology in the veterinary medicines and vaccines. The important medicinal plant species showed the highest fidelity such as: *Rumex nepalensis*, *Primula denticulata*, (100%) used for dysuria, red urination, *Skimmia laureola* (100%), *Swertia paniculata* (99%), and *Angelica glauca* (97%), used for ague, cold, shivering, gastric ailments, *Melia azedarach* (100%), used to reduce intestinal worm load in cattle showing the conformity of knowledge on these species (**Table 2**). Plant communities have been largely disturbed due to deforestation for fuel, over consumption of medicinal resources for the population explosion, treatment of diseases, increased tourism and lack of awareness. Vulnerable species include *Sorghum halepense, Acacia modesta* and *Solanum nigrum.* Medicinal species like *Cissus carnosa, Butea monosperma Ajugabracteosa, Mallottus philippinensis*

	Botanical name	Family	Common name	Traditional uses
1.	*Ajuga bracteosa* Wall.ex. Benth.	Lamiaceae	Heri-booti	A decoction is used for curing intestinal ulcer, jaundice, throat infection and high blood pressure
2.	*Argemone mexicana* L.	Papaveraceae	Dudhli kandyari	Seeds are analgesic and laxative.
3.	*Alstonia scholaris* (L.) R.Br.	Apocynaceae	—	The bark is used to treat malaria, fever, asthma and tumors.
4.	*Amaranthus viridis* L.	Amaranthaceae	Ganar	Leaves are used on scorpion sting and snake bite. Root juice is used to treat constipation and inflammation during urination
5.	*Alternanthera pungens* Kunth	Amaranthaceae	Itsit	Roots and leaves are blood purifier and diuretic.
6.	*Anisomeles indica* (L.) S. Kurz.	Lamiaceae	—	Decoction of leaves is anti-rheumatic and used in stomachic and toothache.
7.	*Achyranthes aspera* L.	Amaranthaceae	Puth kanda	Leaves are used in pneumonia and asthma.
8.	*Albizzia lebbeck* (L.) Benth.	Mimosaceae	Sreeia	Seeds are used for curing kidney infection
9.	*Bauhinia veriegata* L.	Caesalpiniaceae	Katchnar	Fruit is edible and useful for leprosy and skin diseases.
10.	*Butea monosperma* (Lam.) Taubert	Papilionaceae	Chechra	Gum is tonic given for backache after birth.
11.	*Buddleja asiatica* Lour.	Buddlejaceae	Batta	Used for skin disease, and as a cure for loss of weight
12.	*Barleria cristata* L.	Acanthaceae	—	Seeds are antidote for snake bites and for serious catarrhal infections
13.	*Boerhavia diffusa* L.	Amaranthaceae	Sanati	Improve eyesight, diuretic and useful in controlling blood sugar levels
14.	*Buglossoides arvensis* (L.) Johnston	Boraginaceae	Kalu	Leaves infusion is sedative
15.	*Croton bonplandianus* Baill	Euphorbiaceae	—	Leaves control blood pressure
16.	*Cissampelos pareira* L.	Menispermaceae	Batrarr	A rhizome decoction and pounded leaves are externally applied as a febrifuge and stomachic, cough and snake bite
17.	*Carissa opaca* Staplf ex. Haines	Apocynaceae	Grunda	Fruit is edible and blood purifier
18.	*Cassia fistula* L.	Caesalpiniaceae	Amaltas	The root helps in reliving the symptoms of fever, asthma, leprosy and heart related diseases

	Botanical name	Family	Common name	Traditional uses
19.	*Chenopodium album* L.	Chenopodiaceae	Bathwa	This plant is laxative
20.	*Cissus carnosa* (L.) Lamk	Vitaceae	Dakh	Fruit is good for abdominal diseases
21.	*Calotropis procera*	Asclepiadaceae	Desi akk	Plant extract is applied on dog bite. Latex is used for skin diseases and ring worm.
22.	*Cannabis sativa* L. (Ait.)	Cannabinaceae	Bhang	Root is used for liver disorders. Leaves and flowers are analgesic, sedative, narcotic, laxative and aphrodisiac.
23.	*Cascuta reflexa* Roxb	Cuscutaceae	Neel Dhari	Its infusion is anti-lice. It is also used in skin diseases and weaknesses of children.
24.	*Chenopodium album* L.	Chenopodiaceae	Bathwa	Leaves are anthelmintic and laxative
25.	*Convolvulus arvensis* L.	Convolvulaceae	Rawari	Root is diuretic and purgative
26.	*Diplocyclos palmatus* (L.) C. Jeffery	Cucurbitaceae		Plant is used for skin diseases and cough
27.	*Dodonea viscosa* (L.) Jacq.	Sapindaceae	Sanatha	Decoction of wood is used as febrifuge and skin diseases
28.	*Dalbergia sissoo* Roxb.	Papilionaceae	Tali	Branches kill worms in sthe teeth
29.	*Eugenia jambolana* Lam.	Myrtaceae	Jaman	It is used for the treatment of cancer
30.	*Fumaria indica* (Hausskn.)Pugsley	Fumariaceae	Papra	Its infusion is used as diaphoretic, blood purifier and antipyretic
31.	*Ficus palmate* Forssk	Moraceae	Phugwara	Fruit is laxative, soothes bee sting
32.	*Gymnosporia royleana* (Wall.ex Lawson) Cuf	Celastraceae		It is used for treatment of cough, asthma, tonic and abdominal pain
33.	*Galium aparine* L.	Rubiaceae	Lahndara	Plant extract is diuretic
34.	*Hedera nepalensis* K.Koch.	Araliaceae		Leaves are used for treatment of diabetes
35.	*Justicia adhatoda* L.	Acanthaceae	Bhakar	It is used to treat colds, coughs, asthma, fevers, skin infections and inflammations
36.	*Juglans regia* L.	Juglandaceae	Khor	Root and leaves are antiseptic. Fruit is aphrodisiac, remove stones in gall bladder
37.	*Malva parviflora* L.	Malvaceae	Sonchul	Leaves extract is anthelmintic

	Botanical name	Family	Common name	Traditional uses
38.	*Mallotus philippinensis* (Lam.) Muell.	Euphorbiaceae	Kamella	Fruit is purgative and anthelmintic
39.	*Medicago polymorpha* L.	Papilionaceae	Sriri	Leaves are helpful in digestive disorders
40.	*Melia azadarach* (L.)	Meliaceae	Draik	Leaves and fruit are blood purifier, antipyretic and antidiabetic
41.	*Malvastrum Coromandelianum* (L.) Garcke	Malvaceae		Leaves paste relieve pain. Flowers are diaphoretic
42.	*Morus nigra* L.	Moraceae	Kala Toot	Fruit is tonic and used for throat irritation and cough
43.	*Mentha longifolia* Benth	Lamiaceae	Jangli podina	Leaves are carminative and stimulant. Leaves are antispasmodic
44.	*Nasturtium officinale* R.Br.	Brassicaceae	Chooch	The leaves and stem are used for internal tumors, scurvy and anemia
45.	*Nerium indicum* L.	Apocynaceae	Gandeera	Leaves decoction is applied on skin diseases
46.	*Oxalis corniculata* L.	Oxalidaceae	Khati Buti	Leaves decoction is used in dysentery and fever
47.	*Plantago lanceolata* L.	Plantaginaceae		Leaf extract is applied to wounds, sores and bruises; seeds are purgative
48.	*Pinus roxburghii* Sargent	Pinaceae	Cheer	Resin is used for bleeding wounds and tumors and cough. Leaves and bark powder is useful for dysentery
49.	*Papaver dubium* L.	Papaveraceae	Jangli post	Its infusion is blood purifier, antipyretic, and diaphoretic
50.	*Periploca aphylla* Decne	Asclepidiaceae	Bata	It is used for treatment of swellings and tumors
51.	*Rumex dentatus* L.	Polygonaceae	Herfli	Leaves are astringent and diuretic
52.	*Rhamnus triquetra* (Wall.) Brandis	Rhamnaceae	Clader	Fruit and leaves are used in hemorrhagic septicemia
53.	*Ranunculus muricatus* L.	Ranunculaceae	Kor-Kandoli	Fruits and leaves are useful on bursts and tumor

Table 1.
Important medicinal plant species with traditional uses in Azad Jammu and Kashmir.

and *Zanthoxylum armatum* are critically endangered. Among endangered species, *Juglans regia, Olean ferruginaea, Phyllanthus emblica, Viola canescens* are the notable species. Some medicinal plants like *G. wallichianum, J. dolomiaea, A. bracteosa,*

	Scientific name	Family	Common name	Ethno veterinary practices	Ailments
1.	*Amaranthus viridis* L.	Amaranthaceae	Safed kannar	Decoctions	Malaria
2.	*Arisaema flavum* (Forssk.) Schoot	Araceae	Toosh	Infusion	Mouth and foot disease of cattle
3.	*Arisaema jacquemontii* Blume	Araceae	Tooshganda	As whole plant	Inflammation, cholera, dysentery, flu, dyspepsia and snake bite
4.	*Senecio chrysanthemoides* DC.	Asteraceae	Bghoo	Decoction	Antiscorbutic, anthelmintic, and diaphoretic
5.	*Lactuca brunoniana* (DC.) Wall. ex C.B. Clarke	Asteraceae	Korijari	The whole plant	Pinworms
6.	*Aesculus indica* (Wall. ex Cambess.) Hook.	Sapindaceae	Bankhorr	Decoction/dried plant powder is mixed with gurr and flour	Indigestion and constipation
7.	*Melia azedarach* L.	Meliaceae	Drek	Powder	Anthelmintic
8.	*Ficus palmata* Forssk.	Moraceae	Pagaaar	Whole	Anorexia
9.	*Bergenia ciliata* (Haw.) Sternb.	Saxifragaceae	Butpeewa	Powder	Stomachic and intestinal troubles
10.	*Fagopyrum acutatum* (Lehm.) Mansf. ex K.	Polygonaceae	Khattra	As fodder	Antimicrobial, bactericidal and diuretic
11.	*Primula denticulata* Sm.	Primulaceae	Chiatpatra	Decoction	Dysuria, hepatic fever and hemoglobinuria
12.	*Sorbaria sorbifolia* (L.) A. Braun	Rosaceae	Karleee	As fodder	Stimulant

Table 2.
Ethno-veterinary practices of important plants in Azad Jammu and Kashmir.

B. amplexicaule, S. lappa, A. heterophyllum, and *B. lyceum* are on the edge of extinction due to high rate of intake [25].

7. Wild mushrooms

Morel collection is an important activity during spring season and villagers take keen interest in collection of morels as it provides them a source of income. Mushroom flora and species diversity as important component of the natural environment in Azad Jammu and Kashmir. Wild mushrooms are sources of edible proteins, dietary fiber, essential amino acids, carbohydrates, and are an important source of food, livelihood, and traditional ethnobotanical health care. Kashmir region is rich with unknown macro fungal wealth. Among total morel population of Pakistan, 90% was reported from the Himalayan mountain ranges of Northern Pakistan. Wild edible fungi dominate the global morel trade, with an estimated value of more than US$2 billion. Ullah et al. [26] reported 56 wild edible mushrooms in Pakistan, of which 44 are from the Kashmir region. Important species include *Agaricus campestris, Hydnum imbricatum, Sparassis crispa, Morchella esculenta, M. crassipes, M. elata, M. conica, Pleurotus ostreatus, Lycoperdon gemmatum, Helvella crispa, Tricholoma megnivelare* and *Gyrometra esculenta*. The local communities of valley well recognize the habitats, morphological features, and qualities of these mushrooms. Ethno mycological data were collected through the use of a questionnaire and found that these species have great medicinal value against different ailments. Four species (*A. campestris, H. imbricatum, P. ostreatus*, and *S. crispa*) are highly recommended for their frequent use as food based on nutritional analysis (proteins, fats, fiber, and moisture). The major identified species from AJK are *Agaricus arvensi, Amanita vaginata, A. fulva, Cantharellus cinereus, Coprinus micaceus, C. comatus, C. domesticus, Cycoperdon perlatum, Daedalea quercina, Helvella crispa, Hygrophuorus melizeus, Lepista nuda, Lactarius turpis, Marasmius alliaceus, Panaeolus campanulatus, Pleurotus ostreatus, Trametes versicolor,* and *Tricholoma ustaloides.*

Although Azad Jammu and Kashmir (AJK) have ample of medicinal plants to treat broad spectrum of ailments, there are many factors which are contributing for loss of ethnic flora e.g. over grazing, over exploitation, fire, deforestation etc. Lack of concern in the present generation has wiped out many rich wild flora of the area. It is necessary to create awareness about the usefulness of the flora. Cultivation of threatened medicinal plants should be encouraged by the local community in order to relieve pressure on wild plants. People should spread useful information on conservation and sustainable use of the natural resources of the area. There must be correct documentation, conservation of plants samples in herbarium of research institutes, and growing plants in gardens.

Author details

Fozia Abasi[1], Muhammad Shoaib Amjad[2*] and Huma Qureshi[3]

1 Department of Botany, Faculty of Basics and Applied Sciences, University of the Poonch Rawalakot, Pakistan

2 Department of Botany, Faculty of Sciences, Women University of Azad Jammu and Kashmir, Bagh, Pakistan

3 Institute of Biological Sciences (Botany Program), Gomal University, Dera Ismail Khan, Pakistan

*Address all correspondence to: malikshoaib1165@yahoo.com

References

[1] Pei, S. J. (1992). Mountain culture and forest resource management of Himalayas. Himalayan Ecosystem. Intel. Book. Distributors, Dehra Dun, India.

[2] Jee, V., U. Dhar and P. Kachroo. (1989). Cytogeography of some endemic taxa of Kasmir Himalaya. Proc. Indian Nat. Sci. Acad., 55 (3): 177-184

[3] Ali, S. I. (2008). Significance of flora with special reference to Pakistan. Pak. J. Bot., 40(30): 967-971.

[4] Ali, S. I. and M. Qaiser. (1986). A Phytogeographic analysis of the phanerogams of Pakistan and Kashmir. Proc. Roy. Soc. Edinb., 89: 89-101.

[5] Pie, S.J. and N.P. Manadhar (1987). Source of some local medicines in the Himalayan Regions. Himalayan Ecosystems, 77-112.

[6] Bokhari, A.H. (1994). Ethnobotanical survey abd vegetation analysis of Machyara National Park Azad Kashmir, Pakistan. M.Sc. Thesis, University of Azad Kashhmir.

[7] Zandial, R. (1994). Ethnobotanical studies and population analysis of Machyara National Park, Azad Kashmir, M. Sc. Thesis University of Azad Kashmir.

[8] Hussain, F. and A. Khaliq. (1996). Ethnobotanical studies on some plants of Dabargai Hills. Swat. Proceedings of first training workshop on Ethnobotany and its application to conservation. NARC, Islamabad. pp. 207-215.

[9] Zaidi, S. H. (2001). Existing indigenous medicinal plant resources of Pakistan and their prospects for utilization. Medicinal Plants of Pakistan. 53 pp.

[10] Mahmood, A., R.N. Malik, Z.K. Shinwari and A. Mahmood. (2011). Ethnobotanical survey of plants from Neelum, Azad Jammu & Kashmir, Pakistan. Pak. J. Bot., 43: 105-110.

[11] Ishtiaq, M., He, Q., Wang, Y. and Cheng, Y.Y. (2010). A Comparative Study of Chemometric and Numerical Taxonomic Approaches in Identification and Classification of Traditional Chinese Medicines (TCMs) of Genus Clematis species. J. Plant Biosyst. 144(2): 288-297.

[12] Saghir, I.A., A.A. Awan, S. Majid, M.A. Khan and S.J. Qureshi. (2001). Ethnobotanical studies of Chikar and its allied area of District Muzaffarabad. Online Journal of Biological Sci., 1(12): 1165-1170.

[13] Gorsi, M.S. and R. Shahzad. (2002). Medicinal uses of plants with particular reference to the people of Dhirkot, Azad Jammu and Kashmir. Asian J. Plant Sci., 1: 222-223.

[14] Ishtiaq, C.M., M.A. Khan and W. Hanif. (2006). Ethnoveterinary medicinal uses of plants from Samahni valley, Bhimber, Azad Kashmir, Pakistan. Asian J. Plant Sci., 5: 390-396.

[15] Khan, M.A., M.A. Khan, M. Hussain, and G. Mujtaba. (2010). An Ethnobotanical Inventory of Himalayan Region Poonch Valley Azad Kashmir (Pakistan). Ethnobotany Research & Applications. 8: 107-123.

[16] Ajaib, M., Z. Khan, N. Khan and M. Wahab. (2010). Ethnobotanical studies on useful shrubs of District kotli, Azad Jammu & Kashmir, Pakistan. Pak. J. Bot., 42(3): 1407-1415.

[17] Saqib, Z., R.N. Malik, M.I. Shinwari and Z.K. Shinwari. (2011). Species Richness, Ethnobotanical Species Richness and Human Settlements along a Himalayan Altitudinal Gradient: Prioritizing Plant Conservation in Palas

Valley, Pakistan. Pak. J. Bot., 43 (Special Issue): 129-133.

[18] Shaheen H., Z.K. Shinwari, R.A. Qureshi and Z. Ullah. (2012). Indigenous Plant Resources and their Utilization Practices in Village Populations of Kashmir Himalayas. Pak. J. Bot., 44(2): 739-745.

[19] Shaheen H, Shinwari ZK (2012). Phyto diversity and Endemic richness of Karambar Lake Vegetation from Chitral, Hindukush- Himalayas. Pak. J. Bot. 44(1):17-21.

[20] Alam, N., Z.K. Shinwari, M. Ilyas and Z. Ullah. (2011). Indigenous knowledge of medicinal plants of Chagharzai Valley, District Buner, Pakistan. Pak. J. Bot., 43: 773-780.

[21] Shinwari, Z.K. (2010). Medicinal plants research in Pakistan. Journal of Medicianal Plants Research, 4(3): 161-176.

[22] Dar, M.E.I. (2003). Ethnobotanical uses of plants of Lawat District Muzaffarabad, Azad Jammu and Kashmir. Asian J. Plant Sci., 2(9): 680-682.

[23] Ishtiaq, M. (2007). An Ethnomedicinal Survey and Documentation of Important Medicinal Folklore Food Phytonims of Flora of Samahni Valley, (A.K) Pakistan. Pakistan Journal of Biological Sciences, 10(13): 2241-2256.

[24] Ajaib, M., Ashraf, Z., Abid, A., Ishtiaq. M and Siddiqui, M.F. (2016). Ethnobotanical studies of wild plant resources of Puna hills, district Bhimber, AJK. AJAIB ET AL FUUAST J. BIOL., 6(2): 257-264.

[25] Khan, M.A., S.A. Khan, M.A. Qureshi, G. Ahmed, M.A. Khan, M. Hussain and G. Mujtaba. (2011). Ethnobotany of some useful plants of Poonch Valley Azad Kashmir. J. Med. Plants Res., 5(26): 6140-6151.

[26] Ullah, T.S., Firdous S.S., Mehmood, A., Shaheen, H., Dar, M. E. I. 2017. Ethnomycological and Nutritional Analyses of Some Wild Edible Mushrooms from Western Himalayas, Azad Jammu and Kashmir (Pakistan). International Journal of Medicinal Mushrooms, 19(10): 10.1615/IntJMedMushrooms.2017024383.